Advances in
APPLIED MICROBIOLOGY

VOLUME **64**

Advances in
APPLIED MICROBIOLOGY

VOLUME **64**

Edited by

ALLEN I. LASKIN
Somerset, New Jersey, USA

SIMA SARIASLANI
Wilmington, Delaware, USA

GEOFFREY M. GADD
Dundee, Scotland, UK

AMSTERDAM • BOSTON • HEIDELBERG • LONDON
NEW YORK • OXFORD • PARIS • SAN DIEGO
SAN FRANCISCO • SINGAPORE • SYDNEY • TOKYO
Academic Press is an imprint of Elsevier

Academic Press is an imprint of Elsevier
84 Theobald's Road, London WC1X 8RR, UK
30 Corporate Drive, Suite 400, Burlington, MA 01803, USA

First edition 2008

Copyright © 2008 Elsevier Inc. All rights reserved

No part of this publication may be reproduced, stored in a retrieval system or transmitted in any form or by any means electronic, mechanical, photocopying, recording or otherwise without the prior written permission of the publisher

Permissions may be sought directly from Elsevier's Science & Technology Rights Department in Oxford, UK: phone (+44) (0) 1865 843830; fax (+44) (0) 1865 853333; email: permissions@elsevier.com. Alternatively you can submit your request online by visiting the Elsevier web site at http://elsevier.com/locate/permissions, and selecting, *Obtaining permission to use Elsevier material*

Notice
No responsibility is assumed by the publisher for any injury and/or damage to persons or property as a matter of products liability, negligence or otherwise, or from any use or operation of any methods, products, instructions or ideas contained in the material herein. Because of rapid advances in the medical sciences, in particular, independent verification of diagnoses and drug dosages should be made

ISBN: 978-0-12-374338-1
ISSN: 0065-2164

For information on all Academic Press publications
visit our website at books.elsevier.com

Printed and bound in USA
08 09 10 11 12   10 9 8 7 6 5 4 3 2 1

Working together to grow
libraries in developing countries

www.elsevier.com | www.bookaid.org | www.sabre.org

ELSEVIER   BOOK AID International   Sabre Foundation

CONTENTS

Contributors ix

1. Diversity of Microbial Toluene Degradation Pathways

R. E. Parales, J. V. Parales, D. A. Pelletier, and J. L. Ditty

I. Introduction	2
II. Dioxygenase Mediated Pathway: *Pseudomonas putida* F1	4
A. Pathway details and enzymology	4
B. Genetics and regulation	8
III. Toluene 2-Monooxygenase Pathway: *Burkholderia vietnamiensis* G4	12
A. Pathway details and enzymology	12
B. Genetics and pathway regulation	14
IV. Toluene-3-monooxygenase Pathway: *Ralstonia pickettii* PKO1	15
V. Toluene-4-monooxygenase Pathway: *Pseudomonas mendocina* KR1	20
VI. A Strain with Multiple Toluene Degradation Pathways: *Burkholderia* sp. Strain JS150	25
VII. A Nonspecific Toluene Monooxygenase Pathway: *Pseudomonas stutzeri* OX1	28
A. Pathway details and enzymology	28
B. Pathway regulation	32
VIII. TOL Pathway: *Pseudomonas putida* mt-2	33
A. Pathway and enzymology	34
B. Genetics and regulation	40
IX. A Fungal Toluene Degradation Pathway	43
X. Anaerobic Toluene Pathway: *Thauera aromatica* K172	47
A. Pathway details and enzymology	47
B. Genetics and regulation	52
XI. Conclusions	52
Acknowledgments	55
References	55

2. Microbial Endocrinology: Experimental Design Issues in the Study of Interkingdom Signalling in Infectious Disease

Primrose P. E. Freestone and Mark Lyte

| I. Microbial Endocrinology as a New Scientific Discipline | 76 |
| A. Overview and goal of review | 76 |

  B. The stress response and its influence on the immune system and the infectious agent 77
  C. Mechanisms by which stress-related neuroendocrine hormones can enhance bacterial growth and virulence 79
  D. The spectrum of catecholamine responsive bacteria 84
  E. Catecholamine specificity in enteric bacteria 85
  F. Molecular analyses of bacterial catecholamine responsiveness 88
II. Experimental Design Issues in the Study of Microbial Endocrinology 89
  A. Medium comparability 90
  B. Bacterial inoculum size and its influence on stress neurohormone responsiveness 92
  C. The importance of neuroendocrine hormone concentration 94
  D. Neuroendocrine stress neurohormone assay methodologies 97
III. Conclusions and Future Directions 101
References 102

3. Molecular Genetics of Selenate Reduction by *Enterobacter cloacae* SLD1a-1

Nathan Yee and Donald Y. Kobayashi

I. Introduction 108
II. Physiology and Biochemistry of *E. cloacae* SLD1a-1 108
III. Basic Systems of Molecular Genetics in Facultative Se-reducing Bacteria 110
  A. Growth of *E. cloacae* SLD1a-1 on selenate-containing agar plates 110
  B. Direct cloning experiments 110
  C. Gene specific mutagenesis 112
  D. Random mutagenesis using transposons 113
IV. Genetic Analysis of Selenate Reduction in *E. cloacae* SLD1a-1 115
V. A Molecular Model for Selenate Reduction in *E. cloacae* SLD1a-1 118
VI. Conclusions and Future Prospects 120
Acknowledgments 120
References 120

4. Metagenomics of Dental Biofilms

Peter Mullany, Stephanie Hunter, and Elaine Allan

I. Introduction 126
II. Isolation of Metagenomic DNA 126
III. Sample Collection and Processing 126
IV. DNA Extraction 128
V. Preparation of the Insert DNA for Cloning 128
  A. Partial restriction digestion 129
  B. Mechanical shearing of the DNA 129
VI. Removal of Human DNA 129

VII. Metagenomic Library Construction	130
A. Construction of large insert metagenomic libraries	130
B. Construction of small insert libraries	132
C. Phage display	132
VIII. Limitations of Metagenomics	133
IX. Conclusions and Future Perspectives	134
References	134

5. Biosensors for Ligand Detection

Alison K. East, Tim H. Mauchline, and Philip S. Poole

I. Introduction	137
II. Induction Biosensors	139
A. General considerations for use of induction biosensors	139
B. Reporter genes	140
III. Molecular Biosensors	150
A. Optical biosensors	150
B. Electrical biosensors	159
IV. Conclusions and Future Prospects	160
Acknowledgments	160
References	160

6. Islands Shaping Thought in Microbial Ecology

Christopher J. van der Gast

I. Introduction	167
II. The Importance of Islands	169
A. Biogeographic islands for studies of bacterial diversity	169
B. Island biogeography	170
III. Species–Area Relationships	170
A. Microbial biogeography	171
B. Island size and bacterial diversity	172
IV. Beta Diversity	174
A. Smaller islands are less stable than larger ones	175
B. Species–time relationships	175
C. Distance–decay relationships	177
V. Opposing Perspectives on Community Assembly	177
VI. Conclusions	179
Acknowledgments	180
References	180

7. Human Pathogens and the Phyllosphere

John M. Whipps, Paul Hand, David A. C. Pink, and Gary D. Bending

I. Introduction	183

II.	Food Poisoning Outbreaks Associated with Consumption of Fresh and Minimally Processed Fresh Vegetables, Salads, and Fruit	185
III.	Sources of Human Pathogens on Plants	186
IV.	Ecology of Human Pathogens in Relation to Phyllosphere Contamination	189
	A. Ecology and survival in sewage, manure, soil, and water	189
	B. Ecology and survival in association with plants	190
	C. Ecology and survival during processing	203
V.	Conclusions and Future	208
	Acknowledgments	209
	References	209

8. Microbial Retention on Open Food Contact Surfaces and Implications for Food Contamination

Joanna Verran, Paul Airey, Adele Packer, and Kathryn A. Whitehead

I.	Introduction	224
II.	Microbial Attachment, Biofilm Formation, and Cell Retention	225
III.	Surfaces (substrata) Encountered	226
IV.	Factors Affecting Retention	227
	A. Surface topography	227
	B. Surface chemistry	230
	C. Presence of organic material	231
V.	Characterization of Surfaces	232
	A. Topography	232
	B. Chemistry	234
VI.	Measuring Retention and Assessing Cleaning and Disinfection	234
	A. Substratum preparation	234
	B. Amount of retention	235
	C. Strength of retention	238
	D. Quantifying organic material	240
VII.	Conclusions	241
	References	241

Index — 247
Contents of Previous Volumes — 255
Color Plate Section

CONTRIBUTORS

Paul Airey
School of Biology, Chemistry and Health Science, Manchester Metropolitan University, Manchester M1 5GD, United Kingdom.

Elaine Allan
Division of Microbial Diseases, Eastman Dental Institute, University College London, London WC1X 8LD, United Kingdom.

Gary D. Bending
Warwick HRI, University of Warwick, Wellesbourne, Warwick, CV35 9EF, United Kingdom.

J. L. Ditty
Department of Biology, University of St. Thomas, St. Paul, Minnesota 55105.

Alison K. East
Molecular Microbiology, John Innes Centre, Colney Lane, Norwich NR4 7UH, United Kingdom.

Primrose P. E. Freestone
Department of Infection, Immunity and Inflammation, University of Leicester School of Medicine, Leicester, United Kingdom.

Paul Hand
Warwick HRI, University of Warwick, Wellesbourne, Warwick, CV35 9EF, United Kingdom.

Stephanie Hunter
Institute of Structural and Molecular Biology, University College London, London, WC1E 6BT, United Kingdom.

Donald Y. Kobayashi
Department of Plant Biology and Pathology, Rutgers, The State University of New Jersey, New Brunswick, New Jersey 08901.

Mark Lyte
Department of Pharmacy Practice, School of Pharmacy, Texas Tech University Health Sciences Center, Lubbock, Texas 79430-8162.

Tim H. Mauchline
School of Biological Sciences, University of Reading, Berkshire RG6 6AJ, United Kingdom.

Peter Mullany
Division of Microbial Diseases, Eastman Dental Institute, University College London, London WC1X 8LD, United Kingdom.

Adele Packer
School of Biology, Chemistry and Health Science, Manchester Metropolitan University, Manchester M1 5GD, United Kingdom.

J. V. Parales
Department of Microbiology, University of California, Davis, California 95616.

R. E. Parales
Department of Microbiology, University of California, Davis, California 95616.

D. A. Pelletier
Biosciences Division, Oak Ridge National Laboratory, Oak Ridge, Tennessee 37831.

David A. C. Pink
Warwick HRI, University of Warwick, Wellesbourne, Warwick, CV35 9EF, United Kingdom.

Philip S. Poole
Molecular Microbiology, John Innes Centre, Colney Lane, Norwich NR4 7UH, United Kingdom.

Christopher J. van der Gast
NERC Centre for Ecology and Hydrology, Oxford, OX1 3SR, United Kingdom.

Joanna Verran
School of Biology, Chemistry and Health Science, Manchester Metropolitan University, Manchester M1 5GD, United Kingdom.

John M. Whipps
Warwick HRI, University of Warwick, Wellesbourne, Warwick, CV35 9EF, United Kingdom.

Kathryn A. Whitehead
School of Biology, Chemistry and Health Science, Manchester Metropolitan University, Manchester M1 5GD, United Kingdom.

Nathan Yee
Department of Environmental Sciences, Rutgers, The State University of New Jersey, New Brunswick, New Jersey 08901.

CHAPTER 1

Diversity of Microbial Toluene Degradation Pathways

R. E. Parales,[*] **J. V. Parales,**[*] **D. A. Pelletier,**[†] **and J. L. Ditty**[‡]

Contents			
	I.	Introduction	2
	II.	Dioxygenase Mediated Pathway: *Pseudomonas putida* F1	4
		A. Pathway details and enzymology	4
		B. Genetics and regulation	8
	III.	Toluene 2-Monooxygenase Pathway: *Burkholderia vietnamiensis* G4	12
		A. Pathway details and enzymology	12
		B. Genetics and pathway regulation	14
	IV.	Toluene-3-monooxygenase Pathway: *Ralstonia pickettii* PKO1	15
	V.	Toluene-4-monooxygenase Pathway: *Pseudomonas mendocina* KR1	20
	VI.	A Strain with Multiple Toluene Degradation Pathways: *Burkholderia* sp. Strain JS150	25
	VII.	A Nonspecific Toluene Monooxygenase Pathway: *Pseudomonas stutzeri* OX1	28
		A. Pathway details and enzymology	28
		B. Pathway regulation	32
	VIII.	TOL Pathway: *Pseudomonas putida* mt-2	33
		A. Pathway and enzymology	34
		B. Genetics and regulation	40
	IX.	A Fungal Toluene Degradation Pathway	43

[*] Department of Microbiology, University of California, Davis, California 95616
[†] Biosciences Division, Oak Ridge National Laboratory, Oak Ridge, Tennessee 37831
[‡] Department of Biology, University of St. Thomas, St. Paul, Minnesota 55105

X.	Anaerobic Toluene Pathway:	
	Thauera aromatica K172	47
	A. Pathway details and enzymology	47
	B. Genetics and regulation	52
XI.	Conclusions	52
	Acknowledgments	55
	References	55

I. INTRODUCTION

Both natural and anthropogenic sources contribute to the presence of toluene in the environment. Toluene is found naturally in petroleum and coal, and is a major component of gasoline; it is also produced by the burning of organic materials and has been detected in cigarette smoke (Cline *et al.*, 1991; Darrall *et al.*, 1998; Koppmann *et al.*, 1997; Sinninghe Damste *et al.*, 1992). Toluene is also produced industrially for use as a solvent and in the production of various chemicals. In addition, toluene is produced and emitted by plants (Heiden *et al.*, 1999; Holzinger *et al.*, 2000; Vrkocova *et al.*, 2000). Benzene, toluene, ethylbenzene, and xylenes (BTEX), which together represent a significant fraction of the aromatic hydrocarbons in gasoline, are known or suspected carcinogens (Dean, 1985; Snyder, 2002) and present a serious threat to drinking water supplies because of their relatively high aqueous solubilities. A significant amount of effort has gone into the study of toluene fate in the environment and the contribution of biodegradation to its removal.

Because of the stability of the benzene ring, the initial step in toluene degradation is the most difficult. Aromatic hydrocarbons such as toluene are significantly more reduced than cellular organic material and thus must be oxidized prior to assimilation. Bacteria and fungi have evolved a variety of interesting pathways for the degradation of toluene. An overview of known toluene degradation pathways is shown in Fig. 1.1. Under aerobic conditions, oxygen is used directly as a substrate, destabilizing the aromatic ring and preparing the molecule for further metabolism. Some bacteria use a dioxygenase to catalyze the formation of a *cis*-dihydrodiol product, which is then subjected to dehydrogenation to form 3-methylcatechol. Other bacteria have monooxygenases that catalyze the initial hydroxylation at the ortho, meta, or para position of toluene, which is then followed by a second hydroxylation by the same or a different hydroxylase to form catecholic products. Alternatively, the methyl group of toluene may undergo oxidation prior to conversion to a catechol as in the TOL pathway (Fig. 1.1). A recently described fungal pathway uses a variation of the TOL pathway. Thus, in the aerobic pathways,

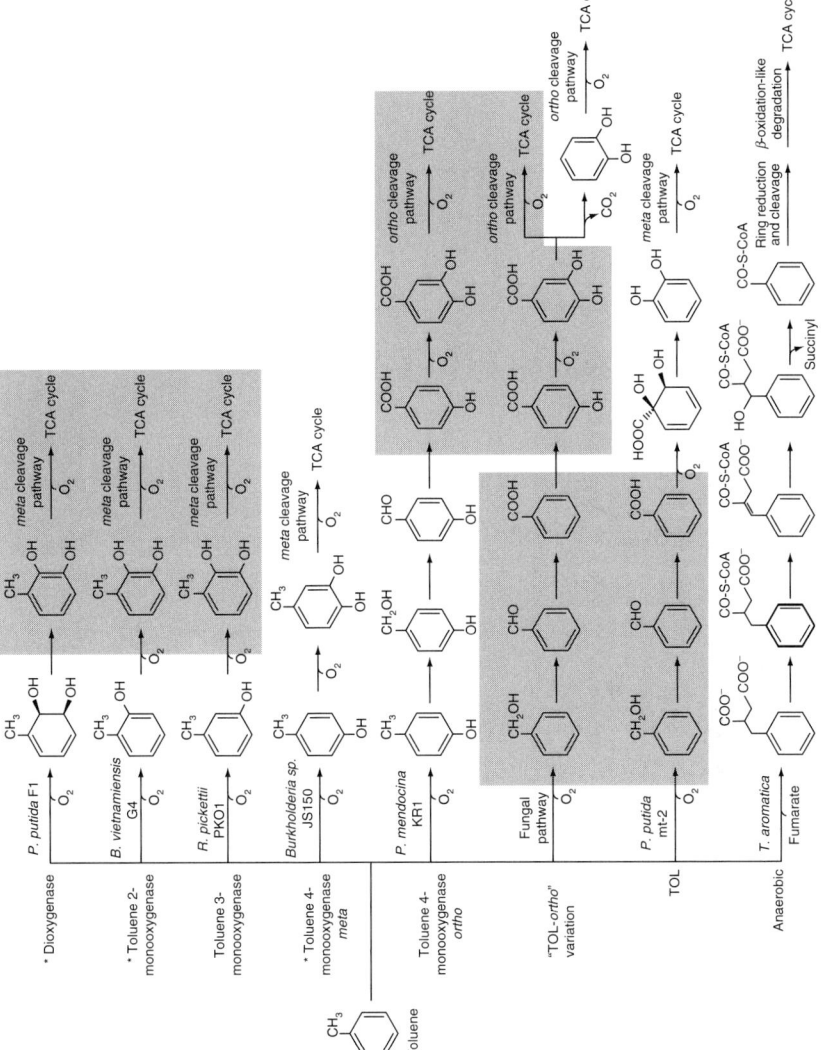

FIGURE 1.1 Overview of known toluene degradation pathways in bacteria and fungi. Shaded boxes indicate similar pathway segments. All pathways require oxygen as a substrate except the anaerobic pathway, which is present in *T. aromatica* and a few other bacterial isolates. *Burkholderia* sp. strain JS150 carries multiple toluene degradation pathways (indicated by asterisks).

catechol and its derivatives generally serve as central metabolites in aerobic toluene degradation, and several of the pathways converge at a common intermediate (Fig. 1.1). Catechols are then cleaved by either an extradiol (meta) or intradiol (ortho) ring fission reaction, again with oxygen participating as a substrate. In the absence of available oxygen, a completely different degradation strategy must be used. To initiate toluene degradation, anaerobic toluene degraders carry out a unique reaction that adds fumarate to toluene (Fig. 1.1). The product of this reaction, benzylsuccinate, is further metabolized by a series of β-oxidation-like reactions to yield benzoyl-CoA, a central intermediate of anaerobic aromatic degradation. Benzoyl-CoA is then metabolized to TCA cycle intermediates by a pathway that involves ring reduction prior to a series of reactions similar to β-oxidation and ring cleavage. This review compares the details of the known bacterial pathways and a fungal pathway for the complete degradation of toluene.

II. DIOXYGENASE MEDIATED PATHWAY: *PSEUDOMONAS PUTIDA* F1

Pseudomonas putida F1 was isolated based on its ability to grow with ethylbenzene as sole carbon and energy source, although the strain has been mainly studied for its ability to degrade toluene (Gibson *et al.*, 1968a). The complete pathway for toluene degradation in *P. putida* F1 (Fig. 1.2A) has been characterized using a combination of enzymology, analysis of mutants, identification of intermediates, and molecular cloning and sequencing of the structural and regulatory genes. Benzene, toluene, and ethylbenzene serve as growth substrates for *P. putida* F1 (Gibson *et al.*, 1968a, 1970a), and all three substrates were oxidized equally rapidly by ethylbenzene-induced cell suspensions.

A. Pathway details and enzymology

Several possible metabolic intermediates, including catechol, 3-methylcatechol, and benzene *cis*-dihydrodiol were oxidized at high rates, while related compounds (phenol, *o*-, *m*-, and *p*-cresol, 4-methylcatechol, and benzene *trans*-dihydrodiol) were oxidized much more slowly (Gibson *et al.*, 1968a). These results led to a proposed pathway with a *cis*-dihydrodiol as the first intermediate. This idea was supported by the identification of *cis*-2,3-dihydroxy-2,3-dihydro-4-chlorotoluene, which was formed by the dioxygenation of *p*-chlorotoluene by induced *P. putida* F1 cells (Gibson *et al.*, 1968b). The isolation of a *cis*-dihydrodiol dehydrogenase mutant of *P. putida* F1 (strain F39/D), made possible the identification of the initial metabolite of toluene

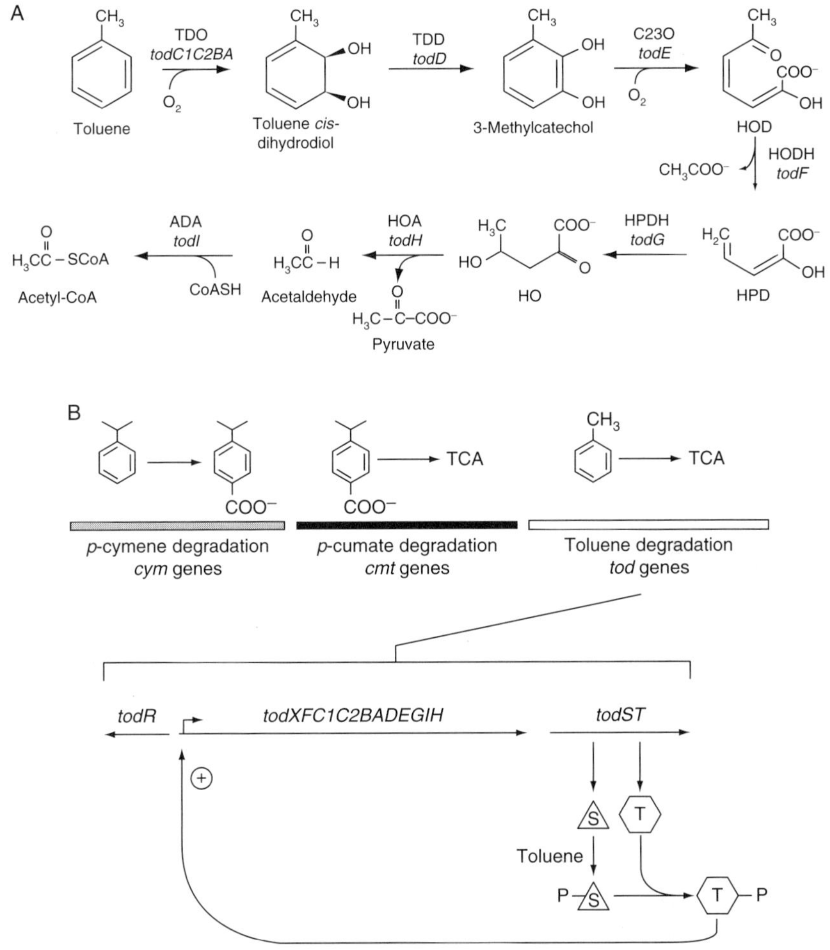

FIGURE 1.2 Dioxygenase-mediated toluene degradation pathway in *P. putida* F1. (A) Pathway sequence indicating intermediates, enzymes, and structural genes. Intermediates: HOD, 2-hydroxy-6-oxohepta-2,4-dienoate; HPD, 2-hydroxypenta-2,4-dienoate; HO, 4-hydroxy-2-oxovalerate. Enzymes: TDO, toluene 2,3-dioxygenase; TDD toluene *cis*-dihydrodiol dehydrogenase; C23O, catechol 2,3-dioxygenase; HODH, 2-hydroxy-6-oxohepta-2,4-dienoate hydrolase (also known as HMSH, 2-hydroxymuconic semialdehyde hydrolase); HPDH, 2-hydroxypent-2,4-dienoate hydratase (also known as OEH, 2-oxopent-4-enoate hydratase); HOA, 4-hydroxy-2-oxovalerate aldolase; ADA, acetaldehyde dehydrogenase (acylating). *P. putida* F1 does not appear to carry genes for the oxalocrotonate branch of the meta pathway (see Section VIII and *xylGHI* in Fig. 1.8A). (B) Gene organization and regulation. The genes for toluene degradation are located in a large catabolic island with genes for *p*-cymene and *p*-cumate degradation. Arrows below

degradation, (+)-cis-1,2-dihydroxy-3-methylcyclohexa-3,5-diene (toluene cis-dihydrodiol), and confirmed the initial reaction in the pathway (Gibson et al., 1970b).

cis-Dihydroxylation of the aromatic ring of toluene and related substrates is the hallmark reaction catalyzed by Rieske nonheme iron oxygenases. Toluene dioxygenase (TDO), which is encoded by the todC1C2BA genes (Zylstra and Gibson, 1989; Zylstra et al., 1988), catalyzes the initial cis-dihydroxylation in the P. putida F1 toluene degradation pathway. The enzyme is a member of a large family of multicomponent Rieske nonheme iron-containing oxygenases (for reviews, see Butler and Mason, 1997; Coulter and Ballou, 1999; Gibson and Parales, 2000; Wackett, 2002), and is composed of three protein components: a flavoprotein reductase (encoded by todA), a Rieske iron–sulfur center-containing ferredoxin (encoded by todB) and the catalytic oxygenase (encoded by todC1C2) (Fig. 1.2). Initial purification of the oxygenase component was by a two-step procedure in which p-toluic acid was used as a ligand for affinity chromatography (Subramanian et al., 1979). Subsequently, the enzyme was purified by standard chromatographic methods (Lee, 1995), and then a monoclonal antibody specific for the TDO β-subunit was used for immunoaffinity purification of both the intact oxygenase and the β-subunit (Lynch et al., 1996). The oxygenase is composed of α- and β-subunits with molecular weights of 52,500 and 20,800, respectively (Subramanian et al., 1979). These values are comparable to those deduced from the sequences of the cloned genes encoding the two subunits (todC1 and todC2; 50,900 and 22,000) (Zylstra and Gibson, 1989). Although the oxygenase was originally predicted to have an $\alpha_2\beta_2$ conformation (Subramanian et al., 1979), it is likely that the enzyme has an $\alpha_3\beta_3$ hexameric structure based on the crystal structures of other related dioxygenases (Dong et al., 2005; Ferraro et al., 2007; Friemann et al., 2005; Furusawa et al., 2004; Gakhar et al., 2005; Kauppi et al., 1998). Absorption and electron paramagnetic resonance spectra, iron and sulfur contents, and conserved sequence motifs indicated the presence of a Rieske [2Fe–2S] center and iron at the active site (Jiang et al., 1996, 1999; Lynch et al., 1996; Subramanian et al., 1979).

---

gene designations indicate the direction of transcription. The expression of the structural genes in the todXFC1C2BADEGIH operon is controlled by the two-component regulatory system TodS (triangle) and TodT (hexagon). TodS is autophosphorylated (the (–P) symbol designates the phosphorylated state of each respective protein) in the presence of toluene and related inducers and transfers its phosphate to TodT. When phosphorylated, TodT binds the tod box upstream of the tod promoter (indicated by ↦) to activate transcription. The ⊕ symbol represents positive regulation of the tod promoter. todR encodes a truncated and presumably nonfunctional LysR-type protein.

The ferredoxin and ferredoxin reductase components of TDO have also been purified and characterized. The reductase is a 42.9 kDa monomer that contains FAD and accepts electrons from NADH (Subramanian et al., 1981). This protein transfers electrons to the ferredoxin component (Subramanian et al., 1981; Zylstra and Gibson, 1989), which is an 11.9 kDa monomer containing a Rieske [2Fe–2S] center (Subramanian et al., 1985; Zylstra and Gibson, 1989). Cross-linking experiments indicated that a ternary complex does not form between the soluble reductase, ferredoxin, and oxygenase components. Rather, the ferredoxin forms a one-to-one complex with the reductase and accepts an electron; it then dissociates and interacts with the oxygenase to transfer the electron to one of its Rieske centers (Lee, 1998).

TDO is capable of oxidizing over 100 substrates, and many of the products are chiral compounds in high enantiomeric purity (reviewed in Boyd and Sheldrake, 1998; Hudlicky et al., 1999). The enzyme has therefore been particularly useful for the chemoenzymatic synthesis of a wide range of complex chiral chemicals of pharmaceutical and industrial interest (reviewed in Brown and Hudlicky, 1993; Carless, 1992; Hudlicky et al., 1999; Sheldrake, 1992). In addition, TDO is one of several bacterial oxygenases with the ability to convert indole to the blue dye indigo (Gibson et al., 1990), and it also oxidizes trichloroethylene (TCE) to formic acid and glyoxylic acid, although TCE is not used as a growth substrate by strain F1 (Li and Wackett, 1992; Zylstra et al., 1989).

Toluene *cis*-dihydrodiol dehydrogenase (TDD; encoded by *todD*), catalyzes the second reaction in the toluene degradation pathway, converting toluene *cis*-dihydrodiol to 3-methylcatechol (Gibson et al., 1970b). TDD is a homotetramer composed of 27 kDa subunits that requires NAD^+. It has a wide substrate range, catalyzing the formation of the corresponding catechols from *cis*-dihydrodiols of toluene, benzene, biphenyl, phenanthrene, and anthracene (Rogers and Gibson, 1977). The enzyme did not appear to differentiate between the optical isomers of naphthalene *cis*-dihydrodiol (Rogers and Gibson, 1977), but it was specific for the (+)-enantiomers of biphenyl 2,3- and biphenyl 3,4-dihydrodiol (Parales et al., 2000). *P. putida* F39/D, a mutant strain that accumulates toluene *cis*-dihydrodiol, was shown to be specifically defective in TDD activity (Finette and Gibson, 1988).

Aromatic ring cleavage proceeds via a meta cleavage pathway in *P. putida* F1 (Fig. 1.2A). 3-Methylcatechol is oxidized by catechol 2,3-dioxygenase (C23O; encoded by *todE*), forming 2-hydroxy-6-oxohepta-2,4-dienoate (Klecka and Gibson, 1981). The mutant strain *P. putida* F107 was specifically defective in this extradiol dioxygenase activity, and accumulated 3-methylcatechol when grown in the presence of toluene (Finette and Gibson, 1988). The substrate range of TodE also included catechol, 4-methylcatechol and 4-fluorocatechol, but the enzyme was inhibited by

3-chlorocatechol and 4-chlorocatechol (Klecka and Gibson, 1981). TodE is a member of the I.3 Family of extradiol dioxygenases, which are two-domain iron-containing enzymes that generally oxidize bicyclic aromatic substrates (Eltis and Bolin, 1996). Therefore, TodE is different in its substrate preferences compared with closely related meta cleavage dioxygenases.

The next enzyme in the pathway, 2-hydroxy-6-oxohepta-2,4-dienoate hydrolase (HODH; encoded by *todF*), was purified and the N-terminal sequence of the protein was used to clone the *todF* gene (Menn et al., 1991), which had been localized upstream of *todC1* based on transposon mutagenesis studies (Zylstra et al., 1988). The remaining enzymes of the meta cleavage pathway from P. putida F1 have not been characterized in detail. The *todGIH* genes, which are located downstream of *todE*, were cloned and sequenced (Lau et al., 1994). Functions were assigned based on sequence compared with the enzymes of the phenol/dimethylphenol (*dmp*) pathway in *Pseudomonas putida* CF600 (Shingler et al., 1992), with *todG* encoding 2-hydroxypenta-2,4-dienoate hydratase, *todH* encoding 4-hydroxy-2-oxovalerate aldolase, and *todI* encoding acetaldehyde dehydrogenase (acylating) (Fig. 1.2A). The reactions catalyzed appear to be the same as those determined for the TOL plasmid lower pathway in *P. putida* mt-2, although genes for the oxalocrotoanate branch of the pathway (*xylGHI*) are not present (Fig. 1.3; see Section VIII.A for a discussion of the role of the two meta pathway branches).

B. Genetics and regulation

The toluene degradation (*tod*) genes from *P. putida* F1 have been cloned and sequenced (Lau et al., 1994, 1997; Menn et al., 1991; Wang et al., 1995; Zylstra and Gibson, 1989, 1991; Zylstra et al., 1988). No plasmids have been detected in *P. putida* F1, and the genes encoding the toluene degradation pathway were expected to be located on the chromosome (Finette and Gibson, 1988; Finette et al., 1984). This assumption was supported by the complete genome sequence of *P. putida* F1, which confirmed the absence of plasmids (http://genome.jgi-psf.org/finished_microbes/psepu/psepu.home.html). The *tod* gene cluster (Figs. 1.2B and 1.3) is located within a 46-kb catabolic island adjacent to genes for the degradation of *p*-cymene (*cym*) and *p*-cumate (*cmt*) (Eaton, 1996; Eaton, 1997). A phage-like integrase gene (*int-F1*) was found downstream of the *tod* genes, which may indicate that an insertion event occurred in this region of the *P. putida* F1 genome (Phoenix et al., 2003). The GC-rich repeat sequences located upstream and downstream of the *p*-cymene/*p*-cumate degradation gene cluster (Eaton, 1997) and the presence of the truncated *todR* gene upstream (Figs. 1.2B and 1.3) of the *tod* operon (Wang et al., 1995) provide

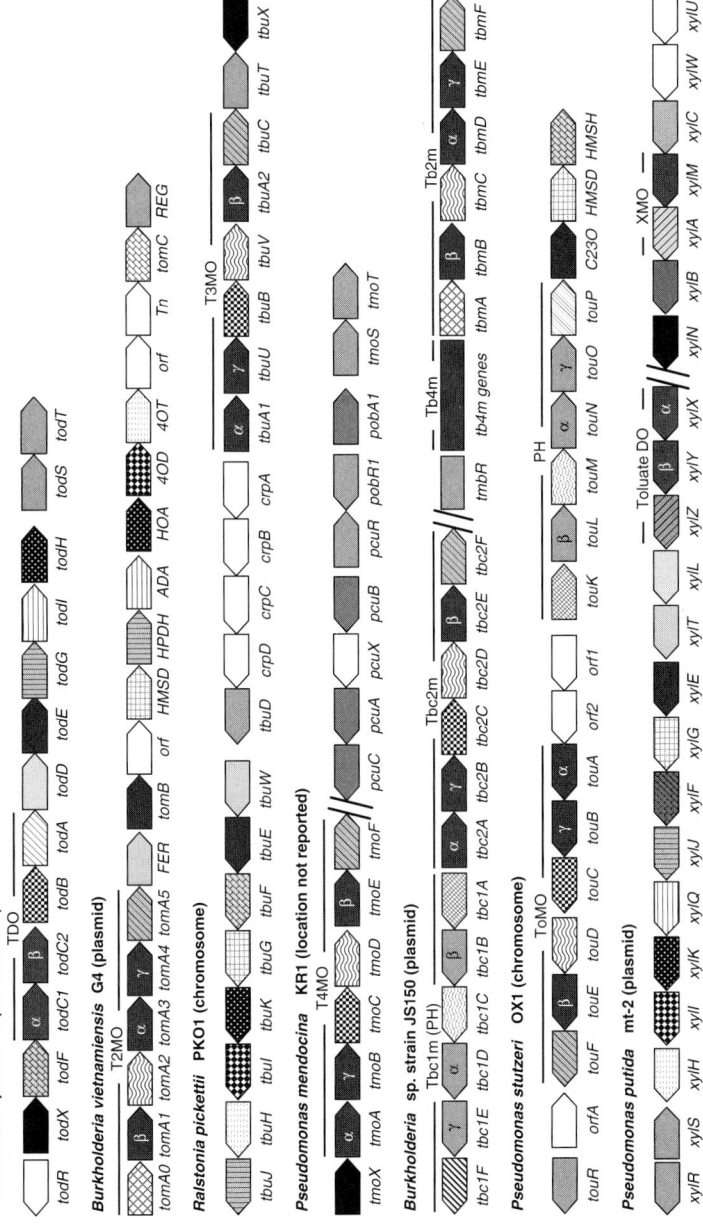

FIGURE 1.3 Comparison of toluene degradation gene clusters in aerobic bacteria. Gene locations (plasmid or chromosome) are indicated. White boxes indicate open reading frames (*orf*) without known function, transposons (Tn), or genes not required for the pathway. Colored boxes indicate orfs involved in toluene degradation pathways. Genes are not drawn to scale. Arrowheads indicate the direction of transcription; genes drawn as rectangles indicate that the direction of transcription is not known. Two diagonal black lines indicate that gene

additional suggestive evidence that insertion or recombination events contributed to the evolution of this large catabolic island.

The *todFC1C2BADEGIH* cluster encodes all of the steps in toluene degradation as described above, and the genes appear to be cotranscribed in a single operon (Wang et al., 1995). Two additional genes are present upstream of *todF*: *todX*, the first gene in the *tod* operon, and the divergently transcribed *todR* (Figs. 1.2B and 1.3). The *todR* gene apparently encodes a truncated and nonfunctional LysR-type regulatory protein that plays no role in toluene degradation (Wang et al., 1995). TodX, a 453 amino acid protein with a putative 20-amino acid N-terminal signal sequence and a single transmembrane domain, was shown to associate with the membrane fraction when expressed in *Escherichia coli*. TodX has homology with the *E. coli* outer membrane protein FadL, which is involved in uptake of long-chain fatty acids (Black, 1991). Inactivation of *todX* resulted in a strain that grew more slowly than wild type at low toluene concentrations, suggesting that TodX may facilitate entry of toluene into the cell (Wang et al., 1995). Homologues of *todX* have been identified in other aromatic hydrocarbon-degradation gene clusters from Gram-negative bacteria, suggesting that genes encoding specialized

---

clusters are not linked. The color scheme is as follows: Blue, dioxygenase genes; light blue, *cis*-dihydrodiol dehydrogenase genes; red, multicomponent toluene monooxygenase genes; grey, phenol/cresol hydroxylase genes; brown, xylene monooxygenase genes; tan, alcohol dehydrogenase gene; light brown, aldehyde dehydrogenase gene; orange, *p*-cresol utilization genes; purple, *meta* pathway genes; lavender, plant-like ferredoxin genes; black, outer membrane channel genes; green, regulatory genes. The pattern scheme for the upper pathway genes is as follows: solid, catalytic oxygenase or hydroxylase subunits (labeled α,β,γ as appropriate); diagonal hashed, reductase components; checkerboard, ferredoxin components; zigzag, effector components; open diamonds, assembly proteins. The genes *tomA0*, *touK*, *tbmA*, and *tbc1A* were designated as assembly proteins based on sequence homology to *dmpK* from *P. putida* CF600 (Powlowski et al., 1997). Upper pathway enzyme designations are indicated above the corresponding genes. XMO, xylene monooxygenase; toluate DO, toluate dioxygenase. The pattern scheme for the *meta* pathway genes is as follows: solid, C23O; diagonal bricks, HMSH (also known as HODH); vertical lines, hydratase (OEH/HPDH); dotted, aldolase (HOA); grid, HMSD; closed diamonds, decarboxylase (4OD); horizontal lines, acetaldehye dehydrogenase (acylating) (ADA); horizontal dashed lines, tautomerase (4OT). Putative functions of the *B. vietnamiensis* G4 *meta* pathway genes were deduced by BLAST searches with the genome sequence data (http://img.jgi.doe.gov/cgi-bin/pub/main.cgi?section=TaxonDetail&page=taxonDetail&taxon_oid=640069307) and are labeled with the enzyme abbreviation as gene name have not been assigned. The gene labeled "FER" appears to encode a plant-type ferredoxin similar to XylT from *P. putida* mt-2; the gene labeled "REG" encodes a XylR/NtrC family regulator with similarity to TbuT from *R. pickettii* PKO1. (See Color Plate Section in the back of the book.)

outer membrane proteins are coordinately expressed with degradation genes. For example, TbuX from *Ralstonia pickettii* PKO1 (see Section IV) is 34% identical to TodX, and it appears to be involved in toluene entry into *R. pickettii* PKO1 cells (Kahng *et al.*, 2000). XylN, encoded by the last gene in the upper operon of the TOL plasmid (Harayama *et al.*, 1989b), shares 38% identity in amino acid sequence with TodX and was localized to the outer membrane of *P. putida* mt-2 (see Section VIII).

Initial experiments indicated that the *tod* genes in *P. putida* F1 were coordinately induced in the presence of toluene and other volatile aromatic compounds, and surprisingly, TCE (Finette and Gibson, 1988; Heald and Jenkins, 1994; Shingleton *et al.*, 1998; Wackett, 1984). Transcriptional *tod–lacZ* and *tod–lux* fusions, and C23O assays have identified a broad range of compounds capable of inducing the *tod* pathway (Applegate *et al.*, 1997, 1998; Cho *et al.*, 2000; Lacal *et al.*, 2006). Two proteins, TodS and TodT, form a two-component sensor and response regulator signal transduction system that is required for activation of the *tod* operon in the presence of toluene. The *todST* operon was identified by analysis of sequence downstream of the *todXFC1C2BADEGIH* genes (Fig. 1.2B). TodS and TodT mutants were unable to grow with toluene and lacked TDO or C23O activity (Lau *et al.*, 1997). The TodS–TodT two-component systems from both *P. putida* F1 and *P. putida* DOT–T1E (Ramos *et al.*, 1995), a solvent resistant strain with an identical toluene degradation pathway, have been studied (Lacal *et al.*, 2006; Lau *et al.*, 1997). TodS is a large (978 amino acid) and complex sensor kinase that contains an N-terminal basic leucine zipper motif, two histidine kinase domains, two PAS/PAC domains, and a receiver domain of a response regulator (Lau *et al.*, 1997). The sequence of TodS indicates that it does not contain a transmembrane domain and it appears to be localized to the cytoplasm. TodT, a response regulator with a C-terminal DNA-binding domain, was shown to bind to two *tod* boxes, which are conserved inverted repeats centered 107 and 85 bases upstream of the *todX* transcription start site (Lacal *et al.*, 2006; Lau *et al.*, 1997). Toluene was shown to bind with high affinity in a 1:1 ratio to the N-terminal sensor domain of TodS, but not the C-terminal sensor histidine kinase domain. In addition, TodS was shown to undergo autophosphorylation, which was stimulated in the presence of toluene, and its phosphate could be transferred to TodT *in vitro* (Lacal *et al.*, 2006). When the conserved aspartate residue at position 56 in TodT (which was predicted to be the site of phosphorylation) was changed to asparagine or alanine, the mutant proteins were still able to bind the *tod* box but did not stimulate transcription from the *todX* promoter (Lacal *et al.*, 2006; Lau *et al.*, 1997). The participation of integration host factor (IHF) in stimulating transcription and the distant location of the *tod* box relative to the *todX* promoter suggested that DNA bending occurs during transcription activation (Lacal et al., 2006). The role of the second input domain and the reason for the extreme complexity of TodS are not yet understood.

III. TOLUENE 2-MONOOXYGENASE PATHWAY: *BURKHOLDERIA VIETNAMIENSIS* G4

Burkholderia vietnamiensis G4 (formerly *Pseudomonas cepacia*; *Burkholderia cepacia*) was originally isolated from a holding pond at an industrial waste treatment facility because of its ability to oxidize TCE (Nelson et al., 1987). Although TCE was mineralized by strain G4, it did not serve as a growth substrate. Growth substrates for strain G4 included toluene, phenol, *o*-, *m*-, and *p*-cresol, and benzoate.

A. Pathway details and enzymology

It was originally hypothesized that toluene degradation in *B. vietnamiensis* G4 was similar to that in *P. putida* F1 (Nelson et al., 1987). However, analysis of mutants that could not metabolize toluene made it clear that a monooxygenase catalyzed the two initial oxidation steps in toluene degradation in *B. vietnamiensis* G4 (Fig. 1.4) (Shields et al., 1989; Shields et al., 1991).

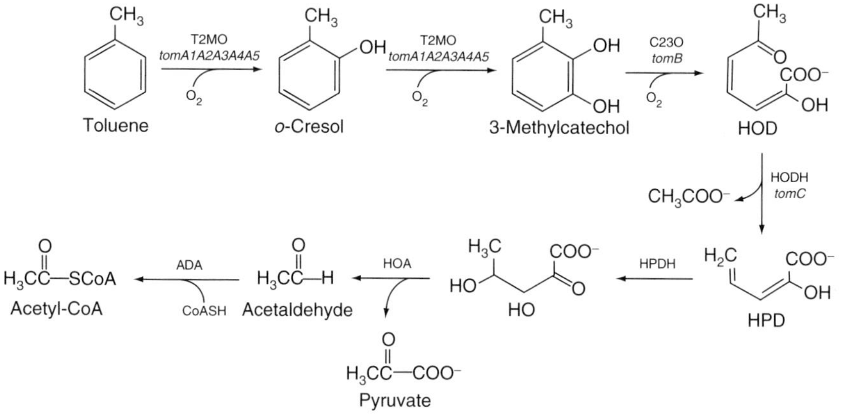

FIGURE 1.4 Toluene 2-monooxygenase-mediated toluene degradation pathway in *B. vietnamiensis* G4. Pathway sequence indicating intermediates, enzymes, and genes. Intermediates: HOD, 2-hydroxy-6-oxohepta-2,4-dienoate; HPD, 2-hydroxypenta-2, 4-dienoate; HO, 4-hydroxy-2-oxovalerate. Enzymes: T2MO, toluene 2-monooxygenase; C23O, catechol 2,3-dioxygenase; HODH, 2-hydroxy-6-oxohepta-2,4-dienoate hydrolase (also known as HMSH); HPDH, 2-hydroxypent-2,4-dienoate hydratase (also known as OEH); HOA, 4-hydroxy-2-oxovalerate aldolase; ADA, acetaldehyde dehydrogenase (acylating).

Three classes of *B. vietnamiensis* G4 mutants were isolated by nitrosoguanidine mutagenesis. The TOMA⁻ class (represented by strain G4 100) lacked monooxygenase (T2MO) activity but could still catabolize catechol or 3-methylcatechol. The TOMB⁻ class (represented by strain G4 102) lacked C23O activity but retained T2MO and downstream meta cleavage pathway activities. The TOMC⁻ class (represented by strain G4 103) was defective in the step following meta cleavage (HODH) (Shields *et al.*, 1991).

T2MO is a three-component enzyme system that is comprised of the catalytic hydroxylase, a flavo-iron–sulfur reductase, and a small (10.5 kDa) protein (Newman and Wackett, 1995). Denaturing gel electrophoresis demonstrated that the hydroxylase holoenzyme (total weight approximately 211 kDa) was comprised of three subunits of 54.5, 37.7, and 13.5 kDa each, suggesting a $\alpha_2\beta_2\gamma_2$ quaternary structure. The hydroxylase component is encoded by the genes *tomA3*, *tomA1*, and *tomA4*, respectively (Shields and Francesconi, 1996). Iron and zinc were present at ratios of 4.9 and 2.0 nmol/nmol protein respectively, and the hydroxylase was determined to contain two diiron centers based on EPR spectroscopy. The C23O mutant G4 102 was found to accumulate 3-methylcatechol, but *o*-cresol was detected as a transient product, indicating that 3-methylcatechol was produced by sequential hydroxylations of the aromatic ring (Shields *et al.*, 1989). T2MO was proposed to catalyze the formation of both of these products, first by an ortho and then a meta hydroxylation as determined by ^{18}O isotope incorporation experiments (Shields *et al.*, 1989). Single turn-over experiments confirmed that *o*-cresol was the first intermediate produced prior to the second reaction that converted *o*-cresol to 3-methylcatechol. *o*-Cresol strongly inhibited the oxidation of radiolabeled toluene, suggesting that the same active site bound both substrates. These data, along with tight binding of the hydroxylase component to a toluene affinity column, H_2O_2-dependent toluene oxidation activity and oxidation of toluene by chemically reduced purified enzyme provided evidence that the hydroxylase contained the monooxygenase active site and was responsible for the sequential oxidations of toluene to *o*-cresol and *o*-cresol to 3-methylcatechol (Newman and Wackett, 1995).

The 10.5 kDa component of T2MO, encoded by *tomA2* (Shields and Francesconi, 1996), was shown to dimerize based on gel filtration chromatography. The protein lacked any identifiable organic cofactors or metals, suggesting that it was not responsible for the shuttling of electrons between the reductase and hydroxylase components of the enzyme system. Interestingly, this effector protein was not required for the oxidation of toluene, but its presence increased the rate of toluene oxidation by about 10-fold (Newman and Wackett, 1995).

The reductase component, which is encoded by the gene *tomA5* (Shields and Francesconi, 1996), is a 40-kDa protein that contains 1.2 mol of FAD per mol protein and 2.2 and 2.9 moles of iron and sulfur per mole protein respectively, suggesting one [2Fe–2S] center per protein. The reductase was shown to oxidize NADH and transfer electrons to the hydroxylase component, and was present in purified extracts at approximately half the molar ratio of the hydroxylase component.

This monooxygenase was very different from other multicomponent oxygenases that had been previously described, such as TDO, which consists of a Rieske-type oxygenase that receives electrons from reductase and ferredoxin components (see Section II.A). This T2MO enzyme has more similarity to the diiron protein family and methane monooxygenase (MMO) systems in various methanotrophs (Newman and Wackett, 1995; Wilkins et al., 1994).

In addition to toluene, o-cresol, and TCE, T2MO catalyzes the oxidation of dichloroethylenes, phenol, chloroform, 1,4-dioxane, aliphatic ethers, and diethyl sulfide (Hur et al., 1997; Mahendra and Alvarez-Cohen, 2006; Shim and Wood, 2000). The enzyme was also shown to catalyze the formation of epoxides from a variety of alkene substrates (McClay et al., 2000).

Ring cleavage in *B. vietnamiensis* G4 proceeds via a meta cleavage pathway, whereby 3-methylcatechol is oxidized by C23O and the ring cleavage product is a substrate for HODH (Fig. 1.4). C23O enzyme assays demonstrated the formation of 2-hydroxy-6-oxohepta-2,4-dienoate from 3-methylcatechol (Shields et al., 1989), and a TOMC$^-$ strain (G4 103) accumulated this compound in cell extracts. Enzyme assays demonstrated that G4 103 lacked HODH activity (Shields et al., 1991).

B. Genetics and pathway regulation

The ability of *B. vietnamiensis* G4 to oxidize TCE was dependent upon the presence of an unknown compound that was present in the water from the original site of isolation (Nelson et al., 1986). That compound was identified as phenol; however, stimulation of the TCE-degrading activity was also demonstrated with toluene, o-cresol, and m-cresol, suggesting that the toluene degradation pathway in *B. vietnamiensis* G4 is inducible (Nelson et al., 1987). All of the toluene degradation genes in *B. vietnamiensis* G4 (Fig. 1.3) are located on a large, self-transmissible plasmid (TOM) (Shields and Francesconi, 1996; Shields et al., 1995). G4 variants cured of the TOM plasmid lost the ability to oxidize TCE and use toluene as a source of carbon and energy. Transfer of TOM to a cured strain resulted in transconjugants expressing inducible toluene degradation. Strain PR1$_{23}$ was isolated as a strain that constitutively expressed T2MO and C23O activity (Shields and Reagin, 1992). When a cured strain was mated with the constitutive strain, the transconjugants acquired the ability to degrade toluene constitutively, suggesting that the control elements for toluene degradation are maintained on the TOM plasmid (Shields et al., 1995).

The genome of *B. vietnamiensis* G4 was completely sequenced and the organism was found to carry three chromosomes and five plasmids, including TOM (http://img.jgi.doe.gov/cgi-bin/pub/main.cgi?section= TaxonDetail&page=taxonDetail&taxon_oid=640069307). The complete sequence of the TOM catabolic plasmid (designated pBriE04 in the genome data) revealed the presence of meta cleavage pathway genes downstream of the genes encoding T2MO (Fig. 1.3). A search of the complete genome did not reveal the presence of any other C23O genes, suggesting that a single meta pathway is present in *B. vietnamiensis* G4.

Although T2MO activity was shown to be inducible in *B. vietnamiensis* G4, the *cis*- and *trans*-regulatory elements required for this regulation have not yet been described. However, genome sequence analysis revealed the presence of a gene encoding a XylR/NtrC family regulator downstream of the gene cluster encoding T2MO and the meta cleavage enzymes (Fig. 1.3). The gene encodes a protein of 611 amino acids with 53% sequence identity to the *R. pickettii* PKO1 TbuT protein (see IV), 48% identity to the *Pseudomonas mendocina* KR1 PcuR protein (see V), 45% identity to TouR from *Pseudomonas stutzeri* OX1 (see VII.B), and 44% identity to DmpR from *P. putida* CF600 (O'Neill *et al.*, 1998). This sequence similarity suggests that the putative regulatory protein functions to turn on transcription of the *tom* genes in *B. vietnamiensis* G4 in the presence of toluene and related inducers. C23O activity was found to be expressed in *B. vietnamiensis* G4 under similar inducing conditions as T2MO (Nelson *et al.*, 1987), suggesting that the genes for T2MO and the meta pathway are coordinately regulated, possibly by the same XylR/NtrC family regulatory protein (Fig. 1.3).

IV. TOLUENE-3-MONOOXYGENASE PATHWAY: *RALSTONIA PICKETTII* PKO1

R. pickettii PKO1 (formerly *Pseudomonas pickettii* PKO1) was isolated from a sandy aquifer following incubation with benzene, toluene, and *p*-xylene. The strain used benzene, toluene, ethylbenzene, phenol, and *m*-cresol as carbon and energy sources, but was unable to utilize *o*- or *p*-cresol or any of the isomers of xylene (Kukor and Olsen, 1990a). The genes for toluene degradation were originally isolated from *R. pickettii* PKO1 on a 26-kb chromosomal fragment (pRO1957) that allowed growth of the heterologous host *Pseudomonas aeruginosa* PAO1 on phenol, toluene, and benzene as sole carbon sources (Kukor and Olsen, 1990b). Understanding the initial attack of toluene in *R. pickettii* PKO1 came from expression experiments with pRO1957 subclones in *P. aeruginosa* PAO1 (Kukor and Olsen, 1990b). In these studies, the accumulation of *m*-cresol from toluene was detected by HPLC. $^{18}O_2$ incorporation experiments demonstrated that one atom of oxygen was introduced into the resulting cresol product, indicating a monooxygenase attack (Olsen et al., 1994). Also consistent with this series of reactions (Fig. 1.5A) was the finding that benzene, toluene, ethylbenzene, phenol, and m-cresol induced expression of the monooxygenase, but p-cresol did not (Kukor and Olsen, 1990b; Olsen et al., 1994). This study also reported the formation of 2,4-dimethylphenol from m-xylene by the enzyme, which would be unexpected for a true "meta oxidizing" enzyme (Olsen et al., 1994). In contrast to these early results, recent work has suggested that the T3MO from R. pickettii PKO1 is actually a para hydroxylating enzyme. The formation of p- and m-cresol in a 9:1 ratio from toluene by the recombinant enzyme was demonstrated by gas chromatography and NMR analysis (Fishman et al., 2004c). The recombinant T3MO was also shown to oxidize several other substrates at the para position, and formed a 2:1 mixture of p- and m-nitrophenol from nitrobenzene (Fishman et al., 2004c). A site-directed mutagenesis study identified a T3MO variant with two amino acid substitutions that preferentially formed m-cresol from toluene (Fishman et al., 2004b). At this time the reports are in direct conflict with respect to the initial position of attack on toluene by T3MO from R. pickettii PKO1 and the actual pathway remains controversial (Fig. 1.5B). However, if T3MO preferentially catalyzes the formation of p-cresol, it seems odd that this isomer of cresol is not a growth substrate for R. pickettii PKO1, but m-cresol is readily utilized. Similarly, the observation that m-cresol but not p-cresol is capable of inducing the monooxygenase genes is somewhat troubling. Regardless, the pathway in R. pickettii PKO1 is unique with respect to the T2MO and T4MO pathways in B. vietnamiensis G4 (see Section III.A) and P. mendocina KR1 (see Section V), as G4 clearly oxidizes toluene at the ortho position and uses the meta ring-cleavage pathway (Kukor and Olsen, 1991), while KR1 oxidizes toluene at the para position and uses the β-ketoadipate (ortho) pathway for ring cleavage (Fig. 1.1). In contrast, as described in more detail later, R. pickettii PKO1 utilizes a monooxygenase attack on the ring (either at the meta or para position) followed by meta cleavage (Figs. 1.1 and 1.5B).

Substrates for T3MO besides toluene included benzene (converted to phenol), ethylbenzene (converted to 3- and 4-ethylphenol), o-xylene (converted to 3,4-dimethylphenol), m-xylene (converted to 2,4-dimethylphenol), and p-xylene (converted to 2,5-dimethylphenol); however, p-xylene and ethylbenzene were transformed with about one-tenth of the activity (Olsen et al., 1994). T3MO was later shown to catalyze the formation of epoxides from a variety of alkenes (McClay et al., 2000), and oxidized other diverse substrates including TCE, 1,4-dioxane, nitrobenzene, and N-nitrosodimethylamine (Haigler and Spain, 1991; Leahy et al., 1996; Mahendra and Alvarez-Cohen, 2006; Sharp et al., 2005).

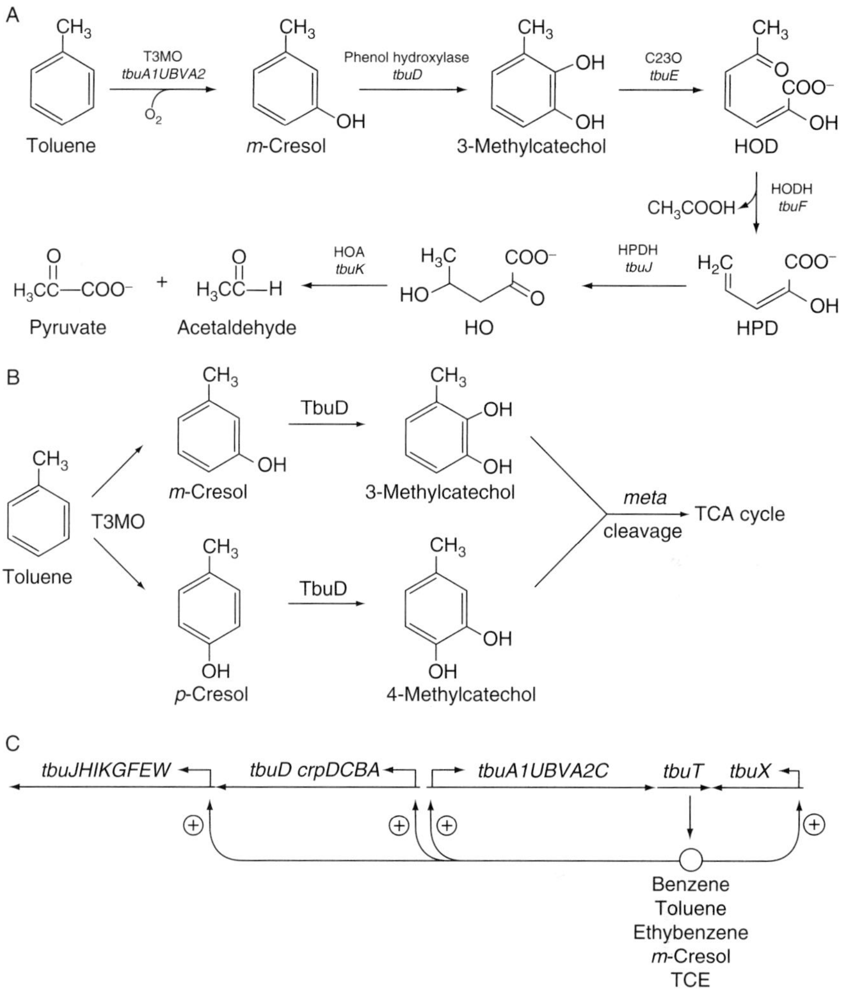

FIGURE 1.5 Toluene 3-monooxygenase-mediated toluene degradation pathway in R. pickettii PKO1. (A) Pathway sequence indicating intermediates, enzymes, and genes. Intermediates: HOD, 2-hydroxy-6-oxohepta-2,4-dienoate; HPD, 2-hydroxypenta-2, 4-dienoate; HO, 4-hydroxy-2-oxovalerate. Enzymes: T3MO, toluene 3-monooxygenase; C23O, catechol 2,3-dioxygenase; HODH, 2-hydroxy-6-oxohepta-2,4-dienoate hydrolase (also known as HMSH); HPDH, 2-hydroxypent-2,4-dienoate hydratase (also known as OEH); HOA, 4-hydroxy-2-oxovalerate aldolase. (B) R. pickettii PKO1 pathway controversy. At this time it is not clear whether the initial reaction is hydroxylation at the 3-position or the 4-position of toluene. See text for details. (C) Gene organization and regulation. Arrows below gene designations represent the direction of transcription. The ⊕ symbol represents positive regulation of the respective promoters (indicated by ↦ or ↤) by TbuT (open circle) in the presence of effector molecules.

Protein extracts from recombinant *P. aeruginosa* PAO1 expressing T3MO activity showed the presence of three unique protein bands compared to the control (Olsen *et al.*, 1994). Subsequent sequence analysis of the T3MO-encoding region identified the presence of a total of six structural genes, *tbuA1UBVA2C* (Byrne *et al.*, 1995) (Figs. 1.3 and 1.5C). The conserved sequence motif responsible for diiron binding was found in TbuA1, and both TbuA1 and TbuA2 were shown to have homology to the oxygenase components of the T4MO multicomponent monooxygenase from *P. mendocina* KR1. Therefore, TbuA1 and TbuA2 were assigned as the hydroxylase subunits for T3MO. TbuB was found to contain the conserved cysteine and histidine residues present in Rieske-type iron–sulfur centers, and TbuC contained sequences homologous with FAD binding and a [2Fe–2S] cluster, suggesting these two proteins functioned as the ferredoxin and reductase components of the T3MO respectively. TbuU and TbuV shared homology with the TmoB and TmoD proteins of *P. mendocina* KR1 (see Section V) (Byrne *et al.*, 1995). Based upon sequence homology, *tbuU* appears to encode the γ subunit of T3MO and TbuV most likely functions as an effector protein to stimulate T3MO activity.

The gene encoding phenol/cresol hydroxylase (*tbuD*; formerly *phlA*) was identified by expressing subclones of pRO1957 in the presence of a *xylE* clone (encoding C23O from the TOL plasmid pWWO). *P. aeruginosa* PAO1 cells containing both *tbuD* and *xylE* constructs grown in the presence of phenol accumulated a yellow product, indicating that 2-hydroxymuconate semialdehyde was produced from catechol (Kukor and Olsen, 1990b). TbuD has similarity to FAD-containing aromatic hydroxylases (Kukor and Olsen, 1992). Purified TbuD protein from recombinant *P. aeruginosa* PAO1 was used in *in vitro* experiments that identified the preferred electron donor for this enzyme as NADPH. The substrates for purified TbuD included *m*-, *p*-, and *o*-cresol (in order of highest to lowest activity), while the enzyme had limited activity with catechol, resorcinol, and all isomers of fluoro-, chloro-, and aminophenols. In general, *m*- and *p*-substituted phenols were most readily attacked over the *ortho* isomers. The products of TbuD oxidation of the cresol isomers were mixtures of 3- and 4-methylcatechol for *m*-cresol, 3-methylcatechol for *o*-cresol, and 4-methylcatechol for *p*-cresol (Kukor and Olsen, 1992).

Genes for the meta cleavage pathway were identified on fragments of pRO1957 downstream of *tbuD* (Kukor and Olsen, 1991). Deletion analysis and enzyme assays identified the order of the *tbu* genes for meta cleavage as *tbuEFGKIHJ*, encoding C23O, 2-hydroxymuconate semialdehyde hydrolase, 2-hydroxymuconate semialdehyde dehydrogenase, 4-hydroxy-2-oxovalerate aldolase, 4-oxalocrotonate decarboxylase, 4-oxalocrotonate tautomerase, and 2-hydroxypent-2,4-dienoate hydratase, respectively (Figs. 1.3 and 1.5C). These genes were in a single operon driven by a promoter upstream of *tbuE* (Kukor and Olsen, 1991).

The genes that encode T3MO were shown to be regulated by the product of the *tbuT* locus, which is located just downstream of the *tbuA1UBVA2C* gene cluster (Figs. 1.3 and 1.5C) (Olsen *et al.*, 1994). TbuT showed homology with the NtrC family regulators XylR and DmpR (Byrne and Olsen, 1996), and a promoter with homology to σ^{54}-regulated promoters was identified just upstream of *tbuA1UBVA2C*, which is consistent with regulation by TbuT (Byrne *et al.*, 1995). Expression of genes encoding T3MO in *P. aeruginosa* PAO1 was induced in the presence of toluene, benzene, ethylbenzene, TCE, phenol, and *o*- and *m*-cresol (Byrne and Olsen, 1996). Interestingly, the *tbuT* gene was shown to lack its own promoter, and a transcriptional terminator was identified just downstream of *tbuT*, suggesting that *tbuT* is transcribed from the *tbuA1* promoter (*PtbuA1*) (Byrne and Olsen, 1996). It was determined that transcription of *tbuT* was due to read-through from *PtbuA1* based on β-galactosidase reporter fusions. Therefore, toluene-induced expression of *PtbuA1* leads to increased expression of T3MO and TbuT, resulting in a positive feedback circuit. Although the regiospecificities of T3MO of *R. pickettii* PKO1 and the toluene–benzene-2-monooxygenase (Tb2m) of *Burkholderia* sp. strain JS150 differ (see Section VI), the genetic organization of these two operons is similar in that the regulatory genes (encoding TbuT and TbmR respectively) are each located at the end of the monooxygenase operon (Fig. 1.3) (Leahy *et al.*, 1997). In assays conducted in *P. aeruginosa* PAO1, each regulator was shown to activate its own and the reciprocal promoter sequence with the same set of effector molecules, suggesting that distinct monooxygenase pathways can be regulated by structurally similar regulators with the same range of effector molecules (Leahy *et al.*, 1997).

Initial genetic analyses of the *tbuD* and *tbuEFGKIHJ* genes (encoding phenol/cresol monooxygenase and enzymes of the meta cleavage pathway, respectively) suggested that two additional trans-acting regulatory factors (*tbuR* and *tbuS*) functioned to control the expression of these genes. Expression of *tbuD* and *tbuEFGKIHJ* appeared to be induced by phenol when a DNA fragment located between the *tbuD* and *tbuA1* genes was provided *in trans* in a heterologous *P. aeruginosa* host (Kukor and Olsen, 1990b; Kukor and Olsen, 1991). On the basis of these results, a complex regulatory cascade mechanism involving three different regulatory proteins controlling the three *tbu* operons was proposed (Olsen *et al.*, 1994). However, subsequent sequence analysis revealed the absence of any obvious regulatory genes in the predicted region, and results of *lacZ* fusion analyses indicated that *tbuD* is transcribed from a promoter located ~2.5 kb upstream of the translational start site (Fig. 1.5C) (Olsen *et al.*, 1997). This unusual promoter location can partially explain the requirement for the upstream region for inducible expression of *tbuD* (Kukor and Olsen, 1990b), but how it could function *in trans* is not clear. In addition, the 1997 study (Olsen *et al.*, 1997) implicated TbuT as the activator of the *tbuD* and *tbuE* promoters, a conclusion that is difficult to reconcile with the earlier genetic data. The 2.5-kb region between *tbuD* and *tbuA1* appeared to encode cryptic genes from an ancestral multicomponent monooxygenase very similar to the toluene 2-monooxygenase (Tb2m) from *Burkholderia* sp. strain JS150 (see Section VI), and led to the hypothesis that the single component phenol/cresol monooxygenase TbuD evolved from a multicomponent toluene monooxygenase (Olsen *et al.*, 1997). Several gaps remain in our understanding of the control of gene expression in the *tbu* cluster in *R. pickettii* PKO1, but the current model is shown in Fig. 1.5C.

In addition to the genes required for toluene degradation, transport of toluene into the cell has been shown to be important for toluene degradation in *R. pickettii* PKO1 (Kahng *et al.*, 2000). Located just downstream of the *tbuA1UBVA2C–tbuT* operon was an open reading frame (*tbuX*; Figs. 1.3 and 1.5C) that had homology to FadL, an outer membrane protein known to be important for long-chain fatty acid transport in *E. coli*. As with the other *tbu* operons, the *tbuX* gene was found to have its own σ^{54}-regulated promoter and a *tbuX–lacZ* fusion was induced in the presence of TbuT and toluene or other effector molecules. The phenotype of a *tbuX* deletion was assessed in the *P. aeruginosa* PAO1 heterologous host system. This strain was significantly impaired in its ability to use toluene as a growth substrate. In addition, T3MO peptides were not synthesized and the formation of *m*-cresol could not be detected, suggesting that toluene did not accumulate, thereby leaving these cells uninduced for T3MO and meta pathway gene expression (Kahng *et al.*, 2000). Together, these results suggest that TbuX plays a role in toluene passage across the outer membrane as proposed for TodX in *P. putida* F1 (see Section II.A).

V. TOLUENE-4-MONOOXYGENASE PATHWAY: *PSEUDOMONAS MENDOCINA* KR1

P. mendocina KR1 was isolated from an algal-bacterial mat from the Colorado River in Austin, TX following incubation in the presence of toluene and chlorocatechol. This strategy took advantage of the observation that 3-chlorocatechol inhibits C23O (Klecka and Gibson, 1981), and thus organisms that degraded toluene through a meta cleavage pathway would be inhibited, but strains that used an ortho pathway would survive (D. T. Gibson, personal communication). The resulting isolate carried a unique toluene degradation pathway in which toluene is first oxidized at the para position to form *p*-cresol, followed by oxidation of the methyl group to form 4-hydroxybenzoate. A second ring hydroxylation generates protocatechuate, which is a substrate for ortho cleavage (Fig. 1.6A).

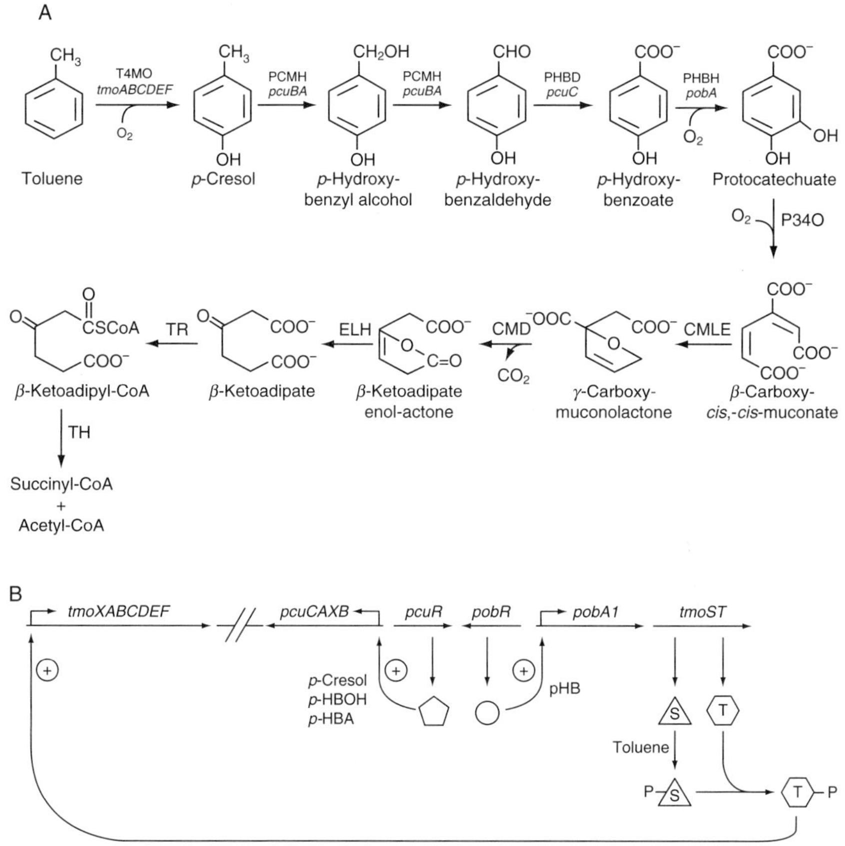

FIGURE 1.6 Toluene 4-monooxygenase-mediated toluene degradation pathway in P. mendocina KR1. (A) Pathway sequence indicating intermediates, enzymes, and genes. Enzymes: T4MO, toluene 4-monooxygenase; PCMH, p-cresol methylhydroxylase; PHBD, p-hydroxybenzaldehyde dehydrogenase; PHBH, p-hydroxybenzoate hydroxylase; P34O, protocatechuate 3,4-dioxygenase; CMLE, β-carboxy-cis,cis-muconate lactonizing enzyme; CMD, γ-carboxymuconolactone decarboxylase; ELH, β-ketoadipate enol-lactone hydrolase; TR, β-ketoadipate succinyl-CoA transferase; TH, β-ketoadipyl-CoA thiolase. (B) Gene organization and regulation. Arrows below gene designations indicate the direction of transcription. The ⊕ symbol represents positive regulation of the *tmoX* promoter (represented by ⟶) via the two-component regulatory system driven by TmoS (triangle) and TmoT (hexagon) in response to toluene. The (−P) symbol designates the phosphorylated state of each respective protein. The ⊕ symbol indicates positive regulation of *pobA1* and the *pcuCAXB* operon (represented by ⟶ or ⟵) by PcuR (pentagon) or PobR (circle), respectively, in the presence of the indicated effector molecules. p-HBOH, p-hydroxybenzyl alcohol; p-HBA, p-hydroxybenzaldehyde; pHB, p-hydroxybenzoate.

A three-component enzyme system designated toluene 4-monooxygenase (T4MO), which required NAD(P)H, Fe^{2+} and FAD for optimal activity, catalyzed the formation of p-cresol from toluene in P. mendocina KR1 (Whited and Gibson, 1991b). Subsequent detailed analyses indicated that the enzyme actually consisted of four protein components (Pikus et al., 1996). The hydroxyl group of p-cresol was derived from oxygen based on $^{18}O_2$ incorporation analysis. Electrons were transferred from NAD(P)H to a flavin-containing oxidoreductase to a 12-kDa Rieske iron–sulfur ferredoxin and finally to the catalytic hydroxylase (Pikus et al., 1996; Whited and Gibson, 1991b). Oxidation of deuterated toluene and p-xylene indicated that the reaction proceeded through an epoxide intermediate (Mitchell et al., 2003; Whited and Gibson, 1991b), and the ability of T4MO to catalyze the formation of epoxides was confirmed with a variety of alkene substrates (McClay et al., 2000).

The genes encoding T4MO were cloned by complementation of a P. mendocina KR1 mutant that was capable of growth on p-cresol but not toluene (Yen et al., 1991). A cluster of five genes, tmoABCDE was identified (Figs. 1.3 and 1.6B), and all were required for T4MO activity. Significant similarities of the tmo products were found with the products of the dmp genes from P. putida CF600, which catalyze the oxidation of phenol and methylphenols to catechols (Nordlund et al., 1990; Yen et al., 1991). T4MO hydroxylase was initially found to contain at least two subunits of ~50 and 32 kDa (Whited and Gibson, 1991b). Based on protein characteristics and sequence homologies, tmoA and tmoE were predicted to encode the large and small subunits of the hydroxylase component (Yen et al., 1991); a third subunit of 9.6 kDa (encoded by tmoB) was later identified (Pikus et al., 1996). The tmoC gene was predicted to encode the ferredoxin (Yen et al., 1991). Subsequent studies identified a downstream gene, tmoF, which significantly enhanced the activity of recombinant T4MO in E. coli and P. putida; the gene was found to encode the NADH:ferredoxin oxidoreductase component (Yen and Karl, 1992). TmoF was a 37-kDa protein with homology to several characterized reductases with both a plant-type ferredoxin domain and a flavin-binding domain. TmoF could be replaced by the reductase from the naphthalene dioxygenase system (NahAa; Yen and Karl, 1992), spinach ferredoxin reductase (Pikus et al., 1996), or by unidentified reductase(s) present in E. coli (Yen et al., 1991). The tmoD gene was later shown to encode a monomeric 11.6-kDa effector protein that stimulated T4MO activity in vitro and affected regiospecificity with various substrates (Mitchell et al., 2002; Pikus et al., 1996). TmoD, which had significant similarity to component B of MMO (Lipscomb, 1994), copurified with the hydroxylase component, but at substoichiometric levels (Pikus et al., 1996).

High levels of the components of T4MO have been individually expressed and purified from E. coli [hydroxylase, TmoABE (Studts et al.,

2000); ferredoxin, TmoC (Xia et al., 1999); effector protein, TmoD (Studts and Fox, 1999)] and detailed mechanistic and structure–function studies have been carried out as described briefly here. The hydroxylase component was demonstrated to have a $(\alpha\beta\epsilon)_2$ subunit structure with variable amounts of TmoD bound (Pikus et al., 1996). Like other members of this family of monooxygenases, the hydroxylase contained a diiron center at the site of catalysis (Pikus et al., 1996), and a mechanism involving both radical and cation intermediates was proposed based on the application of diagnostic probes (Moe and Fox, 2005; Moe et al., 2004). T4MO has a wide substrate specificity, catalyzing oxidation of alkenes, chloroform, TCE, N-nitrosodimethylamine, various polycyclic aromatic substrates, and substituted benzenes (McClay et al., 1996, 2000; Oppenheim et al., 2001; Pikus et al., 1997; Sharp et al., 2005; Winter et al., 1989). T4MO could be converted to an ortho-oxidizing enzyme by mutagenesis of key positions in the active site (Mitchell et al., 2002; Tao et al., 2004), and several other studies have resulted in the generation of T4MO variants with altered product formation profiles, thus identifying residues important for controlling regiospecificity (Fishman et al., 2004a; McClay et al., 2005; Mitchell et al., 2002; Pikus et al., 1997, 2000; Tao et al., 2004, 2005).

Solution and crystal structures of TmoD confirmed the absence of any cofactors or metal ions, providing additional support to the idea that TmoD interacts with the hydroxylase to generate conformational changes required for efficient catalysis (Hemmi et al., 2001; Lountos et al., 2005). The corresponding effector protein from the R. pickettii PKO1 T3MO could effectively substitute for TmoD while maintaining the regiospecificity of T4MO (Hemmi et al., 2001; Mitchell et al., 2002). In contrast, the more distantly related T2MO effector proteins from Burkholderia sp. strain JS150 and from B. vietnamiensis G4 did not form functional complexes (Hemmi et al., 2001).

On the basis of sequence analysis, the TmoC ferredoxin was predicted to carry a Rieske-type iron–sulfur center (Yen et al., 1991); this was later confirmed biochemically with the purified protein (Pikus et al., 1996), and subsequently by structural analyses (Moe et al., 2006a; Skjeldal et al., 2004; Xia et al., 1999). The interaction between TmoC and the hydroxylase was quite specific, as only TbuB from the R. pickettii T3MO enzyme was able to reconstitute activity. Other ferredoxins, including the Burkholderia xenovorans LB400 BphF and Thermus thermophilus Tt Rieske protein, were not capable of reconstituting any detectable activity even when provided at high concentrations (Elsen et al., 2007). Amino acid substitutions on two different faces of TmoC affected catalysis, suggesting that TmoC is involved in two different types of protein–protein interactions (Elsen et al., 2007). TmoC is known to interact with the hydroxylase during electron transfer, and there is now some evidence that TmoC also interacts with TmoD (Moe et al., 2006b).

Analysis of the downstream pathway enzymes indicated that *p*-cresol was oxidized at the methyl group by *p*-cresol methylhydroxylase (PCMH; Fig. 1.6A), an enzyme that was previously identified in a strain of *P. putida* (Hopper, 1976; Hopper and Taylor, 1977; McIntire *et al.*, 1985). PCMH catalyzed the oxidation of both *p*-cresol and *p*-hydroxybenzyl alcohol to *p*-hydroxybenzaldehyde, suggesting that two sequential oxidations occurred at the methyl group of *p*-cresol (Whited and Gibson, 1991a). PCMH was shown to be a homodimer of 56-kDa subunits that contained one heme and one FAD per enzyme (Whited and Gibson, 1991a). *p*-Hydroxybenzaldehyde dehydrogenase catalyzed the formation of *p*-hydroxybenzoate, which was then oxidized by *p*-hydroxybenzoate hydroxylase to protocatechuate (Fig. 1.6A). Unlike the other bacterial toluene degradation pathways, ring cleavage proceeded via the β-ketoadipate (ortho) pathway in *P. mendocina* KR1 (Whited and Gibson, 1991b). This conclusion was based on the presence of protocatechuate 3,4-dioxygenase (P34O) activity in toluene-grown cells, and the formation of the key intermediate of the β-ketoadipate pathway, β-carboxy-*cis*, *cis*-muconate (Harwood and Parales, 1996).

All downstream pathway genes are located in a cluster (Figs. 1.6B and 1.3) (Wright and Olsen, 1994). Sequence analysis revealed that *pcuCAXB*, which encodes the *p*-cresol utilization enzymes, was divergently transcribed from the regulatory gene *pcuR* (Ben-Bassat *et al.*, 2003). The product of *pcuBA* was PCMH; apparently the enzyme may be a heterodimer, although the purified protein was reported to be a homodimer (Whited and Gibson, 1991a). The *pcuC* gene encodes *p*-hydroxybenzaldehyde dehydrogenase. No function for PcuX was proposed, but the protein was predicted to be located in the inner membrane (Ben-Bassat *et al.*, 2003). PcuR, which has homology with TbuT from *R. pickettii* PKO1 (see Section IV), controls expression of *pcuCBXA*. A *lacZ* fusion was used to show that the *pcuC* promoter was induced in the presence of *p*-cresol, *p*-hydroxybenzyl alcohol, and *p*-hydroxybenzaldehyde in a recombinant *E. coli* strain expressing *pcuR* (Ben-Bassat *et al.*, 2003). Just downstream of the *pcu* gene cluster is a pair of divergently transcribed genes, *pobR1–pobA1*, which code for a regulatory protein and *p*-hydroxybenzoate hydroxylase, respectively (Figs. 1.3 and 1.6B) (Ramos-Gonzalez *et al.*, 2002).

The *pcu–pob* gene cluster is apparently not linked to the *tmoABCDEF* genes in *P. mendocina* KR1, but two genes (*tmoST*) required for expression of the *tmo* operon were located 714 bp downstream of *pobA1* (Figs. 1.3 and 1.6B). Interestingly, *tmoST* encodes a two-component signal transduction system that is ~85% identical in amino acid sequence to the *todST* products from *P. putida* F1 (see Section II.B) and *P. putida* DOT–T1E (Ramos-Gonzalez *et al.*, 2002). The *tmoST* genes complemented a *P. putida* DOT–T1E *todST* mutant, demonstrating a conserved function. In order to identify and study the *tmo* promoter, the region upstream of *tmoA* was sequenced. An additional open reading frame was located upstream of *tmoABCDEF* (Figs. 1.3 and 1.6B). The product was found to be 83% identical to TodX from *P. putida* F1 and DOT–T1E, and the gene was therefore designated *tmoX* (Ramos-Gonzalez *et al.*, 2002). The function of TmoX in *P. mendocina* KR1 was not investigated, but it is presumed to play a role in toluene passage across the outer membrane by analogy with functional analyses of TodX (see Section II.B). The *tmoX* promoter was very similar to the *todX* promoter, and a conserved *tod* box (presumably the site of TmoT binding) was located ~100 bases upstream the *tmoX* promoter. Expression from the *tmoX* promoter was induced in the presence of toluene and *p*-cresol (Ramos-Gonzalez *et al.*, 2002). Toluene oxidizing activity in *P. mendocina* KR1 was also shown to be induced by the chlorinated solvents TCE, perchloroethylene, *cis*-1,2-dichloroethylene, and chloroethene as well as linear alkanes (McClay *et al.*, 1995). The similarities in the putative outer membrane channel proteins (TmoX/ TodX) and the two-component regulation systems (TmoST/TodST) between *P. mendocina* KR1 and *P. putida* F1 are striking. These observations, together with the functional cross-regulation demonstrated for TodST/TmoST suggest that these components were recruited independently for use in two very different toluene degradation pathways (Ramos-Gonzalez *et al.*, 2002).

VI. A STRAIN WITH MULTIPLE TOLUENE DEGRADATION PATHWAYS: *BURKHOLDERIA* SP. STRAIN JS150

Burkholderia (formerly *Pseudomonas*) sp. strain JS150 is an unencapsulated mutant of the 1,4-dichlorobenzene degrading isolate JS1, which was obtained from combined sewage from Tyndall Air Force Base and Panama City, FL (Spain and Nishino, 1987). The strain is capable of growth on a large number of aromatic hydrocarbons and chlorinated aromatic compounds, including benzene, toluene, ethylbenzene, naphthalene, chlorobenzene, and 1,4-dichlorobenzene (Haigler *et al.*, 1992). In addition, spontaneous mutants that grew with other aromatic compounds were easily isolated. The strain was shown to differentially express at least three different dioxygenases (including one for toluene that is apparently very similar to the *P. putida* F1 toluene dioxygenase; Haigler and Spain, 1991) and four different ring cleavage pathways, including both ortho and meta cleavage pathways (Fig. 1.1) (Haigler *et al.*, 1992). *Burkholderia* sp. strain JS150 also carries multiple monooxygenase-mediated pathways for toluene degradation. Therefore, in strain JS150, toluene can be initially converted to toluene *cis*-dihydrodiol, *o*-cresol, or *p*-cresol (explained later). All three initial intermediates are then converted to 3- or 4-methylcatechol, and degradation is expected to proceed via the meta cleavage pathway based on the presence of C23O activity (but not C12O activity) in toluene-induced cells (Haigler et al., 1992).

Genes for a *toluene/benzene 2-monooxygenase* (Tb2m) were cloned from *Burkholderia* sp. strain JS150 by selecting for recombinant *P. aeruginosa* clones that could grow with phenol or benzene as sole carbon source (Johnson and Olsen, 1995). This selection resulted in the identification of the *tbmABCDEF* genes (Fig. 1.3), which had significant sequence homology with the *dmp* cluster encoding the multicomponent phenol hydroxylase (32–65% amino acid sequence identity) from *P. putida* CF600 (Shingler et al., 1992). The gene order was identical to that of the *dmp* and other phenol hydroxylase gene clusters and also the *B. vietnamiensis* G4 T2MO gene cluster (Fig. 1.3). The genes were localized to a plasmid native to strain JS150 by Southern hybridizations. On the basis of the sequence analysis, the enzyme is expected to contain three components. The recombinant enzyme catalyzed the conversion of toluene to *o*-cresol and *o*-cresol to 3-methylcatechol (Johnson and Olsen, 1995), which indicates that the enzyme is similar to T2MO from *B. vietnamiensis* G4 (see Section III.A). ^{18}O incorporation analysis demonstrated that the oxygen atom in *o*-cresol was derived from molecular oxygen.

A regulatory gene (*tbmR*) was located approximately 5 kb upstream of the *tmbABCDEF* cluster (Fig. 1.3) (Johnson and Olsen, 1995). A putative σ^{54}-dependent -12, -24 sequence was identified upstream of the *tbmA* gene, and TbmR showed homology to the NtrC family regulators XylR from the TOL pathway in *P. putida* mt-2 and DmpR from the dimethylphenol pathway in *P. putida* CF600 (Leahy et al., 1997). Expression from a P*tbmA-lacZ* fusion demonstrated that toluene, benzene, and chlorobenzene were recognized as effectors; *o*- and *m*-cresols were weak inducers. Interestingly, the TbuT protein from *R. pickettii* PKO1 was capable of controlling expression from the *tbmA* promoter and the opposite (TbmR control of the *tbuA1* promoter) was also demonstrated (Leahy et al., 1997). This finding was surprising because the T3MO from *R. pickettii* PKO1 differs biochemically and genetically from the Tb2m enzyme from *Burkholderia* sp. strain JS150, but it provides additional evidence for recruitment of regulatory genes by distinct catabolic clusters.

The same recombinant plasmid that harbored the genes encoding Tb2m was found to carry genes for a toluene 4-monooxygenase (Tb4m) pathway based on the formation of both *o*-cresol and *p*-cresol from toluene by the *P. aeruginosa* host carrying the plasmid (Johnson and Olsen, 1997). The genes encoding Tb4m were localized to a region between *tbmR* and *tbmABCDEF* (Fig. 1.3), and the regulatory protein TbmR was shown to control expression of the genes encoding both Tb2m and Tb4m in response to toluene. Using a subclone containing only the Tb4m gene region, ^{18}O incorporation analysis into toluene showed that the oxygen atom in *p*-cresol was derived from molecular oxygen. Interestingly, Tb4m was unable to catalyze the formation of 4-methylcatechol from *p*-cresol, indicating that Tb2m or another monooxygenase is required to catalyze the second step in toluene degradation following oxidation by Tb4m (Johnson and Olsen, 1997).

A genetic region encoding a second toluene 4-monooxygenase was cloned from the chromosome of *Burkholderia* sp. strain JS150. Some homology with the Tb2m region was detected in Southern hybridizations, but the genes were reported to be distinct. However, no sequence data was reported for either of the two toluene-4-monooxygenase encoding regions (Johnson and Olsen, 1997).

Genes encoding a second toluene 2-monooxygenase that converted toluene to *o*-cresol, benzene to phenol, and chlorobenzene to 2-chlorophenol (*tbc*) were cloned in *P. aeruginosa*, allowing the strain to grow with toluene as sole carbon source (Kahng *et al.*, 2001). The initial clone was found to carry genes for a multicomponent phenol hydroxylase (*tbc1ABCDEF*; Fig. 1.3) that was similar to the Dmp enzyme in *P. putida* CF600 (Shingler *et al.*, 1992). Interestingly, the deduced amino acid sequences of the *tbc1ABCDF* genes had 84–98% sequence identity with the products of the *Burkholderia* sp. strain JS150 *tbmABCDF* genes (explained later). Products of the *tbc1ABCDEF* genes were shown to catalyze the oxidation of *o*-cresol to 3-methylcatechol, but did not oxidize toluene, indicating that this enzyme is a phenol/cresol hydroxylase (Kahng *et al.*, 2001). A second set of monooxygenase genes was located adjacent to the *tbc1* gene cluster and designated the *tbc2* cluster (*tbc2ABCDEF*; Fig. 1.3). The *tbc2* gene products had the highest sequence identity with the components of T3MO from *R. pickettii* PKO1 (Byrne *et al.*, 1995). The Tbc2 monooxygenase (Tbc2m) catalyzed the conversion of toluene to *o*-cresol as well as the oxidation of *o*-cresol to 3-methylcatechol. Since Tbc2m can catalyze these sequential oxidations, it is not clear whether the Tbc1 monooxygenase (Tbc1m) plays a role in toluene degradation (Kahng *et al.*, 2001). Like the *tbm* cluster, the *tbc1–tbc2* cluster appeared to be located on a large catabolic plasmid native to strain JS150 (Johnson and Olsen, 1995; Kahng *et al.*, 2001).

To date, *Burkholderia* sp. strain JS150 has been shown to have a TDO pathway, at least five toluene/benzene monooxygenases with differing specificities, and at least four ring cleavage pathways. Regardless of which initial oxidation sequence takes place, in all cases 3- or 4-methylcatechol is formed and ring cleavage appears to proceed through a meta pathway based on induction experiments (Haigler *et al.*, 1992). This means that *Burkholderia* sp. strain JS150 carries a previously unreported type of toluene degradation pathway involving an initial toluene 4-monooxygenase followed by meta cleavage (Fig. 1.1). However, since all of the known toluene mono- and dioxygenases are induced during growth with toluene, it appears that multiple pathways function concurrently in *Burkholderia* sp. strain JS150, and toluene may be metabolized by several routes at the same time.

VII. A NONSPECIFIC TOLUENE MONOOXYGENASE PATHWAY: *PSEUDOMONAS STUTZERI* OX1

An interesting variation of the previously described pathways has been identified in *P. stutzeri* OX1. *P. stutzeri* OX1 is of interest because although it has a monooxygenase-mediated pathway to convert toluene to methylcatechol, the initial monooxygenase is most similar to T4MO in sequence and structure, but the enzyme is not regiospecific and catalyzes the formation of all three cresol isomers, which are then converted to 3-methylcatechol and 4-methylcatechol by a second monooxygenation reaction (Fig. 1.7A).

A. Pathway details and enzymology

A toluene/*o*-xylene monooxygenase (ToMO) pathway was identified in *P. stutzeri* OX1, a strain that was isolated from a wastewater treatment plant enriched with *o*-xylene (Baggi *et al.*, 1987). The original isolation of this organism was based on growth with *o*-xylene (but not on *m*- or *p*-xylene), and 2,3- and 3,4-dimethylphenol (Baggi *et al.*, 1987). *o*-Xylene was converted into 2,3- (and smaller amounts of 3,4-) dimethylphenol by this strain, and these intermediates were converted to 2,3- or 3,4-dimethylcatechol, suggesting sequential monooxygenations of the aromatic ring. In addition, C23O activity was induced when *P. stutzeri* OX1 was grown on *o*-xylene, indicating that the ring is cleaved by the meta pathway (Baggi *et al.*, 1987). Subsequent studies demonstrated that *m*- and *p*-xylene were growth substrates for mutants of *P. stutzeri* OX1 that lost the ability to grow on *o*-xylene (Barbieri *et al.*, 1993). Exposure of wild-type *P. stutzeri* OX1 to either *m*- or *p*-xylene resulted in high cell lethality due to the accumulation of the toxic intermediates 2,4- and 2,5-dimethylphenol. However, mutants M1, M2 (isolated on *m*-xylene) and P1 (isolated on *p*-xylene), which lost the ability to grow on *o*-xylene, could metabolize *m*- and *p*-xylene without toxic effects, suggesting that two independent pathways are responsible for the oxidation of the methylbenzene isomers; *o*-xylene required oxidation on the aromatic ring itself, whereas *m*- and *p*-xylene were oxidized at the methyl group (via the TOL pathway; see Section VIII.A). All products were subsequently degraded by meta cleavage. It was shown that a chromosomal rearrangement resulted in the selection of the M1, M2, and P1 mutants (Barbieri *et al.*, 1993). Comparison of the chromosomal arrangements in *P. stutzeri* OX1, the M1 mutant, and a M1 revertant (R1) that could grow on all three methylbenzene isomers (Di Lecce *et al.*, 1997) showed that a 3-kb insertion sequence, named ISPs1, was present in the genomic regions for *m*- and *p*-xylene degradation in wild-type OX1, and in the genomic region for *o*-xylene degradation in M1, but absent in the revertant strain R1 (Bolognese *et al.*, 1999).

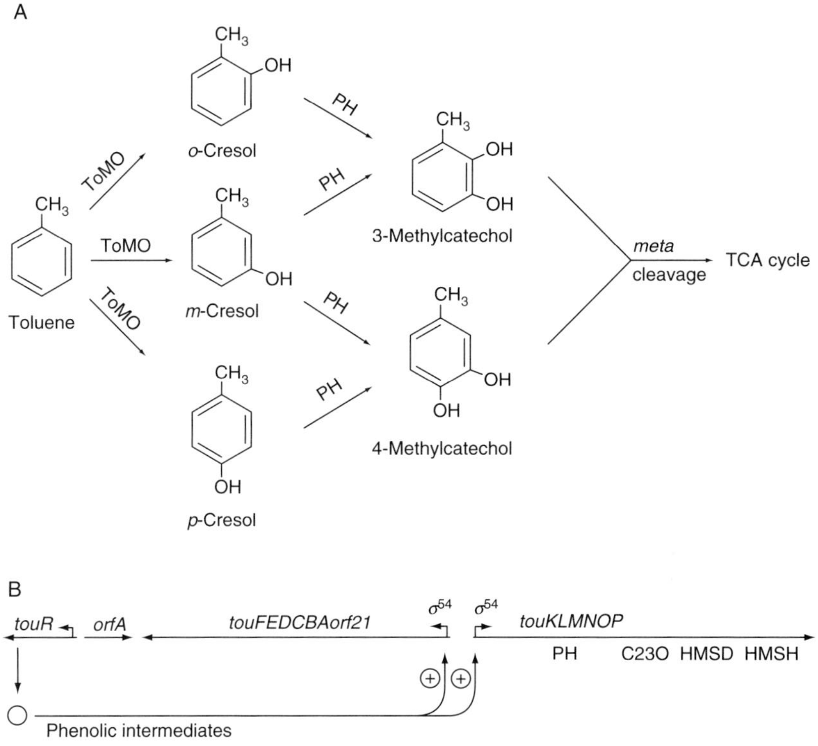

FIGURE 1.7 Nonspecific monooxygenase-mediated pathway in *P. stutzeri* OX1. (A) Upper pathway sequence indicating intermediates and enzymes. Enzymes: ToMO, toluene/o-xylene monooxygenase; PH, phenol/cresol hydroxylase. (B) Gene organization and regulation. Arrows below gene designations represent the direction of transcription. *touKLMNOP* encode a multicomponent phenol/cresol hydroxylase (PH) that catalyses the second hydroxylation in the pathway. *tou* gene designations have not been assigned for the *meta* pathway enzymes catechol 2,3-dioxygenase (C23O), 2-hydroxymuconic semialdehyde dehydrogenase (HMSD), or 2-hydroxymuconic semialdehyde hydrolase (HMSH/HODH). The ⊕ symbol indicates positive regulation of the respective promoters (represented by ⟶ or ⟵) by TouR (circle) in the presence of effector molecules and the RNA polymerase sigma factor 54 (σ^{54}).

Understanding the initial oxidations of *o*-xylene and toluene in *P. stutzeri* OX1 came from experiments in which the genes for the early stages of degradation were cloned and expressed in *P. putida* and *E. coli* (Bertoni et al., 1996). Using a *P. stutzeri* genomic library, *P. putida* PaW340 transconjugants were screened for the ability to convert toluene and *o*-xylene to 3-methylcatechol and 3,4-dimethylcatechol, respectively.

The expression of these genes in PaW340 did not allow this strain to utilize either toluene or *o*-xylene as growth substrates, but did allow a *P. putida* strain that contained the meta cleavage genes from the TOL plasmid to grow with toluene and *o*-xylene as sole carbon and energy sources. A subclone of the complementing plasmid expressed in *E. coli* was used to verify that incubation with toluene resulted in the accumulation of 3- and 4-methylcatechol, and *o*- or *m*- and *p*-cresol. When this strain was incubated with *o*-cresol, 3-methylcatechol was detected, and 2,3- and 3,4-dimethylphenol were produced from *o*-xylene. These results demonstrate that toluene and *o*-xylene are converted into mixtures of all possible methylphenols, which are subsequently converted into the corresponding catechols (Fig. 1.7A), and lends support for the occurrence of two sequential monooxygenations (Bertoni *et al.*, 1996).

The genes for toluene and *o*-xylene degradation are chromosomally encoded. A large plasmid was identified in *P. stutzeri* OX1; however, strains that were cured of this plasmid (which codes for mercury resistance) retained the ability to degrade *o*-xylene (Barbieri *et al.*, 1989). The genetic region for toluene and *o*-xylene degradation was localized to a 6-kb region that contains six open reading frames designated the *tou* (toluene/*o*-xylene utilization) locus (Figs. 1.7B and 1.3) (Bertoni *et al.*, 1998). On the basis of sequence analysis, a ribosome-binding site preceded the *touABCDEF* genes, which were followed by a *rho*-independent terminator sequence suggesting these genes constitute an operon. Based on protein sequence alignments, TouA was shown to have sequences similar to dinuclear iron binding ligand sites, suggesting that it may be the large subunit of the monooxygenase, while TouE and TouB appeared to be the other subunits that are known to be required for monooxygenase activity in other systems. Indeed, independent mutations in the *touA*, *B*, or *E* genes led to a total loss of toluene and *o*-xylene oxidizing activity (Bertoni *et al.*, 1998). When purified from *E. coli*, recombinant ToMOA (TouA), B, and E coeluted in ion-exchange and gel filtration chromatography, suggesting that they form a 206-kDa, $TouB_2E_2A_2$ polypeptide complex named ToMOH. ToMOH was determined to contain 3.4 mol of iron, which is consistent with the presence of a diiron center in each TouA subunit. ToMOH, when incubated with dithionite and methyl viologen, was shown to convert *p*-cresol to 4-methylcatechol, confirming that ToMOH alone contains hydroxylase activity (Cafaro *et al.*, 2002). The crystal structure of ToMOH has been determined (Sazinsky *et al.*, 2004). Interestingly, the TouA subunit contains a large active site pocket ($\sim$30–35 Å long, $\sim$6–10 Å wide) that connects the diiron center to the surface of the protein. This large site is probably important for the wide substrate specificity of the enzyme. Within the active site, ToMOH differs from other T3MO and T4MO monooxygenases in that it contains a glutamate residue at position 103 compared to a glycine residue. In addition, the charged portion of

Glu-103 points away from the active site and the charged carboxylate group is buried, possibly explaining the relaxed regiospecificity of ToMOH (Sazinsky et al., 2004). ToMOH has very broad substrate specificity, and is able to oxidize benzene, ethylbenzene, m- and p-xylene, 2,3- and 3,4-dimethylphenol, cresols, styrene, naphthalene, TCE, 1,1-dichloroethylene, tetrachloroethylene, and chloroform in addition to toluene and o-xylene (Bertoni et al., 1996; Chauhan et al., 1998; Ryoo et al., 2000).

Sequence alignments and biochemical analyses indicated that TouC and TouF had the characteristics of ferredoxin and reductase components that participate in electron transfer (Bertoni et al., 1998; Cafaro et al., 2002). NADH was shown to be the electron donor to the reductase ToMOF, as the specific activity was 100-fold higher when ToMOF was incubated with NADH compared with NADPH. Electron transfer from ToMOF to ToMOC was also demonstrated (Cafaro et al., 2002). The TouD protein sequence had similarity with low molecular weight proteins involved in stimulating the activity of hydroxylases in other aromatic degradation pathways (Bertoni et al., 1998). The surface topology of ToMOD (TouD) was compared with the effector protein component (DmpM) of the phenol hydroxylase from P. putida CF600, whose presence has been shown to be important for hydroxylase activity. Although there is high protein sequence similarity between DmpM and ToMOD, it was found by comparing protein cleavage profiles that ToMOD has a much higher conformational flexibility relative to DmpM, which may contribute to the broad regiospecificity of ToMO (Scognamiglio et al., 2001).

Equal amounts of the recombinant ToMO proteins (H, F, C, and D) were shown to transform p-cresol to 4-methylcatechol in vitro. When ToMOH was incubated with ToMOD alone, a 3.6-fold increase in the conversion of p-cresol to 4-methylcatechol was demonstrated. As ToMOD is unable to transfer electrons, the presence of ToMOD in this complex was attributed to its ability to physically modulate hydroxylase activity. Therefore, the ToMO enzyme is a member of the four-component alkene/aromatic monooxygenase family (reviewed in Leahy et al., 2003). When all of the ToMO proteins were incubated together, the optimal stoichiometry for each protein complex was found to be 1:2:2, (ToMOH: ToMOC:ToMOD) with substoichiometric amounts of ToMOF (0.2 mol ToMOF per mol ToMOH) (Cafaro et al., 2002).

Genes that encode a multicomponent phenol/cresol hydroxylase similar to the enzyme from P. putida CF600 were identified ~2-kb downstream of the genes encoding ToMO (Arenghi et al., 2001b; Cafaro et al., 2004). Detailed analysis of the substrate specificities and kinetic parameters of ToMO and the phenol/cresol hydroxylase with benzene, toluene, o-xylene, cresols and phenol indicated that ToMO was significantly more efficient with unoxidized substrates, while the phenol/cresol hydroxylase was more efficient at oxidation of hydroxylated substrates, although the substrate specificities of the two enzymes overlapped (Cafaro et al., 2004; Cafaro et al., 2005). Interestingly, catechol production was much more efficient when both enzymes were present than when either was present alone, suggesting that both enzymes are required for optimal benzene or toluene degradation. These results suggest that two different monooxygenases catalyze the first and second oxidations in toluene degradation in P. stutzeri OX1 (Fig. 1.7A), as seen in R. pickettii PKO1 (see IV).

Ring cleavage in P. stutzeri OX1 proceeds via a meta cleavage pathway. Subclones were used to measure enzymatic activity, which confirmed the order of the genes encoding C23O, 2-hydroxymuconate semialdehyde dehydrogenase, and 2-hydroxymuconate semialdehyde hydrolase in the chromosomal meta cleavage pathway (Arenghi et al., 2001b).

B. Pathway regulation

When first isolated, it was demonstrated that C23O activity in P. stutzeri OX1 was inducible by growth on o-xylene, and 2,3-, and 3,4-dimethylphenol (Baggi et al., 1987), suggesting that expression of the genes required for toluene and o-xylene degradation are regulated. Using P. putida PaW340 as a heterologous host harboring the touABCDEF gene region, a series of deletions in the tou gene region were screened for phenotypes that lacked inducible expression of ToMO in the presence of toluene, o-xylene, or their cognate phenols (Arenghi et al., 1999). The touR gene was located downstream of the tou operon, just past a divergently transcribed open reading frame (orfA; encoding a putative transposase), and is transcribed in the same direction as the tou genes (Figs. 1.3 and 1.7B). Primer extension analysis determined that touR has its own promoter and transcriptional start site. The TouR protein has high similarity to the NtrC family of transcriptional activators, with highest homology to the DmpR regulator of phenol catabolism in P. putida CF600. In addition, the sequence of the touABCDEF promoter region (Ptou) was found to have homology to σ^{54}-dependent promoters. Interestingly, toluene and o-xylene were not effector molecules for TouR; however, the mono- and diphenols activated Ptou to varying degrees, suggesting that pathway intermediates are responsible for inducing the tou pathway. Indeed, when tou mRNA levels were measured in uninduced cells, low basal levels of tou transcript were detected, implying that low levels of cellular ToMO are responsible for converting toluene or o-xylene to the corresponding phenols, subsequently activating TouR and tou operon transcription by interacting with σ^{54}-RNA polymerase (Arenghi et al., 1999, 2001a,b). TouR was also shown to be responsible for the induction of the phenol hydroxylase and meta pathway genes in P. stutzeri OX1 (Arenghi et al., 2001b). While o-cresol was the most effective effector molecule for TouR *in vitro*

(Arenghi *et al.*, 2001a), the *tou* promoter was also regulated by gratuitous activation, whereby P*tou* transcription was activated by TouR in a growth-phase dependent manner in the absence of genuine effector molecules (Solera *et al.*, 2004). In uninduced stationary phase cultures, TouR was able to initiate transcription of *tou* transcripts, as *tou* mRNA was detected at levels much higher than were evident in exponentially growing cells, although overall levels were lower than in *o*-cresol induced cells. TouR was essential for this gratuitous activation based on experiments in *P. putida* PaW340. When present with a P*tou*–*lacZ* fusion in this heterologous host, β-galactosidase activity was detected in stationary phase without *o*-cresol. β-Galactosidase was not detectable in the control strain lacking TouR (Solera *et al.*, 2004).

The gene cluster encoding ToMO was suggested to have been acquired recently by *P. stutzeri* OX1 based on its position, regulation, low GC content, and proximity to a transposase gene (Barbieri *et al.*, 2001). This acquisition effectively increased the substrate range of the organism. On the basis of studies to date, it appears that, unlike the toluene monooxygenases from strains *B. vietnamiensis* G4, *R. pickettii* PKO1, and *P. mendocina* KR1, ToMO from *P. stutzeri* OX1 is not regiospecific and has no clear preference for the position of oxidation on the toluene ring. A multicomponent phenol/cresol hydroxylase with a similar substrate range but different regiospecificity and substrate preferences relative to ToMO contributes to the efficient degradation of toluene and other aromatic hydrocarbon substrates by efficiently processing the hydroxylated products. Thus *P. stutzeri* OX1 appears to have a mixed pathway in which all three isomers of cresol are produced from toluene and both 3- and 4-methylcatechol are formed from the cresols (Fig. 1.7A). The catechols are then cleaved by a common extradiol dioxygenase. The lack of regiospecificity of ToMO and the presence of the phenol hydroxylase may contribute to the ability of *P. stutzeri* OX1 to grow with *o*-xylene, which appears to be a relatively rare trait (Cafaro *et al.*, 2005).

VIII. TOL PATHWAY: *PSEUDOMONAS PUTIDA* MT-2

The use of *Pseudomonas putida* mt-2 as a model system for the degradation of aromatic compounds has a long history that has involved the work of many different international investigators (Nakazawa, 2002). *P. putida* mt-2 [formerly *Pseudomonas putida* (*arvilla*)] was originally isolated for growth on *o*-, *m*-, and *p*-toluate from a cultivated soil sample in Japan (Nakazawa, 2002; Nozaki *et al.*, 1963). More recently, 1,2,4-trimethylbenzene, 3-ethyltoluene, 3,4-dimethylbenzyl alcohol, 3,4-dimethylbenzoic acid, and 3-ethylbenzoic acid were shown to be growth substrates for this strain (Kunz and Chapman, 1981). Early studies demonstrated that the toluene degradation pathway was maintained on a large transmissible plasmid designated TOL or pWW0 in *P. putida* mt-2 (Williams and Murray, 1974; Wong and Dunn, 1974). Research on the "TOL pathway" as it has become known, has focused not only on the pathway itself, but also on the catabolic plasmid.

A. Pathway and enzymology

The TOL plasmid-encoded upper pathway for toluene degradation converts toluene to benzoate by the action of three enzymes, xylene monooxygenase, benzyl alcohol dehydrogenase, and benzaldehyde dehydrogenase (Fig. 1.8A). Xylene monooxygenase is a two-component enzyme consisting of XylM, a membrane-bound catalytic component with ferrous iron at the active site, and XylA, a NADH ferredoxin reductase that has a plant-type [2Fe–2S] cluster and contains FAD (Shaw and Harayama, 1992; Suzuki *et al.*, 1991). Purified XylA was shown to be a 40-kDa monomer with one C-terminal FAD and one [2Fe–2S] cluster in the N-terminus (Shaw and Harayama, 1992). The protein has similarity to the toluate dioxygenase reductase from the same strain and the *P. putida* CF600 phenol hydroxylase reductase (Suzuki *et al.*, 1991). XylM, which receives electrons from XylA, is a 42-kDa protein with hydroxylase activity. XylM shares homology with the AlkB alkane hydroxylase and eukaryotic fatty acid desaturases. This family of enzymes contains eight conserved histidines that are essential for catalysis (Shanklin *et al.*, 1994). Based on homology with AlkB, XylM is expected to be a membrane-bound enzyme with a diiron center at the active site for oxygen activation (Shanklin *et al.*, 1997). The eight-histidine motif in the membrane-bound diiron enzymes differs from the consensus for the soluble diiron monooxygenases.

The substrate range of xylene monooxygenase includes 1,2,4-trimethylbenzene, 3-ethyltoluene, toluene, and *m*- and *p*-xylene, which all serve as growth substrates for *P. putida* mt-2 (Kunz and Chapman, 1981). The recombinant enzyme expressed in *E. coli* oxidized substituted toluenes to the corresponding alcohols, and also converted styrene to styrene epoxide (Panke *et al.*, 1999; Wubbolts *et al.*, 1994). Nitroaromatic compounds were also substrates for xylene monooxygenase; when the genes were expressed in *P. putida* and *E. coli*, *m*- and *p*-nitrotoluene were converted to *m*- and *p*-nitrobenzyl alcohol and subsequently to *m*- and *p*-nitrobenzaldehyde, which accumulated in the growth medium (Delgado *et al.*, 1992). Attempts to solubilize and purify XylM resulted in loss of activity, but the enzyme was assayed *in vitro* and shown to have activity with *p*-xylene, *m*-xylene, toluene, and *o*-xylene (in order of best to worst substrate), and activity required the presence of ferrous iron (Shaw and Harayama, 1995). Evidence for a radical intermediate in the

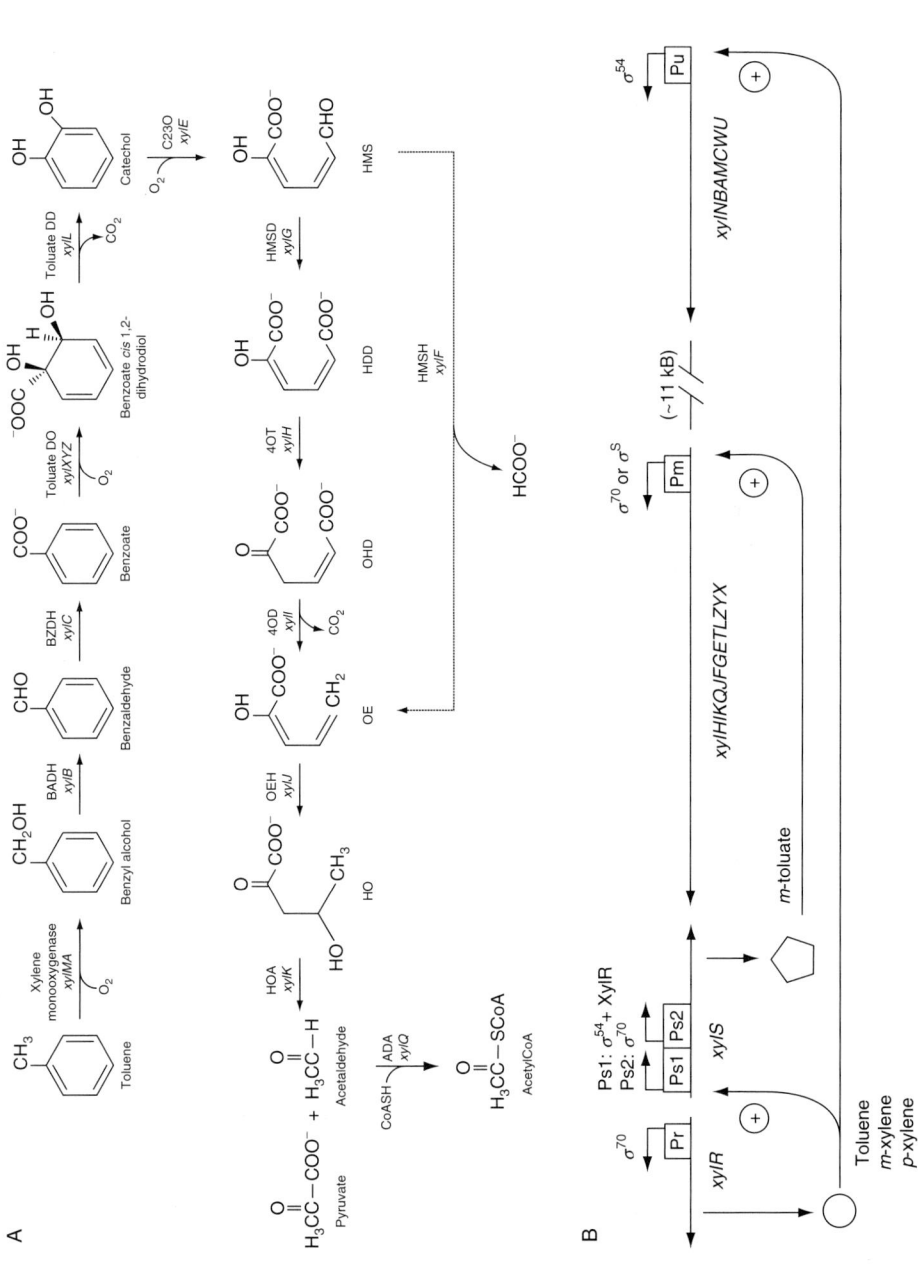

monooxygenase mechanism was obtained by incubating whole cells expressing xylene monooxygenase in the presence of the radical clock substrate norcarane (Austin et al., 2003), which is consistent with results of similar studies with AlkB (Austin et al., 2000).

The TOL plasmid carries genes encoding benzyl alcohol dehydrogenase (*xylB*) and benzaldehyde dehydrogenase (*xylC*) (Harayama et al., 1989b). Both enzymes were purified from *P. putida* carrying the TOL plasmid, and were shown to have the expected NAD^+-dependent dehydrogenase activity with various benzyl alcohols and benzaldehydes, respectively (Shaw and Harayama, 1990; Shaw et al., 1992). The benzyl alcohol dehydrogenase was a homodimer of 42-kDa subunits and the benzaldehyde dehydrogenase was a homodimer of 57-kDa subunits (Shaw and Harayama, 1990). The sequence of *xylB* indicated that benzyl alcohol dehydrogenase was a member of the long-chain zinc-containing alcohol dehydrogenase family and as expected, two atoms of zinc were found per subunit (Shaw et al., 1993). The 57-kDa protein with benzaldehyde dehydrogenase activity was later shown to be the product of the *xylG* gene based on its N-terminal amino acid sequence. The enzyme was significantly more active with 2-hydroxymuconic semialdehyde than benzaldehyde, indicating that it functioned in the lower (meta cleavage) pathway rather than in the upper pathway (Inoue et al., 1995). The *xylC* product was therefore purified from a $\Delta xylG$ mutant. The enzyme, a homotetramer of 52.5-kDa subunits, had activity with benzaldehyde and related substituted benzaldehydes, but no activity with ring cleavage products of catechols, confirming its role in the upper pathway (Inoue et al., 1995).

FIGURE 1.8 TOL pathway in *P. putida* mt-2. (A) Pathway sequence indicating intermediates, enzymes, and genes. Intermediates: HMS, 2-hydroxymuconic semialdehyde; HDD, 2-hydroxyhexa-2,4-diene-1,6-dioate; OHD, 2-oxohex-4-ene-1,6-dioate; OE, 2-oxopent-4-eneoate; HO, 4-hydroxy-2-oxovalerate. Enzymes: BADH, benzylalcohol dehydrogenase; BZDH, benzaldehyde dehydrogenase; Toluate DO, toluate dioxygenase; Toluate DD, toluate *cis*-dihydrodiol dehydrogenase; C23O, catechol 2,3-dioxygenase; HMSH (HODH), 2-hydroxymuconic semialdehyde hydrolase; HMSD, 2-hydroxymuconic semialdehyde dehydrogenase; 4OT, 4-oxalocrotonate tautomerase; 4OD, 4-oxalocrotonate decarboxylase; OEH (HPDH), 2-oxopent-4-enoate hydratase; HOA, 4-hydroxy-2-oxovalerate aldolase; ADA, acetaldehyde dehydrogenase (acylating). (B) Gene organization and regulation. Arrows above gene designations indicate the direction of transcription. Open boxes represent promoter sequences for each operon and regulation by various sigma factors is indicated. The ⊕ symbol represents positive regulation of the respective promoters (represented by ⟼ or ⟵) by XylR (circle) or XylS (pentagon) in the presence of effector molecules (see text for a complete list of effector molecules). Promoters are differentially regulated by different sigma factors (σ^{70}, σ^{54}, σ^{S}); see text for details.

Initial studies of cloned xylene monooxygenase in *E. coli* indicated that the enzyme catalyzed not only the oxidation of toluene to benzyl alcohol, but also the conversion of benzyl alcohol to benzaldehyde (Harayama *et al.*, 1986a); however, the oxidation of benzyl alcohol by extracts containing xylene monooxygenases could not be demonstrated (Shaw and Harayama, 1995), leaving validity of this activity an open question. Subsequently, ^{18}O incorporation strongly supported the xylene monooxygenase-catalyzed sequential oxidation of toluene to benzyl alcohol, benzaldehyde, and benzoate with the proposed formation of a *gem*-diol intermediate (Bühler *et al.*, 2000). The presence of toluene was found to inhibit the xylene monooxygenase-catalyzed conversions of benzyl alcohol to benzaldehyde and benzaldehyde to benzoate, suggesting that benzyl alcohol and benzaldehyde dehydrogenases catalyze these reactions in the wild-type strain (Fig. 1.8A) (Bühler *et al.*, 2002). The use of dehydrogenases for these steps in the pathway would be energetically more favorable, as less NADH would be consumed in the process. It has been proposed that the ancestral strain may have carried only xylene monooxygenase for the conversion of toluene to benzoate, and later acquired the dehydrogenases, which further optimized the pathway (Bühler *et al.*, 2002). The benzyl alcohol dehydrogenase-catalyzed reaction was reversible (Harayama *et al.*, 1989b), and in recombinant strains carrying XylMA and XylB, benzyl alcohol rather than benzaldehyde was formed, indicating that the reaction in the reverse direction is favored, possibly as a means to prevent the accumulation of toxic aldehydes (Bühler *et al.*, 2000).

The TOL plasmid lower pathway converts benzoate to catechol, which is then metabolized to TCA cycle intermediates by the enzymes of the meta cleavage pathway (Fig. 1.8A). Benzoate formed through the oxidation of the toluene methyl group by toluate 1,2-dioxygenase (Harayama *et al.*, 1986b), a two-component enzyme that is a member of the Rieske nonheme iron oxygenase family (Gibson and Parales, 2000). Toluate dioxygenase is encoded by the *xylXYZ* genes and is closely related to benzoate dioxygenase (Harayama *et al.*, 1991). The oxygenase, which was expressed from a recombinant *P. putida* strain, was purified under anaerobic conditions using standard chromatographic techniques (Ge *et al.*, 2002). This component consisted of α- and β-subunits in an $\alpha_3\beta_3$ configuration, and contained ~6 mol sulfur per mol of protein but larger than expected amounts of iron to account for the typical Rieske [2Fe–2S] cluster and iron at the active site; excess iron may have been due to nonspecific binding. The enzyme had an absorption spectrum typical of reduced Rieske proteins (absorbance maxima at 280, 323, and 455 nm). The reductase was purified as the His-tagged form and contained FAD and approximately 2 mol of iron and sulfur each per mol of protein, suggesting the presence of one [2Fe–2S] cluster. The enzyme had a broader substrate range than benzoate dioxygenase, catalyzing the oxidation of benzoate, *o*-, *m*-, and *p*-toluate, and 3- and 4-chlorobenzoate to the corresponding 1,2-*cis*-dihydrodiols (Ge *et al.*, 2002). However, the oxidation of ortho-substituted benzoates was strongly uncoupled, and the affinity of the enzyme for these substrates was much lower than for benzoate or meta- or para-substituted benzoates. Analysis of purified hybrid toluate and benzoate dioxygenases demonstrated that the α-subunit plays the major role in controlling substrate specificity (Subramanya *et al.*, 1996), but the β-subunit can modulate activity and regioselectivity (Ge and Eltis, 2003), which is consistent with results of an earlier genetic analysis of toluate dioxygenase (Harayama *et al.*, 1986b). Most studies with other dioxygenases have shown that the α-subunit alone controls specificity (Barriault *et al.*, 2001; Mondello *et al.*, 1997; Parales *et al.*, 1998a,b; Zielinski *et al.*, 2002); however, a role for β has been reported with certain hybrid biphenyl dioxygenases (Chebrou *et al.*, 1999; Furukawa *et al.*, 1994; Hurtubise *et al.*, 1998).

Benzoate *cis*-dihydrodiol is oxidized to catechol by the product of the *xylL* gene, toluate *cis*-dihydrodiol dehydrogenase (Fig. 1.7A) (Neidle *et al.*, 1992). The protein is 58% identical to benzoate *cis*-dihydrodiol dehydrogenase (BenD) from *Acinetobacter baylyi* ADP1, and both are members of the short chain alcohol dehydrogenase family (Neidle *et al.*, 1992). Presumably XylL has a broader substrate range than BenD due to the funneling of alkybenzoate substrates through the TOL pathway, and in fact substrates for XylL included *cis*-dihydrodiols of benzoate, chlorobenzoates, toluates, benzene, and toluene (Lee *et al.*, 1999; Lehrbach *et al.*, 1984). XylL was purified and shown to have an unexpected activity, that of NAD^+-dependent methyl transfer to adenosylcobalamin from toluene *cis*-dihydrodiol (Lee *et al.*, 1999). This reaction resulted in the production of catechol; in the absence of adenosylcobalamin, the expected dehydrogenation product, 3-methylcatechol was formed. Methyl transfer was not catalyzed when toluate *cis*-dihydrodiol was provided as the substrate, and 4-methylcatechol was formed both in the presence and absence of adenosylcobalamin.

C23O, the product of the TOL plasmid-encoded gene *xylE*, catalyzes the key step in the meta cleavage pathway (Fig. 1.8A). The enzyme was first purified from *P. putida* mt-2 in 1963 (Nozaki *et al.*, 1963), and the expression, purification, and assay conditions were later optimized (Nakai *et al.*, 1983a). The enzyme contained approximately one iron per subunit and was a homotetramer of 35-kDa subunits, which matched well with the calculations based on the deduced amino acid sequence of the protein (Nakai *et al.*, 1983b). Substrates oxidized included catechol, and 3- and 4-methylcatechol (Nozaki *et al.*, 1970). The crystal structure of the enzyme revealed the presence of a narrow hydrophobic channel through which the substrates and oxygen are proposed to enter (Kita *et al.*, 1999). C23O was found to be inactivated to varying degrees by certain substrates, including 4-ethylcatechol, and 3- and 4-methylcatechol (Cerdan *et al.*, 1994; Polissi and Harayama, 1993). Reactivation of the enzyme was shown to require ferrous iron or the presence of the XylT product, a plant-type ferredoxin whose function has been proposed to be the reduction of oxidized iron in C23O (Cerdan *et al.*, 1994; Polissi and Harayama, 1993). The *xylT* gene is located in the meta cleavage operon of the TOL plasmid (Fig. 1.8B).

Following ring cleavage, the meta pathway branches in order to accommodate a wider range of substrates (Harayama *et al.*, 1987). 3-Methylcatechol is metabolized through the hydrolase branch (Fig. 1.8A) as its ring-cleavage product is not a substrate for hydroxymuconate semialdehyde dehydrogenase (Murray *et al.*, 1972). The product of *xylF*, which has been called 2-hydroxymuconic semialdehyde hydrolase (HMSH) as well as HODH, was purified and shown to be a homodimer of 65-kDa subunits (Duggleby and Williams, 1986). The enzyme preferred ketone substrates (cleavage products of 3-substituted catechols) over aldehyde substrates (cleavage products of 4-substituted catechols), which is consistent with the functions of the two branches of the pathway (Duggleby and Williams, 1986). Studies with mutant strains and cloned genes demonstrated that catechol and 4-methylcatechol are mainly metabolized through the 4-oxalocrotonate branch (*xylGHI*; Fig. 1.8A) (Harayama *et al.*, 1987). Therefore, in *P. putida* mt-2, metabolites of toluene are preferentially routed through the 4-oxalocrotoanate pathway as toluene is degraded through catechol by the TOL pathway. In the 4-oxalocrotonate branch, *xylG*, *xylH*, and *xylI* encode 2-hydroxymuconic semialdehyde dehydrogenase (HMSD), 4-oxalocrotonate tautomerase (4OT; sometimes called 4-oxalocrotonate isomerase), and 4-oxalocrotonate decarboxylase (4OD), respectively, which convert ring cleavage products to the same common intermediate formed by the activity of HMSH (Fig. 1.8A). The next common step in the pathway is catalyzed by the product of *xylJ*, 2-oxopent-4-enoate hydratase (OEH) (Harayama and Rekik, 1990). 4OT was purified and shown to be a homohexamer of 62-amino acid subunits that stimulates the conversion of 4-oxalocrotonate to its keto form, which is the substrate for 4OD (Chen *et al.*, 1992; Harayama *et al.*, 1989a; Taylor *et al.*, 1998). 4OD and OEH copurified as a complex with both decarboxylase and hydratase activity. However, active hydratase could be obtained in the absence of the decarboxylase when purified from a recombinant *E. coli* strain; the opposite was not the case (Harayama *et al.*, 1989a). The products of the *xylK* and *xylQ* genes, 4-hydroxy-2-ketovalerate aldolase and acetaldehyde dehydrogenase (acylating) catalyze the final two steps in the meta pathway (Fig. 1.8A), and have not been studied in detail from the TOL system. However, homologues (DmpG and DmpF) from the phenol degradation pathway in *P. putida* CF600 have been well characterized (Powlowski *et al.*, 1993). These two proteins were purified and found to be physically associated in a complex containing two subunits of each polypeptide (Powlowski et al., 1993); this finding was confirmed by determination of the crystal structure (Manjasetty et al., 2003).

B. Genetics and regulation

Investigation of P. putida mt-2 showed that the meta cleavage pathway was plasmid-encoded, as strains cured of the TOL plasmid (pWWO) lost all meta pathway enzyme activities. In addition, meta pathway enzyme activity was shown to be transferable by conjugation from the wild type to various mutant strains, further demonstrating that these genes are plasmid encoded (Nakazawa and Yokota, 1973; Williams and Murray, 1974; Wong and Dunn, 1974). The work of many investigators has shown that the 117-kb TOL plasmid pWWO is a broad-host-range plasmid belonging to the IncP-9 incompatibility group that can be transferred to other Gram-negative microorganisms besides pseudomonads, although at lower frequencies (Benson and Shapiro, 1978; Downing and Broda, 1979; Duggleby et al., 1977; Ramos-Gonzales et al., 1991; White and Dunn, 1978). In addition to the general ability of this plasmid to mobilize between bacterial species, the *xyl* genes themselves are highly mobile, as they are located between IS*1246* insertion sequences, on the nested transposable elements Tn*4651* and Tn*4653* (Meulien et al., 1981; Reddy et al., 1994; Tsuda and Iino, 1987, 1988). The Tn*4651* and Tn*4653* transposons and the *xyl* meta pathway genes have GC contents of 59%, 57%, and 60% respectively, which are similar to the GC content of P. putida mt-2 (~60%). Interestingly, the *xyl* upper pathway genes have a GC content of 49%, suggesting a different origin for these genes (Harayama, 1994).

Physical and functional mapping of the *xyl* genes and cloning of the entire meta cleavage pathway revealed that there were two regions required for toluene degradation, one containing the upper pathway genes for the conversion of toluene and xylenes to their respective carboxylic acids and the other containing the lower pathway genes required for ring cleavage and conversion to TCA cycle intermediates (Franklin et al., 1981; Inouye et al., 1981b; Nakazawa et al., 1980). The complete sequence of the pWWO plasmid (http://img.jgi.doe.gov/cgi-bin/pub/main.cgi?section=TaxonDetail&page=taxonDetail&taxon_oid=637000222) and numerous genetic studies have since led to a detailed understanding of *xyl* gene organization, expression, and evolution (Greated et al., 2002; Harayama and Rekik, 1990; Harayama et al., 1984). The genes for toluene degradation on pWWO are organized into two operons, the upper pathway operon, *xylUWCMABN*, which codes for the conversion of toluene to benzoate and the lower pathway operon, *xylXYZLTEGFJQKIH*, which codes for conversion of benzoate to TCA cycle intermediates (Figs. 1.3 and 1.8B). Other genes are present in addition to the structural genes encoding the enzymatic steps in degradation (described in Section VIII.A). The first two genes in the upper pathway operon, *xylU* and *xylW* (Fig. 1.8B), are not required for toluene or xylene metabolism in *P. putida* mt-2 (Williams *et al.*, 1996). *xylU* encodes a protein of unknown function and *xylW* encodes a benzyl alcohol dehydrogenase. The last gene in the upper pathway operon, *xylN*, encodes an outer membrane protein with similarity to *todX* from *P. putida* F1 (see Section II), and the *xylN* product has been implicated in *m*-xylene uptake in *P. putida* mt-2 (Kasai *et al.*, 2001).

Because *P. putida* mt-2 contains both ortho and meta ring cleavage pathways, the question of which pathway was preferentially used for toluene metabolism was addressed by assessing enzyme activity in the presence of various inducers. The chromosomally-encoded C12O was shown to be inducible by *cis,cis*-muconate (a product of benzoate degradation), whereas the plasmid-encoded C23O was inducible by benzoate itself, which is an intermediate in toluene metabolism. This data indicated that the meta cleavage pathway is induced first, and is therefore preferentially utilized by *P. putida* mt-2 during growth with toluene (Murray *et al.*, 1972; Worsey and Williams, 1977). Additional studies showed that benzyl alcohol dehydrogenase, benzaldehyde dehydrogenase, and C23O activities were induced by *m*-xylene. However, C23O activity was also induced by *m*-toluate without the concurrent induction of benzyl alcohol dehydrogenase and benzyl aldehyde dehydrogenase, suggesting a multifaceted regulatory mechanism (Worsey and Williams, 1975, 1977).

Two putative regulatory proteins were proposed based upon phenotypic analysis of various strains. The putative XylR protein was shown to respond to toluene, *m*- and *p*-xylene and their respective alcohols, *p*-chlorobenzaldehyde, and mono- and disubstituted methyl-, ethyl-, and chlorotoluenes (Abril *et al.*, 1989; Worsey *et al.*, 1978). The XylR protein was hypothesized to regulate both the upper and lower pathways as growth with these compounds induced the activity of all of the enzymes. XylR mutants had a noninducible phenotype, indicating that XylR functions as a positive regulator (Worsey *et al.*, 1978). A second putative regulatory protein, XylS, was hypothesized to regulate the lower pathway genes, as growth on *m*-toluate induced the activity of only *xylXYZ, E, F,* and *G* (Worsey *et al.*, 1978). Cloning of the *xylS* gene and expression in *E. coli* along with the genes for the meta pathway showed that inducible degradation of *m*-toluate occurred in this strain (Inouye *et al.*, 1981a). C23O, HMSD, and HMSH activity could only be detected in *E. coli* strains containing *xylS* in the presence of *m*-toluate; effectors for the upper pathway did not induce this activity (Inouye *et al.*, 1981a). Tn5 mutagenesis of both regulatory elements and the cloning and expression of *xylR* demonstrated that XylR is required for the expression of both the upper pathway and *xylS* (Fig. 1.8B) (Franklin et al., 1983; Inouye et al., 1983).

xylS and *xylR* are located downstream of the meta pathway operon (Figs. 1.3 and 1.8B). The 67-kDa XylR protein has significant homology with the NtrC family of transcriptional regulators (Inouye et al., 1985, 1988); the 36-kDa XylS is homologous to the AraC family of DNA-binding proteins (Ramos et al., 1990; Spooner et al., 1987). The *xylS* gene is transcribed from its own promoter (designated Ps) and is oriented in the opposite direction from the meta pathway operon, while the *xylR* gene (Figs. 1.3 and 1.8B) is transcribed in the same direction as the meta pathway genes but from its own promoter (Pr) (Greated et al., 2002; Inouye et al., 1983; Ramos et al., 1987; Spooner et al., 1986). The *xylR* gene was shown to be constitutively expressed from Pr (which is recognized by σ^{70}) in the absence of inducers in *P. putida* mt-2 and *E. coli* (Inouye et al., 1985). *xylS* transcripts were found to initiate from two different promoters. The expression of *xylS* from the Ps2 promoter (a σ^{70}-type promoter) was at a low constitutive level, while expression from the Ps1 promoter was induced in the presence of XylR and *m*-xylene or *m*-cresol (Gallegos et al., 1996; Inouye et al., 1987; Marques et al., 1994). The Ps1 promoter of *xylS* and the *xylUWCMABN* operon promoter (designated Pu or OP1) have high homology to σ^{54}-regulated promoters, which is consistent with XylR having homology to NtrC transcriptional activators (Ramos et al., 1987). By measuring β-galactosidase activity from a *xylA–lacZ* reporter fusion in *E. coli*, the upper pathway was shown to be regulated by XylR in the presence of RpoN (σ^{54}) and inducer molecules. In addition, the *rpoN* gene was required for XylR-mediated induction of gene expression in *E. coli* (Dixon, 1986; Inouye et al., 1989; Ramos et al., 1987).

While promoters for both *xylS* (Ps1) and the upper pathway (Pu) had strong homology to σ^{54}-dependent promoters, the promoter for the meta pathway (Pm or OP2) had little similarity to this class of promoters (Inouye et al., 1984). Experiments in *E. coli* showed that XylS binds the inducer *m*-toluate and is responsible for expression from Pm (Inouye et al., 1987; Ramos et al., 1986; Spooner et al., 1987). Therefore, the regulation by XylR and XylS is hierarchical depending on the growth substrate. The lower pathway promoter Pm is controlled by XylS produced from the Ps2 σ^{70} promoter (Fig. 1.8B). In the presence of *m*-toluate, the meta pathway is induced. This cascade form of regulation is controlled by the master regulator XylR in order to coordinate the regulation of both the upper and meta pathways. XylR becomes activated in the presence of inducer molecules (e.g., toluene) and activates the σ^{54} promoters of the upper pathway genes *xylUWCMABN* (Pu) and *xylS* (Ps1), to turn on the entire aromatic degradation pathway (Fig. 1.8B).

There are additional levels of regulation of toluene and *m*-xylene catabolism in *P. putida* mt-2. In addition to being dependent upon σ^{70}

for expression, the Pm promoter is regulated by σ^S, the stationary phase sigma factor encoded by *rpoS* (Hugouvieux-Cotte-Pattat *et al.*, 1990; Marques *et al.*, 1995; Miura *et al.*, 1998). The Pu and Ps1 promoters have IHF-binding sites, and when expressed in an *E. coli* IHF mutant, expression from the Pu promoter was only 25% of the level in wild type. However, the IHF-binding sequences of the Ps1 promoter did not play a significant role in expression (de Lorenzo *et al.*, 1991; Holtel *et al.*, 1995). The TOL pathway has also been shown to be regulated depending upon the available carbon source. When *P. putida* mt-2 is grown in rich medium or with glucose, expression from the Pu and Ps1 promoters was repressed, demonstrating that the system is susceptible to catabolite repression (Duetz *et al.*, 1996; Holtel *et al.*, 1994; Marques *et al.*, 1994; Vercellone-Smith and Herson, 1997). In addition, when a *xylS* mutant was grown on benzoate, the Pm promoter was still activated, suggesting possible cross-talk between the chromosomally-encoded ortho pathway and TOL-plasmid-encoded meta pathway for the catabolism of benzoate (Kessler *et al.*, 1994). For a more detailed description of the *xyl* regulatory mechanism, the reader is referred to detailed reviews and the original papers cited therein (Assinder and Williams, 1990; Ramos *et al.*, 1997).

IX. A FUNGAL TOLUENE DEGRADATION PATHWAY

Until recently, fungi were thought to transform or cooxidize aromatic hydrocarbons, but not utilize them as carbon and energy sources. For example, polycyclic aromatic hydrocarbons (PAHs) are oxidized by cytochrome P-450 enzymes in fungi and other eukaryotic organisms to form arene oxides, which are then converted to *trans*-dihydrodiols or phenols (Cerniglia, 1997). For most fungi, these compounds represent dead-end metabolites that are not degraded further and do not serve as carbon and energy sources. The ligninolytic fungus *Phanerochaete chrysosporium* was shown to degrade all six components of BTEX under nonligninolytic conditions by an unknown mechanism. The organism was apparently not using the aromatic hydrocarbons as carbon and energy sources, although it carried out some mineralization of toluene and benzene (Yadav and Reddy, 1993). In mammals, toluene is oxidized at the methyl group to form benzyl alcohol, further oxidized to benzoate, and then either excreted directly or following enzymatic modification (Bray *et al.*, 1951). It is now clear that certain fungal genera carry out a similar initial monooxygenase attack on the methyl group of toluene and related single-ring aromatic hydrocarbons, but some isolates are capable of complete mineralization, assimilating the carbon and gaining energy in the process (Prenafeta-Boldú *et al.*, 2006).

Several fungal genera were initially shown to degrade various *n*-alkylbenzenes (Fedorak and Westlake, 1986), and clear evidence for toluene utilization and complete degradation was first demonstrated with the fungus *Cladosporium sphaerospermum* (Fig. 1.9) (Weber *et al.*, 1995). This organism was isolated from a biofiltration system for the removal of toluene; similar biofilters now appear to be a ready source of aromatic hydrocarbon-degrading fungi (Cox *et al.*, 1993; Estevez *et al.*, 2005; Woertz *et al.*, 2001). The low moisture content and available solid support seem to favor the enrichment of fungi, even in reactors initially inoculated with bacterial cultures (Prenafeta-Boldú, 2001; Woertz *et al.*, 2001). In general, fungi appear to be particularly well suited for use in biofiltration systems for the removal of volatile alkylbenzenes because of their resistance to dry conditions and low pH (Kennes and Veiga, 2004).

Most of the confirmed toluene-degrading fungal isolates fall into the genera *Cladophialophora* and *Exophiala* and were isolated from toluene biofilters or petroleum-contaminated soils. However, the toluene-degrading ability appears to be strain-specific rather than species-specific (Prenafeta-Boldú *et al.*, 2006), and appears to be relatively rare, as 260 fungal isolates from a culture collection were screened for growth on toluene and none had this ability (Prenafeta-Boldú, 2001).

A toluene degradation pathway for the first toluene-degrading fungal isolate, *Cladosporium sphaerospermum* (recently reclassified as *Cladophialophora* sp. CBS 114326; Prenafeta-Boldú *et al.*, 2006) was proposed based on substrate utilization, oxygen uptake, and enzyme assays (Fig. 1.9) (Weber *et al.*, 1995). This strain grew with ethylbenzene, propylbenzene, and styrene but not benzene, *o*-xylene, or phenol. Of a series of possible toluene degradation intermediates, the strain grew with benzyl alcohol, benzaldehyde, benzoate, and catechol, but not 3-methylcatechol. Consistent with these results, significant oxygen uptake was observed with toluene, benzyl alcohol, benzaldehyde, benzoate, catechol, and *p*-hydroxybenzaldehyde, but not 3-methylcatechol or cresols. Enzyme assays of toluene-grown cell extracts revealed the presence of benzyl alcohol and benzaldehyde dehydrogenases, *p*-hydroxybenzoate hydroxylase, C12O, and protocatechuate 3,4-dioxygenase (Weber *et al.*, 1995). A more recent study with the same strain demonstrated that the initial oxidation of toluene is catalyzed by a cytochrome P-450-dependent toluene monooxygenase (Luykx *et al.*, 2003). This result was perhaps not unexpected, as a cytochrome P-450 enzyme was also shown to catalyze the initial oxidation of the related substrate styrene in the fungus *Exophiala jeanselmei* (Cox *et al.*, 1996), and an earlier study of benzylic hydroxylation of aromatic hydrocarbons by several species of fungi also implicated cytochrome P-450 monooxygenases in the reaction (Holland *et al.*, 1988). The toluene monooxygenase, which preferred NADPH over NADH, was membrane-bound and had activity with toluene, ethylbenzene,

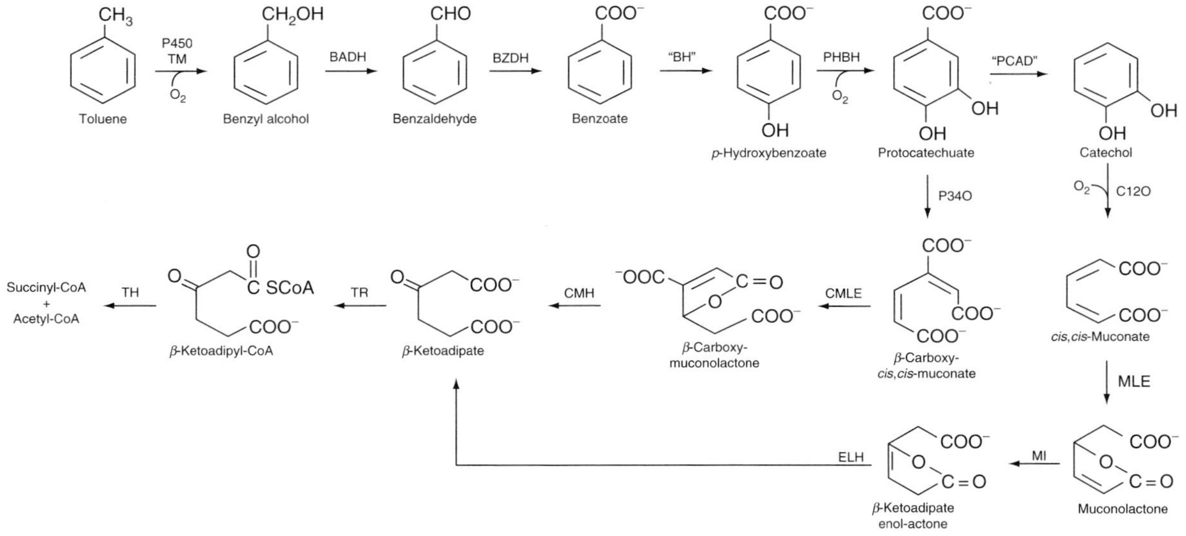

FIGURE 1.9 Toluene degradation pathway in fungi. Pathway sequence indicating intermediates and enzymes. Enzymes: P450 TM, cytochrome P-450 toluene monooxygenase; BADH, benzyl alcohol dehydrogenase; BZDH, benzaldehyde dehydrogenase; "BH", hypothetical benzoate hydroxylase; PHBH, *p*-hydroxybenzoate hydroxylase; P34O, protocatechuate 3,4-dioxygenase; CMLE, β-carboxy-*cis,cis*-muconate lactonizing enzyme; CMH, β-carboxymuconolactone hydrolase; TR, β-ketoadipate succinyl-CoA transferase; TH, β-ketoadipyl-CoA thiolase; "PCAD", hypothetical protocatechuate decarboxylase; C12O, catechol 1,2-dioxygenase; MLE, *cis,cis*-muconate lactonizing enzyme; MI, muconolactone isomerase; ELH, β-ketoadipate enol-lactone hydrolase. "PCAD" and "BH" have not been assayed, and it is not clear at this time whether both proposed branches of the β-ketoadipate pathway are involved in toluene degradation.

propylbenzene, styrene, and all three isomers of xylene, but it did not oxidize benzene (Luykx et al., 2003). The monooxygenase and a NADPH–cytochrome P-450 reductase were purified to homogeneity, but activity could not be reconstituted, suggesting that additional electron donating components might be required. Taken together, these data predict the degradation pathway shown in Fig. 1.9. The initial oxidation on the methyl group of toluene is catalyzed by a cytochrome P-450 monooxygenase to form benzyl alcohol, which is then converted to benzaldehyde and benzoate by the relevant dehydrogenases. Benzoate is then apparently hydroxylated to form p-hydroxybenzoate, which is oxidized by p-hydroxybenzoate hydroxylase to protocatechuate. At this stage there are two options, both of which may be occurring simultaneously. Since the organism has P34O activity (Weber et al., 1995), protocatechuate is likely being metabolized through an ortho-cleavage pathway, which has been seen previously in other fungi (Cain et al., 1968). However, since C12O activity is also present in toluene-grown cells (Weber et al., 1995), protocatechuate may be decarboxylated to catechol, which is then oxidized through the other branch of the β-ketoadipate pathway. The fungal pathway for toluene degradation therefore has some major differences from bacterial pathways, particularly the initial attack by a cytochrome P-450 enzyme, and also some similarities, including the conversion of benzyl alcohol to benzoate (like in the TOL pathway) and the use of the ortho cleavage pathway (as in the later steps in the toluene 4-monooxygenase pathway; Fig. 1.1). Previous studies of the ortho cleavage pathway demonstrated that there was one clear difference between the bacterial and eukaryotic protocatechuate branches. The conversion of β-carboxymuconate to β-ketoadipate involves the participation of three enzymes in bacteria and only two in fungi (Figs. 1.6A and 1.9) (Harwood and Parales, 1996).

Several studies have evaluated the effectiveness of toluene and alkylbenzene degradation by fungal isolates, and optimized conditions for degradation have been reported (Kennes and Veiga, 2004; Prenafeta-Boldú, 2001; Prenafeta-Boldú et al., 2002, 2004; Qi et al., 2002). The degradation of BTEX mixtures revealed that *Cladophialophora* sp. strain T1, which was isolated from a BTEX-polluted site (Prenafeta-Boldú, 2001), was capable of growth with toluene and ethylbenzene, and partial metabolism of o- and m-xylene to phthalates, but it did not metabolize benzene. p-Xylene appeared to be mineralized, but only when cells were grown in the presence of toluene and not in BTEX mixtures (Prenafeta-Boldú et al., 2002). Available evidence suggested that all of the alkylbenzenes were oxidized at the methyl group by the same monooxygenase, suggesting that the pathway in this strain is similar to that in *Cladophialophora* sp. CBS 114326.

X. ANAEROBIC TOLUENE PATHWAY: *THAUERA AROMATICA* K172

Not long ago it was believed that aromatic compounds such as toluene could not be biologically degraded in the absence of molecular oxygen, but we now know that this is not the case. In contrast to the aerobic degradation pathways in which oxygen is introduced into toluene via mono- and dioxygenase-catalyzed reactions, toluene can be metabolized in the complete absence of molecular oxygen by novel biochemical reactions. The anaerobic degradation of toluene has been demonstrated under a variety of electron acceptor conditions, including denitrification, iron-reduction, sulfate-reduction, phototrophic, and methanogenic conditions (Beller and Spormann, 1997a; Biegert *et al.*, 1996; Coschigano *et al.*, 1998; Kane *et al.*, 2002; Rabus and Heider, 1998; Washer and Edwards, 2007; Zengler *et al.*, 1999). Overall, the anaerobic toluene degradation pathway appears to be common to all anaerobic toluene degraders isolated to date (for recent reviews see Chakraborty and Coates, 2004; Spormann and Widdel, 2000; Widdel and Rabus, 2001). The hallmark of the anaerobic toluene degradation pathway is the addition of fumarate to the methyl group of toluene by the glycyl-radical enzyme benzylsuccinate synthase (Bss) to form (*R*)-benzylsuccinate (Fig. 1.10A). (*R*)-Benzylsuccinate is then metabolized via a series of reactions to benzoyl-CoA, a central intermediate in anaerobic aromatic degradation (Boll *et al.*, 2002; Gibson and Harwood, 2002; Harwood *et al.*, 1998; Spormann and Widdel, 2000). Benzoyl-CoA is further metabolized by reduction of the aromatic ring followed by a series of β-oxidation-like reactions to yield acetyl-CoA (Fig. 1.10A). Detailed studies of the genetics and biochemistry of anaerobic toluene metabolism have been carried out in several strains but the overall pathway has been most well studied in *Thauera aromatica* K172. The following section summarizes our current understanding of the anaerobic toluene pathway using *T. aromatica* K172 as the main example.

A. Pathway details and enzymology

T. aromatica K172 was isolated by selection for anaerobic growth on phenol as the sole carbon and energy source. This denitrifying bacterium was subsequently demonstrated to utilize a variety of aromatic compounds including toluene (Tschech and Fuchs, 1987). Experiments with cell suspensions and extracts demonstrated that toluene degradation was induced during growth on toluene and initiated by the addition of fumarate to toluene (Biegert and Fuchs, 1995; Biegert *et al.*, 1996). Anaerobic degradation of toluene is initiated by benzylsuccinate synthase, which catalyzes the stereospecific addition of fumarate to the methyl-group of toluene to yield (*R*)-benzylsuccinate (Fig. 1.10A). Benzylsuccinate synthase

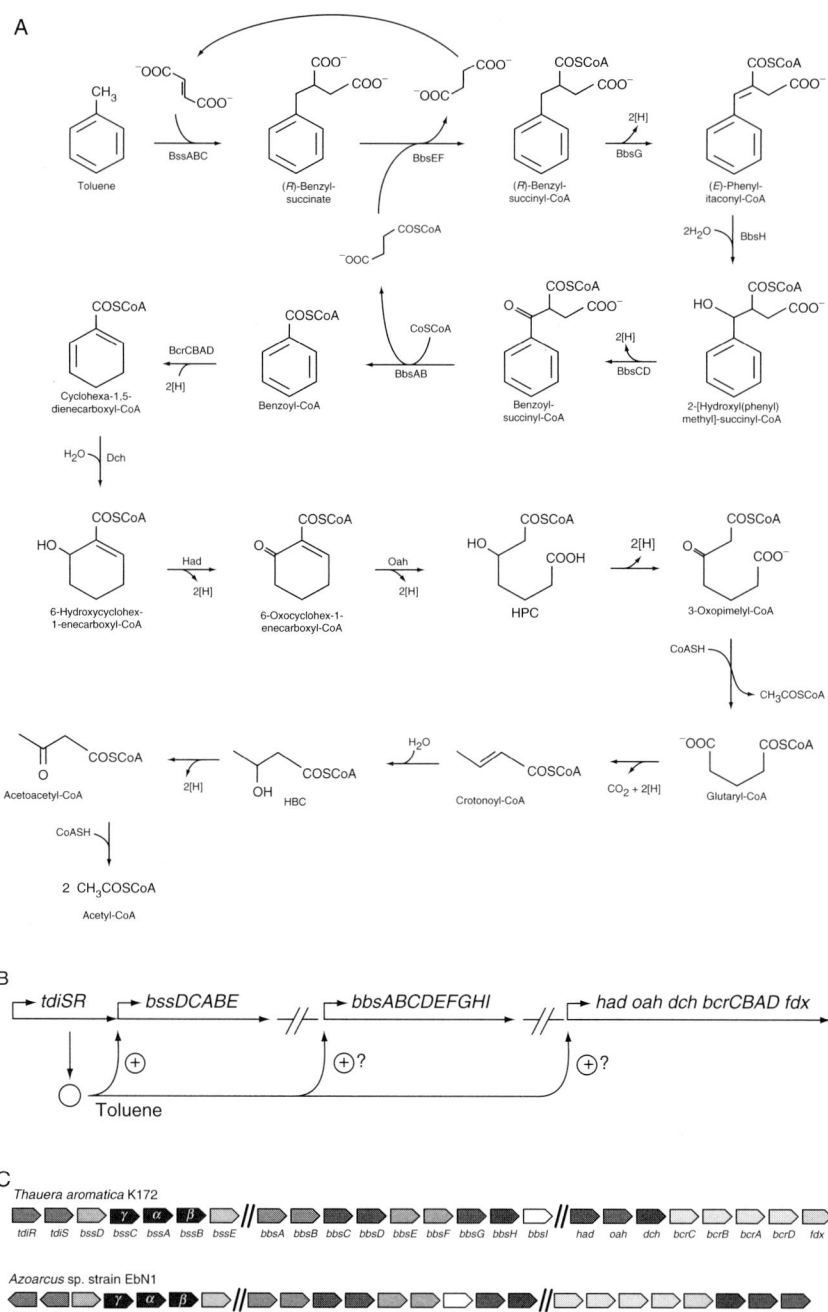

was purified from toluene-grown *T. aromatica* K172 cells (Leuthner *et al.*, 1998). This novel flavin-containing oxygen-sensitive enzyme is composed of three subunits in a $\alpha_2\beta_2\gamma_2$ native conformation. It consists of a glycyl radical-containing large subunit (BssA) of 98 kDa, and two additional structural subunits of unknown function, BssB (8.5 kDa) and BssC (6.5 kDa). The genes encoding benzylsuccinate synthase were cloned and sequenced from strain K172 using a reverse genetic approach (Leuthner *et al.*, 1998), and were shown to be transcribed in an operon with *bssE* in the order *bssDCABE* (Hermuth *et al.*, 2002) (Fig. 1.10B and C). BssA is a member of the pyruvate–formate lyase and anaerobic ribonucleotide reductase glycyl radical enzyme family (Lehtio and Goldman, 2004; Selmer *et al.*, 2005). BssD (36 kDa) is the activase subunit of benzylsuccinate synthase, which is required for adenosylmethionine-dependent free radical formation on a glycine residue of benzylsuccinate synthase. BssE has been hypothesized to be an ATP-dependent chaperone involved in assembly of benzylsuccinate synthase.

Benzylsuccinate is then metabolized by a series of β-oxidation-like reactions to the central intermediate benzoyl-CoA (Leutwein and Heider, 1999) (Fig. 1.10A). These β-oxidation enzymes were identified by 2-dimensional gel electrophoresis analysis of toluene-induced proteins and amino terminal sequence analyses (Leuthner and Heider, 2000).

FIGURE 1.10 Anaerobic toluene degradation pathway in *T. aromatica* K172. (A) Upper (toluene to benzoyl-CoA) and lower (benzoyl-CoA to TCA cycle intermediates) pathway sequence indicating intermediates, enzymes, and genes. Enzymes: BssABC, benzylsuccinate synthase; BbsEF, benzylsuccinate CoA transferase; BbsG, benzylsuccinyl-CoA dehydrogenase; BbsH, phenyl-itaconyl-CoA hydratase; BbsCD, 2-[hydroxyl(phenyl)methyl]-succinyl-CoA dehydrogenase; BbsAB, benzoyl-succinyl-CoA thiolase. BcrCBAD, benzoyl-CoA reductase; Dch, dienoyl-CoA hydratase; Had, 6-hydroxycyclohex-1-enecarboxyl-CoA dehydrogenase; Oah, 6-oxocyclohex-1-enecarboxyl-CoA hydrolase. Intermediates: HPC, 3-hydroxypimelyl-CoA; HBC, 3-hydroxybutyryl-CoA. (B) Gene organization and regulation. Arrows below gene designations indicate direction of transcription. The ⊕ symbol represents positive regulation of the respective promoters (represented by ⟼). Question marks indicate that the proposed regulation mechanism has not been experimentally tested.
(C) Comparison of anaerobic toluene degradation gene organization. Genes are not drawn to scale. Colored boxes indicate the open reading frame for each gene. White boxes indicate open reading frames (*orf*) without known function. Arrowheads indicate the direction of transcription. Two solid black lines indicate that the gene regions are distantly located from one another. The color scheme is as follows: Blue, benzylsuccinate synthase genes; grey, benzylsuccinate synthase activase genes; light blue, potential benzylsuccinate synthase chaperone genes; red, hydratase genes; brown, benzylsuccinate CoA transferase genes; yellow, benzoyl-CoA reductase genes; purple, dehydrogenase genes; orange, benzoylsuccinyl-CoA thiolase genes; green, regulatory genes; light green, ferredoxin genes. (See Color Plate Section in the back of the book.)

The genes encoding the enzymes of the benzylsuccinate β-oxidation-like reactions were cloned and sequenced and were located in one operon on the K172 chromosome in the order *bbsABCDEFGHI* (Fig. 1.10B and C). (*R*)-Benzylsuccinate is activated by benzylsuccinate CoA transferase, which has been purified by standard chromatographic techniques from toluene-grown cells. Benzylsuccinate CoA transferase is composed of BbsE (44 kDa) and BbsF (45 kDa) in a native $\alpha_2\beta_2$ confirmation, and catalyzes the formation of (*R*)-benzylsuccinyl-CoA with succinyl-CoA as the CoA donor (Leutwein and Heider, 2001). (*R*)-Benzylsuccinyl-CoA is then oxidized by the flavin-containing enzyme benzylsuccinyl-CoA dehydrogenase, encoded by BbsG. BbsG was expressed in *E. coli* and active enzyme was purified and characterized. The purified enzyme is a homotetramer of 45-kDa subunits that contains one FAD molecule per subunit. BbsG catalyzes the oxidation of (*R*)-benzylsuccinyl-CoA to form (*E*)-phenylitaconyl-CoA (Leutwein and Heider, 2002). The subsequent β-oxidation enzymes have not been purified but were proposed based on sequence homology of proteins encoded within the *bbs* operon and 2-dimensional gel electrophoresis detection of proteins induced during growth on toluene. These include phenylitaconyl-CoA hydratase, encoded by *bbsH*, which catalyzes the formation of 2-[hydroxy(phenyl)methyl]succinyl-CoA. This compound undergoes a dehydrogenation reaction catalyzed by 2-[hydroxy(phenyl)methyl]succinyl-CoA dehydrogenase, a two-subunit enzyme encoded by *bbsCD*, to give benzoylsuccinyl-CoA. Benzoylsuccinyl-CoA undergoes a thiolytic cleavage catalyzed by a two-subunit thiolase encoded by *bbsAB* to yield succinyl-CoA and benzoyl-CoA (Fig. 1.10A). Succinyl-CoA is then free to be metabolized by succinyl-CoA transferase and a dehydrogenase to regenerate fumarate for the next catalytic cycle.

Benzoyl-CoA is then reduced by another enzyme unique to anaerobic aromatic metabolism, benzoyl-CoA reductase (BcrCBAD) (Fig. 1.10A). This enzyme was purified from *T. aromatica* K172 and is a $\alpha\beta\gamma\delta$ heterotetramer that requires stoichiometric amounts of ATP and contains three cysteine-ligated [4Fe-4S] clusters. A ferredoxin containing two [4Fe-4S] clusters serves as the native electron donor (Boll and Fuchs, 1995; Boll and Fuchs, 1998). Cyclohexa-1,5-diene-1-carboxyl-CoA hydratase (Dch) was purified from *T. aromatica* K172 (Laempe *et al.*, 1998) as were 6-hydroxycyclohex-1-ene-1-carboxyl-CoA dehydrogenase (Had), and 6-oxocyclohex-1-ene-1-carbonyl-CoA hydrolase (Oah) (Laempe *et al.*, 1999). The remaining enzymes in the benzoate pathway have not been studied in detail with the exception of glutaryl-CoA dehydrogenase, which has also been purified from *T. aromatica* K172. The purified enzyme was demonstrated to catalyze the oxidation and decarboxylation of glutaryl-CoA to form crotonoyl-CoA (Hartel *et al.*, 1993). The genes for the benzoyl-CoA pathway were cloned and sequenced and found to be clustered on the *T. aromatica* K172 chromosome in the order *had-oah-dch-bcrCBA-fdx* (Fig. 1.10B and C).

As mentioned previously, anaerobic toluene degradation has been studied in several other bacterial strains. *T. aromatica* strain T1 was isolated for anaerobic utilization of toluene as sole carbon and energy source (Evans *et al.*, 1991). A chemical mutagenesis and complementation approach identified *tutCB*, encoding a two-component regulation system just upstream of genes encoding benzylsuccinate synthase (*tutEFDGH*). These genes are homologous to *bssDCABE* of *T. aromatica* K172 (Coschigano, 2000; Coschigano *et al.*, 1998). Subsequent sequence analysis revealed another two-component system, *tutC1B1*, adjacent to but divergently transcribed from *tutCB*, which is adjacent to but divergently transcribed from *tutEFDGH* (Fig. 1.10C).

Extensive biochemical and mechanistic studies have been carried out in another anaerobic toluene-degrading denitrifier, *Azoarcus* sp. strain T. Benzylsuccinate synthase substrate specificity and enzyme mechanism was determined with partially purified enzyme (Beller and Spormann, 1997a, 1999). The location of glycyl radical in benzylsuccinate synthase was established by EPR studies (Krieger *et al.*, 2001). The genes encoding Bss and putative regulatory proteins were cloned and sequenced and determined to be in the order *tdiRS-bssDCABE* (Fig. 1.10C) (Achong *et al.*, 2001).

Azoarcus sp. strain EbN1 is yet another denitrifying strain isolated and demonstrated to anaerobically utilize toluene (Rabus and Widdel, 1995). Strain EbN1 has the unique ability to utilize toluene and ethylbenzene and investigators established that the two compounds are metabolized by two different pathways based on substrate consumption and 2-dimensional gel electrophoresis analyses (Champion *et al.*, 1999). Genes for the anaerobic toluene pathway were cloned and sequenced (Kube *et al.*, 2004). In addition, the 4.3-Mb genome of this versatile anaerobic aromatic degrading strain has been sequenced and genome sequence analysis identified the genes encoding the *bss* and *bbs* operons (Rabus *et al.*, 2005). Microarray studies with strain EbN1 demonstrated that all 17 genes known to play a role in the toluene degradation pathway were upregulated during growth on toluene, and a subset of these was verified by RT-PCR and proteomics analysis. Interestingly, five additional genes downstream of the *bss* operon were significantly upregulated under toluene-utilization conditions (Kuhner *et al.*, 2005).

In addition to the denitrifying strains, toluene utilization has also been demonstrated under iron-reducing conditions in *Geobacter metallireducens*, and *bss* activity was determined in permeabilized cells. *bssB* and partial *bssA* genes were cloned and sequenced (Kane *et al.*, 2002). Benzylsuccinate synthase genes have also been isolated from methanogenic consortia (Washer and Edwards, 2007) and Bss enzyme activities were also described in the sulfate-reducing bacterial strains PRTOL1 and *Desulfobacula toluolica* (Beller and Spormann, 1997b; Rabus and Heider, 1998) and the phototroph, *Blastochloris sulfovirdis* (Zengler *et al.*, 1999).

B. Genetics and regulation

Relatively few genetic studies have been carried out to analyze the regulation of anaerobic toluene metabolism other than analyses of operon structure and induction of enzyme activities. In *T. aromatica* K172, sequence analysis identified a two-component regulatory system, *tdiSR*, just upstream from the *bss* structural genes and transcribed in the same direction as the *bss* structural genes (Fig. 1.10C) (Leuthner and Heider, 1998). TdiS is a 63.5-kDa protein containing sensor histidine kinase and PAS domains. TdiR is a 25-kDa protein of the LuxR response regulator family, which contains a helix-turn-helix DNA-binding motif. TdiR was shown to bind to a DNA fragment containing the intergenic region between *tdiR* and *bssD* by gel shift assays.

Gene deletion and site-directed mutagenesis experiments in *T. aromatica* T1 have established that disruptions in any of the *tutEFDGH* genes inhibit the ability to utilize toluene (Bhandare *et al.*, 2006; Coschigano, 2002; Coschigano and Bishop, 2004). The authors also demonstrated that benzylsuccinate induced *tutEFDGH* gene expression but toluene only induced expression in strains producing active benzylsuccinate synthase, indicating that benzylsuccinate may be the true inducer in *T. aromatica* T1. This is not the case in *T. aromatica* K172 where toluene has been shown to be the only inducer of *bssDEFGH* (Fig. 1.10C) (Hermuth *et al.*, 2002).

These studies all indicate a common mechanism for the anaerobic activation of toluene by the addition of fumarate to the methyl group of toluene. Additional reports indicate that this Bss type of reaction may be involved anaerobic activation of other hydrocarbon compounds including *m*-cresol, *p*-cresol, *m*-xylene and the alkane *n*-hexane (for a review see Heider, 2007).

XI. CONCLUSIONS

It is clear that microorganisms have evolved a variety of interesting pathways for the degradation of toluene. These pathways can be broadly divided into aerobic and anaerobic pathways. Aerobically, the strategy has been to incorporate oxygen using either dioxygenases or monooxygenases. When monooxygenation is used, attack can either be on the methyl group or the aromatic ring. The aromatic ring-hydroxylating monooxygenases involved in toluene degradation from *B. vietnamiensis*

G4, *R. pickettii* PKO1, *P. mendocina* KR1, *P. stutzeri* OX1 and *Burkholderia* sp. JS150 are all members of a large family of soluble diiron monooxygenases that are related to soluble MMO (Leahy *et al.*, 2003). Analysis of the various monooxygenase enzyme systems (Fig. 1.3), suggests that there were two original phylogenetically-related sources from which all of the enzymes evolved. All of the enzymes have several features in common: all contain a hydroxylase composed of three subunits (α, β, γ), a reductase and an effector protein that is required for optimal monooxygenase activity. The α- and β-subunits of the hydroxylase appear to have originated from an ancient gene duplication followed by loss of the active site in β-subunit. The γ subunits, however, do not appear to be related to each other and may have been independently recruited to these enzyme systems (Leahy *et al.*, 2003; Notomista *et al.*, 2003).

Although the similarities between multicomponent diiron monooxygenases are obvious, clear differences are also apparent. Based on these differences, the diiron monooxygenase enzymes have been divided into four major groups based on the phylogeny of the hydroxyase α- and β-subunits, operon organization, substrate specificity, and the total number of components required for activity (Leahy *et al.*, 2003; Notomista *et al.*, 2003). T3MO from *R. pickettii* G4, T4MO from *P. mendocina* KR1, Tbc2m from *Burkholderia* sp. strain JS150, and ToMO from *P. stutzeri* OX1 are members of a group of four-component monooxygenases, which are comprised of a three-subunit hydroxylase (α, β, γ), reductase and ferredoxin electron transfer components, and an effector protein required for optimal enzymatic activity. The characteristic gene organization for this group of monooxygenases is α-subunit, γ-subunit, ferredoxin, effector protein, β-subunit, reductase (Fig. 1.3). Conversely, T2MO from *B. vietnamiensis* G4, Tbc1m (PH) and Tb2m from *Burkholderia* sp. strain JS150, and PH from *P. stutzeri* OX1 are part of the three-component group that has homology with the phenol monooxygenases represented by the well-studied enzyme from *P. putida* CF600 (Shingler *et al.*, 1992). These monooxygenases are comprised of a three-subunit hydroxylase (α, β, γ), reductase and effector protein (similar to the four-component group), but they lack a ferredoxin in their electron transport chains. In addition, this group contains an assembly protein that is apparently required for incorporating iron into the active site of the hydroxylase α-subunit (Powlowski *et al.*, 1997). The characteristic gene order for the three-component group of monooxygenases is assembly protein, β-subunit, effector protein, α-subunit, γ-subunit, reductase (Fig. 1.3).

Monooxygenase enzymes from both of these groups have been localized to both plasmids and chromosomes, and are found in numerous bacterial lineages. In some cases, evidence of recent gene acquisition is indicated by the non-native GC content of the monooxygenase gene cluster and/or the presence of nearby transposase genes. Most likely the monooxygenase gene clusters were acquired by horizontal transfer and have become associated with appropriate lower pathways for cleavage of hydroxylated products. In fact, with the exception of the T4MO pathway in *P. mendocina* KR1, all of the aerobic bacterial toluene degradation pathways funnel oxidized metabolites through a standard meta cleavage pathway (Fig. 1.3). Each pathway has acquired an appropriate regulation system to control gene expression. In all known instances, the regulation of toluene degradation gene expression is controlled by positive regulatory mechanisms where toluene or relevant toluene metabolites function as the inducers. However, the activator proteins present in the various toluene-degrading strains are members of unrelated protein families, indicating that the type of regulator is less important than its specificity and sensitivity (Cases and de Lorenzo, 2001). Where the regulation of toluene degradation has been studied in detail (e.g., the TOL plasmid), the regulation of toluene degradation genes is extremely complex. This could very well be the case in the less well-studied systems.

In contrast to the variations in aerobic pathways, the single known anaerobic pathway is quite conserved in the available isolated organisms. However, this conserved pathway is broadly distributed in metabolically diverse anaerobes. Interestingly, there is one known isolate, *Thauera* sp. strain DNT-1, which contains both an anaerobic and an aerobic pathway for toluene utilization (Shinoda *et al.*, 2004). This facultative anaerobe was shown to induce both aerobic and anaerobic pathways in the presence of toluene. However, dioxygenase genes similar to the *tod* genes of *P. putida* F1 were only induced in the presence of oxygen, whereas the genes for the classic anaerobic pathway were expressed under both aerobic and anaerobic conditions. Therefore, the transmission and acquisition of toluene degradation genes in the environment goes beyond the conventional aerobic/anaerobic compartmentalization of toluene degradation systems.

In summary, microorganisms have evolved multiple strategies for the initial attack on toluene (dioxygenation, monooxygenation on the ring or methyl group, and anaerobic fumarate addition), and three main strategies for cleavage of the aromatic ring (ortho, meta, and a β-oxidation-like mechanism that follows ring reduction). It is clear that different combinations of upper and lower pathways can be assembled to form complete toluene degradation pathways (Fig. 1.1), and the individual pathways can be envisioned to have evolved by a modular type of pathway evolution (Williams and Sayers, 1994). In several strains, upper pathway genes encoding enzymes that convert aromatic hydrocarbon substrates to catechols appear to have become fused to a lower pathway gene cluster for catechol ring cleavage (Fig. 1.3). These interesting combinations are not limited to the bacterial world, as fungal species have assembled a TOL plasmid-like upper pathway that functions with a β-ketoadipate (ortho) lower pathway. In this case, however, the fungus uses a different type of monooxygenase (a cytochrome P-450) to catalyze the initial oxidation of toluene. Until recently, researchers have focused on the identification of new toluene-degrading *bacteria*, but eukaryotes are now known to play a role in the process. Further enrichment and isolation of eukaryotic toluene degraders should clarify the role and importance of eukaryotes in aromatic hydrocarbon degradation in the environment. In addition, there may be alternative toluene degradation pathways that remain undiscovered. For example, the only strategy for the isolation of toluene-degrading organisms that specifically selected against strains that utilize a meta pathway resulted in the identification of *P. mendocina* KR1, in which a novel pathway involving an initial para-monooxygenation, followed by a methyl-group oxidation and ring cleavage via the β-ketoadipate (ortho) pathway was identified. In enrichments, only the organisms that are best suited to the specific enrichment strategy will predominate; therefore, by designing alternate enrichment strategies additional pathways for toluene attack, cleavage, and mineralization may come to light.

ACKNOWLEDGMENTS

We are grateful to Sol Resnick and Kristina Mahan for helpful comments on portions of the manuscript, and David Gibson for providing the description of the *P. mendocina* KR1 isolation strategy.

REFERENCES

Abril, M.-A., Michan, C., Timmis, K. N., and Ramos, J. L. (1989). Regulator and enzyme specificities of the TOL plasmid-encoded upper pathway for degradation of aromatic hydrocarbons and expansion of the substrate range of the pathway. *J. Bacteriol.* **171,** 6782–6790.

Achong, G. R., Rodriguez, A. M., and Spormann, A. M. (2001). Benzylsuccinate synthase of *Azoarcus* sp. strain T: Cloning, sequencing, transcriptional organization, and its role in anaerobic toluene and *m*-xylene mineralization. *J. Bacteriol.* **183,** 6763–6770.

Applegate, B., Kelly, C., Lackey, L., McPherson, J., Kehrmeyer, S., Menn, F.-M., Bienkowski, P., and Sayler, G. (1997). *Pseudomonas putida* B2: A *tod–lux* bioluminescent reporter for toluene and trichloroethylene co-metabolism. *J. Ind. Microbiol. Biotechnol.* **18,** 4–9.

Applegate, B. M., Kehrmeyer, S. R., and Sayler, G. S. (1998). A chromosomally based *tod–luxCDABE* whole-cell reporter for benzene, toluene, ethybenzene, and xylene (BTEX) sensing. *Appl. Environ. Microbiol.* **64,** 2730–2735.

Arenghi, F. L., Barbieri, P., Bertoni, G., and de Lorenzo, V. (2001a). New insights into the activation of *o*-xylene biodegradation in *Pseudomonas stutzeri* OX1 by pathway substrates. *EMBO Rep.* **2,** 409–414.

Arenghi, F. L., Berlanda, D., Galli, E., Sello, G., and Barbieri, P. (2001b). Organization and regulation of *meta* cleavage pathway genes for toluene and *o*-xylene derivative degradation in *Pseudomonas stutzeri* OX1. *Appl. Environ. Microbiol.* **67,** 3304–3308.

Arenghi, F. L., Pinti, M., Galli, E., and Barbieri, P. (1999). Identification of the *Pseudomonas stutzeri* OX1 toluene–*o*-xylene monooxygenase regulatory gene (*touR*) and of its cognate promoter. *Appl. Environ. Microbiol.* **65,** 4057–4063.

Assinder, S. J., and Williams, P. A. (1990). The TOL plasmids: Determinants of the catabolism of toluene and the xylenes. *Adv. Microbial Physiol.* **31,** 1–69.

Austin, R. N., Buzzi, K., Kim, E., Zylstra, G. J., and Groves, J. T. (2003). Xylene monooxygenase, a membrane-spanning non-heme diiron enzyme that hydroxylates hydrocarbons via a substrate radical intermediate. *J. Biol. Inorg. Chem.* **8,** 733–740.

Austin, R. N., Chang, H.-K., Zylstra, G. J., and Groves, J. T. (2000). The non-heme diiron alkane monooxygenase of *Pseudomonas oleovorans* (AlkB) hydroxylates via a substrate radical intermediate. *J. Am. Chem. Soc.* **122,** 11747–11748.

Baggi, G., Barbieri, P., Galli, E., and Tollari, S. (1987). Isolation of a *Pseudomonas stutzeri* strain that degrades *o*-xylene. *Appl. Environ. Microbiol.* **64,** 2473–2478.

Barbieri, P., Arenghi, F. L., Bertoni, G., Bolognese, F., and Galli, E. (2001). Evolution of catabolic pathways and metabolic versatility in *Pseudomonas stutzeri* OX1. *Antonie Van Leeuwenhoek* **79,** 135–140.

Barbieri, P., Galassi, G., and Galli, E. (1989). Plasmid-encoded mercury resistance in a *Pseudomonas stutzeri* strain that degrades *o*-xylene. *FEMS Microbiol. Lett.* **62,** 375–383.

Barbieri, P., Palladino, L., Gennaro, P., and Galli, E. (1993). Alternative pathways for *o*-xylene or *m*-xylene and *p*-xylene degradation in a *Pseudomonas stutzeri* strain. *Biodegradation* **4,** 71–80.

Barriault, D., Simard, C., Chatel, H., and Sylvestre, M. (2001). Characterization of hybrid biphenyl dioxygenases obtained by recombining *Burkholderia* sp. strain LB400 *bphA* with the homologous gene of *Comamonas testosteroni* B-356. *Can. J. Microbiol.* **47,** 1025–1032.

Beller, H. R., and Spormann, A. M. (1997a). Anaerobic activation of toluene and *o*-xylene by addition to fumarate in denitrifying strain T. *J. Bacteriol.* **179,** 670–676.

Beller, H. R., and Spormann, A. M. (1997b). Benzylsuccinate formation as a means of anaerobic toluene activation by sulfate-reducing strain PRTOL1. *Appl. Environ. Microbiol.* **63,** 3729–3731.

Beller, H. R., and Spormann, A. M. (1999). Substrate range of benzylsuccinate synthase from *Azoarcus* sp. strain T. *FEMS Microbiol. Lett.* **178,** 147–153.

Ben-Bassat, A., Cattermole, M., Gatenby, A. A., Gibson, K. J., Ramos-Gonzalez, M. I., Ramos, J., and Sariaslani, S. (2003). Method for the production of rho-hydroxybenzoate in species of *Pseudomonas* and *Agrobacterium*. US Patent 6,586229.

Benson, S., and Shapiro, J. (1978). TOL is a broad host range plasmid. *J. Bacteriol.* **135,** 278–280.

Bertoni, G., Bolognese, F., Galli, E., and Barbieri, P. (1996). Cloning of the genes for and characterization of the early stages of toluene and *o*-xylene catabolism in *Pseudomonas stutzeri* OX1. *Appl. Environ. Microbiol.* **62,** 3704–3711.

Bertoni, G., Martino, M., Galli, E., and Barbieri, P. (1998). Analysis of the gene cluster encoding toluene/*o*-xylene monooxygenase from *Pseudomonas stutzeri* OX1. *Appl. Environ. Microbiol.* **64,** 3626–3632.

Bhandare, R., Calabro, M., and Coschigano, P. W. (2006). Site-directed mutagenesis of the *Thauera aromatica* strain T1 *tutE tutFDGH* gene cluster. *Biochem. Biophys. Res. Commun.* **346,** 992–998.

Biegert, T., and Fuchs, G. (1995). Anaerobic oxidation of toluene (analogs) to benzoate (analogs) by whole cells and by cell-extracts of a denitrifying *Thauera* sp. *Arch. Microbiol.* **163,** 407–417.

Biegert, T., Fuchs, G., and Heider, J. (1996). Evidence that anaerobic oxidation of toluene in the denitrifying bacterium *Thauera aromatica* is initiated by formation of benzylsuccinate from toluene and fumarate. *Eur. J. Biochem.* **238,** 661–668.

Black, P. N. (1991). Primary sequence of *Escherichia coli fadL* gene encoding an outer membrane protein required for long-chain fatty acid transport. *J. Bacteriol.* **173,** 435–442.

Boll, M., and Fuchs, G. (1995). Benzoyl-coenzyme A reductase (dearomatizing), a key enzyme of anaerobic aromatic metabolism—ATP dependence of the reaction, purification and some properties of the enzyme from *Thauera aromatica* strain K172. *Eur. J. Biochem.* **234,** 921–933.

Boll, M., and Fuchs, G. (1998). Identification and characterization of the natural electron donor ferredoxin and of FAD as a possible prosthetic group of benzoyl-CoA reductase (dearomatizing), a key enzyme of anaerobic metabolism. *Eur. J. Biochem.* **251,** 946–954.

Boll, M., Fuchs, G., and Heider, J. (2002). Anaerobic oxidation of aromatic compounds and hydrocarbons. *Curr. Opin. Chem. Biol.* **6,** 604–611.

Bolognese, F., Di Lecce, C., Galli, E., and Barbieri, P. (1999). Activation and inactivation of *Pseudomonas stutzeri* methylbenzene catabolism pathways mediated by a transposable element. *Appl. Environ. Microbiol.* **65,** 1876–1882.

Boyd, D. R., and Sheldrake, G. N. (1998). The dioxygenase-catalysed formation of vicinal *cis*-diols. *Nat. Prod. Rep.* **15,** 309–324.

Bray, H. G., Thorpe, W. V., and White, K. (1951). Kinetic studies of the metabolism of foreign organic compounds; the formation of benzoic acid from benzamide, toluene, benzyl alcohol and benzaldehyde and its conjugation with glycine and glucuronic acid in the rabbit. *Biochem. J.* **48,** 88–96.

Brown, S. M., and Hudlicky, T. (1993). The use of arene-*cis*-diols in synthesis. In "Organic Synthesis: Theory and Applications" (T. Hudlicky, ed.), Vol. 2, pp. 113–176. JAI Press, Greenwich, CT.

Bühler, B., Schmid, A., Hauer, B., and Witholt, B. (2000). Xylene monooxygenase catalyzes the multistep oxygenation of toluene and pseudocumene to corresponding alcohols, aldehydes, and acids in *Escherichia coli* JM101. *J. Biol. Chem.* **275,** 10085–10092.

Bühler, B., Witholt, B., Hauer, B., and Schmid, A. (2002). Characterization and application of xylene monooxygenase for multistep biocatalysis. *Appl. Environ. Microbiol.* **68,** 560–568.

Butler, C. S., and Mason, J. R. (1997). Structure–function analysis of the bacterial aromatic ring-hydroxylating dioxygenases. *Adv. Microbial. Physiol.* **38,** 47–84.

Byrne, A. M., Kukor, J. J., and Olsen, R. H. (1995). Sequence analysis of the gene cluster encoding toluene-3-monooxygenase from *Pseudomonas picketti* PKO1. *Gene* **154,** 65–70.

Byrne, A. M., and Olsen, R. H. (1996). Cascade regulation of the toluene-3-monooxygenase operon (*tbuA1UBVA2C*) of *Burkholderia pickettii* PKO1: Role of the *tbuA1* promoter (*PtbuA1*) in the expression of its cognate activator, TbuT. *J. Bacteriol.* **178,** 6327–6337.

Cafaro, V., Izzo, V., Scognamiglio, R., Notomista, E., Capasso, P., Casbarra, A., Pucci, P., and Di Donato, A. (2004). Phenol hydroxylase and toluene/*o*-xylene monooxygenase from *Pseudomonas stutzeri* OX1: Interplay between two enzymes. *Appl. Environ. Microbiol.* **70,** 2211–2219.

Cafaro, V., Notomista, E., Capasso, P., and Di Donato, A. (2005). Regiospecificity of two multicomponent monooxygenases from *Pseudomonas stutzeri* OX1: Molecular basis. *Appl. Environ. Microbiol.* **71,** 4736–4743.

Cafaro, V., Scognamiglio, R., Viggiani, A., Izzo, V., Passaro, I., Notomista, E., Piaz, F. D., Amoresano, A., Casbarra, A., Pucci, P., and Di Donato, A. (2002). Expression and purification of the recombinant subunits of toluene/*o*-xylene monooxygenase and reconstitution of the active complex. *Eur. J. Biochem.* **269,** 5689–5699.

Cain, R. B., Bilton, R. F., and Darrah, J. A. (1968). The metabolism of aromatic acids by microorganisms. Metabolic pathways in the fungi. *Biochem. J.* **108,** 797–828.

Carless, H. A. J. (1992). The use of cyclohexa-3,5-diene-1,2-diols in enantiospecific synthesis. *Tetrahedron: Asymmetry* **3,** 795–826.

Cases, I., and de Lorenzo, V. (2001). The black cat/white cat principle of signal integration in bacterial promoters. *EMBO J.* **20,** 1–11.

Cerdan, P., Wasserfallen, A., Rekik, M., Timmis, K. N., and Harayama, S. (1994). Substrate specificity of catechol 2,3-dioxygenase encoded by TOL plasmid pWW0 of *Pseudomonas putida* and its relationship to cell growth. *J. Bacteriol.* **176**, 6074–6081.

Cerniglia, C. E. (1997). Fungal metabolism of polycyclic aromatic hydrocarbons: Past, present and future applications of bioremediation. *J. Ind. Microbiol. Biotechnol.* **19**, 324–333.

Chakraborty, R., and Coates, J. D. (2004). Anaerobic degradation of monoaromatic hydrocarbons. *Appl. Microbiol. Biotechnol.* **64**, 437–446.

Champion, K. M., Zengler, K., and Rabus, R. (1999). Anaerobic degradation of ethylbenzene and toluene in denitrifying strain EbN1 proceeds via independent substrate-induced pathways. *J. Mol. Microbiol. Biotechnol.* **1**, 157–164.

Chauhan, S., Barbieri, P., and Wood, T. K. (1998). Oxidation of trichloroethylene, 1,1-dichloroethylene, and chloroform by toluene/o-xylene monooxygenase from *Pseudomonas stutzeri* OX1. *Appl. Environ. Microbiol.* **64**, 3023–3024.

Chebrou, H., Hurtubise, Y., Barriault, D., and Sylvestre, M. (1999). Catalytic activity toward chlorobiphenyls of purified recombinant His-tagged oxygenase component of *Rhodococcus globerulous* strain P6 biphenyl dioxygenase and of chimeras derived from it and their expression in *Escherichia coli* and in *Pseudomonas putida*. *J. Bacteriol.* **181**, 4805–4811.

Chen, L. H., Kenyon, G. L., Curtin, F., Harayama, S., Bembenek, M. E., Hajipour, G., and Whitman, C. P. (1992). 4-Oxalocrotonate tautomerase, an enzyme composed of 62 amino acid residues per monomer. *J. Biol. Chem.* **267**, 17716–17721.

Cho, M. C., Kang, D.-O., Yoon, B. D., and Lee, K. (2000). Toluene degradation pathway from *Pseudomonas putida* F1: Substrate specificity and gene induction by 1-substituted benzenes. *J. Ind. Microbiol. Biotechnol.* **25**, 163–170.

Cline, P. V., Delfino, J. J., and Rao, P. S. C. (1991). Partitioning of aromatic constituents into water from gasoline and other complex solvent mixtures. *Environ. Sci. Technol.* **25**, 914–920.

Coschigano, P. W. (2000). Transcriptional analysis of the *tutE tutFDGH* gene cluster from *Thauera aromatica* strain T1. *Appl. Environ. Microbiol.* **66**, 1147–1151.

Coschigano, P. W. (2002). Construction and characterization of insertion/deletion mutations of the *tutF*, *tutD*, and *tutG* genes of *Thauera aromatica* strain T1. *FEMS Microbiol. Lett.* **217**, 37–42.

Coschigano, P. W., and Bishop, B. J. (2004). Role of benzylsuccinate in the induction of the *tutE tutFDGH* gene complex of *T. aromatica* strain T1. *FEMS Microbiol. Lett.* **231**, 261–266.

Coschigano, P. W., Wehrman, T. S., and Young, L. Y. (1998). Identification and analysis of genes involved in anaerobic toluene metabolism by strain T1: Putative role of a glycine free radical. *Appl. Environ. Microbiol.* **64**, 1650–1656.

Coulter, E. D., and Ballou, D. P. (1999). Non-haem iron-containing oxygenases involved in the microbial degradation of aromatic hydrocarbons. *Essays Biochem.* **34**, 31–49.

Cox, H. H. J., Faber, B. W., Van Heiningen, W. N., Radhoe, H., Doddema, H. J., and Harder, W. (1996). Styrene metabolism in *Exophiala jeanselmei* and involvement of a cytochrome P-450-dependent styrene monooxygenase. *Appl. Environ. Microbiol.* **62**, 1471–1474.

Cox, H. H. J., Houtman, J. H. M., Doddema, H. J., and Harder, W. (1993). Enrichment of fungi and degradation of styrene in biofilters. *Biotechnol. Lett.* **15**, 737–742.

Darrall, K. G., Figgins, J. A., and Brown, R. D. (1998). Determination of benzene and associated volatile compounds in mainstream cigarette smoke. *Analyst* **123**, 1095–1101.

de Lorenzo, V., Herrero, M., and Timmis, K. N. (1991). An upstream XylR and IHF induced nucleoprotein complex regulates the $\sigma54$-dependent Pu promoter of TOL plasmid. *EMBO J.* **10**, 1159–1167.

Dean, B. J. (1985). Recent findings on the genetic toxicology of benzene, toluene, xylenes, and phenols. *Mutat. Res.* **145**, 153–181.

Delgado, A., Wubbolts, M. G., Abril, M.-A., and Ramos, J. L. (1992). Nitroaromatics are substrates for the TOL plasmid upper pathway enzymes. *Appl. Environ. Microbiol.* **58**, 415–417.

Di Lecce, C., Accarino, M., Bolognese, F., Galli, E., and Barbieri, P. (1997). Isolation and metabolic characterization of a *Pseudomonas stutzeri* mutant able to grow on the three isomers of xylene. *Appl. Environ. Microbiol.* **63**, 3279–3281.

Dixon, R. (1986). The *xylABC* promoter from the *Pseudomonas putida* TOL plasmid is activated by nitrogen regulatory genes in *Escherichia coli*. *Mol. Gen. Genet.* **203**, 129–136.

Dong, X., Fushinobu, S., Fukuda, E., Terada, T., Nakamura, S., Shimizu, K., Nojiri, H., Omori, T., Shoun, H., and Wakagi, T. (2005). Crystal structure of the terminal oxygenase component of cumene dioxygenase from *Pseudomonas fluorescens* IP01. *J. Bacteriol.* **187**, 2483–2490.

Downing, R., and Broda, P. (1979). A cleavage map of the TOL plasmid of *Pseudomonas putida* mt-2. *Mol. Gen. Genet.* **177**, 189–191.

Duetz, W. A., Marques, S., Wind, B., Ramos, J. L., and van Andel, J. G. (1996). Catabolite repression of the toluene degradation pathway in *Pseudomonas putida* harboring pWWO under various conditions of nutrient limitation in chemostat culture. *Appl. Environ. Microbiol.* **62**, 601–606.

Duggleby, C. J., Bayley, S. A., Worsey, M. J., Williams, P. A., and Broda, P. (1977). Molecular sizes and relationships of TOL plasmids in *Pseudomonas*. *J. Bacteriol.* **130**, 1274–1280.

Duggleby, C. J., and Williams, P. A. (1986). Purification and some properties of the 2-hydroxy-6-oxohepta-2,4-dienoate hydrolase (2-hydroxymuconic semialdehyde hydrolase) encoded by the TOL plasmid pWWO from *Pseudomonas putida* mt-2. *J. Gen. Microbiol.* **132**, 717–726.

Eaton, R. W. (1996). *p*-Cumate catabolic pathway in *Pseudomonas putida* F1: Cloning and characterization of DNA carrying the *cmt* operon. *J. Bacteriol.* **178**, 1351–1362.

Eaton, R. W. (1997). *p*-Cymene catabolic pathway in *Pseudomonas putida* F1: Cloning and characterization of DNA encoding conversion of *p*-cymene to *p*-cumate. *J. Bacteriol.* **179**, 3171–3180.

Elsen, N. L., Moe, L. A., McMartin, L. A., and Fox, B. G. (2007). Redox and functional analysis of the Rieske ferredoxin component of the toluene 4-monooxygenase. *Biochemistry* **46**, 976–986.

Eltis, L. D., and Bolin, J. T. (1996). Evolutionary relationships among extradiol dioxygenases. *J. Bacteriol.* **178**, 5930–5937.

Estevez, E., Veiga, M. C., and Kennes, C. (2005). Biodegradation of toluene by the new fungal isolates *Paecilomyces variotii* and *Exophiala oligosperma*. *J. Ind. Microbiol. Biotechnol.* **32**, 33–37.

Evans, P. J., Mang, D. T., Kim, K. S., and Young, L. Y. (1991). Anaerobic degradation of toluene by a denitrifying bacterium. *Appl. Environ. Microbiol.* **57**, 1139–1145.

Fedorak, P. M., and Westlake, D. W. (1986). Fungal metabolism of *n*-alkylbenzenes. *Appl. Environ. Microbiol.* **51**, 435–437.

Ferraro, D. J., Brown, E. N., Yu, C. L., Parales, R. E., Gibson, D. T., and Ramaswamy, S. (2007). Structural investigations of the ferredoxin and terminal oxygenase components of the biphenyl 2,3-dioxygenase from *Sphingobium yanoikuyae* B1. *BMC Struct. Biol.* **7**, 10.

Finette, B. A., and Gibson, D. T. (1988). Initial studies on the regulation of toluene degradation by *Pseudomonas putida* F1. *Biocatalysis* **2**, 29–37.

Finette, B. A., Subramanian, V., and Gibson, D. T. (1984). Isolation and characterization of *Pseudomonas putida* PpF1 mutants defective in the toluene dioxygenase enzyme system. *J. Bacteriol.* **160**, 1003–1009.

Fishman, A., Tao, Y., Bentley, W. E., and Wood, T. K. (2004a). Protein engineering of toluene 4-monooxygenase of *Pseudomonas mendocina* KR1 for synthesizing 4-nitrocatechol from nitrobenzene. *Biotechnol. Bioeng.* **87**, 779–790.

Fishman, A., Tao, Y., Rui, L., and Wood, T. K. (2004b). Controlling the regiospecific oxidation of aromatics via active site engineering of toluene *para*-monooxygenase of *Ralstonia pickettii* PKO1. *J. Biol. Chem.* **280**, 506–514.

Fishman, A., Tao, Y., and Wood, T. K. (2004c). Toluene 3-monooxygenase of *Ralstonia pickettii* PKO1 is a *para*-hydroxylating enzyme. *J. Bacteriol.* **186**, 3117–3123.

Franklin, F. C. H., Bagdasarian, M., Bagdasarian, M. M., and Timmis, K. N. (1981). Molecular and functional analysis of the TOL plasmid pWW0 from *Pseudomonas putida* and cloning of genes for the entire regulated aromatic ring *meta* cleavage pathway. *Proc. Natl. Acad. Sci. USA* **78**, 7458–7462.

Franklin, F. C. H., Lehrbach, P. R., Lurz, R., Rueckert, B., Bagdasarian, M., and Timmis, K. N. (1983). Localization and functional analysis of transposon mutations in regulatory genes of the TOL catabolic plasmid. *J. Bacteriol.* **154**, 676–685.

Friemann, R., Ivkovic-Jensen, M. M., Lessner, D. J., Yu, C.-L., Gibson, D. T., Parales, R. E., Eklund, H., and Ramaswamy, S. (2005). Structural insights into the dioxygenation of nitroarene compounds: The crystal structure of the nitrobenzene dioxygenase. *J. Mol. Biol.* **348**, 1139–1151.

Furukawa, K., Hirose, J., Hayashida, S., and Nakamura, K. (1994). Efficient degradation of trichloroethylene by a hybrid aromatic ring dioxygenase. *J. Bacteriol.* **176**, 2121–2123.

Furusawa, Y., Nagarajan, V., Tanokura, M., Masai, E., Fukuda, M., and Senda, T. (2004). Crystal structure of the terminal oxygenase of biphenyl dioxygenase from *Rhodococcus* sp. strain RHA1. *J. Mol. Biol.* **342**, 1041–1052.

Gakhar, L., Malik, Z. A., Allen, C. C., Lipscomb, D. A., Larkin, M. J., and Ramaswamy, S. (2005). Structure and increased thermostability of *Rhodococcus* sp. naphthalene 1,2-dioxygenase. *J. Bacteriol.* **187**, 7222–7231.

Gallegos, M. T., Marques, S., and Ramos, J. L. (1996). Expression of the TOL plasmid *xylS* gene in *Pseudomonas putida* occurs from a σ70-dependent promoter or from σ70- and σ54-dependent tandem promoters according to the compound used for growth. *J. Bacteriol.* **178**, 2356–2361.

Ge, Y., and Eltis, L. D. (2003). Characterization of hybrid toluate and benzoate dioxygenases. *J. Bacteriol.* **185**, 5333–5341.

Ge, Y., Vaillancourt, F. H., Agar, N. Y. R., and Eltis, L. D. (2002). Reactivity of toluate dioxygenase with substituted benzoates and dioxygen. *J. Bacteriol.* **184**, 4096–4103.

Gibson, D. T., Cardini, G. E., Maseles, F. C., and Kallio, R. E. (1970a). Incorporation of oxygen-18 into benzene by *Pseudomonas putida*. *Biochemistry* **9**, 1631–1635.

Gibson, D. T., Hensley, M., Yoshioka, H., and Mabry, T. J. (1970b). Formation of (+)-*cis*-2,3-dihydroxy-1-methylcyclohexa-4,6-diene from toluene by *Pseudomonas putida*. *Biochemistry* **9**, 1626–1630.

Gibson, D. T., Koch, J. R., and Kallio, R. E. (1968a). Oxidative degradation of aromatic hydrocarbons by microorganisms I. Enzymatic formation of catechol from benzene. *Biochemistry* **7**, 2653–2661.

Gibson, D. T., Koch, J. R., Schuld, C. L., and Kallio, R. E. (1968b). Oxidative degradation of aromatic hydrocarbons by microorganisms II. Metabolism of halogenated aromatic hydrocarbons. *Biochemistry* **7**, 3795–3802.

Gibson, D. T., and Parales, R. E. (2000). Aromatic hydrocarbon dioxygenases in environmental biotechnology. *Curr. Opin. Biotechnol.* **11**, 236–243.

Gibson, D. T., Zylstra, G. J., and Chauhan, S. (1990). Biotransformations catalyzed by toluene dioxygenase from *Pseudomonas putida* F1. In "*Pseudomonas*: Biotransformations, pathogenesis, and evolving biotechnology" (S. Silver, A. M. Chakrabarty, B. Iglewski, and S. Kaplan, Eds.), pp. 121–132. American Society for Microbiology, Washington, DC.

Gibson, J., and Harwood, C. S. (2002). Metabolic diversity in aromatic compound utilization by anaerobic microbes. *Ann. Rev. Microbiol.* **56**, 345–369.

Greated, A., Lambertsen, L., Williams, P. A., and Thomas, C. M. (2002). Complete sequence of the IncP-9 TOL plasmid pWW0 from *Pseudomonas putida*. *Environ. Microbiol.* **4,** 856–871.

Haigler, B. E., Pettigrew, C. A., and Spain, J. C. (1992). Biodegradation of mixtures of substituted benzenes by *Pseudomonas* sp. strain JS150. *Appl. Environ. Microbiol.* **58,** 2237–2244.

Haigler, B. E., and Spain, J. C. (1991). Biotransformation of nitrobenzene by bacteria containing toluene degradative pathways. *Appl. Environ. Microbiol.* **57,** 3156–3162.

Harayama, S. (1994). Codon usage patterns suggest independent evolution of 2 catabolic operons on toluene degradative plasmid pWWO of *Pseudomonas putida*. *J. Mol. Evol.* **38,** 328–335.

Harayama, S., Lehrbach, P. R., and Timmis, K. N. (1984). Transposon mutagenesis analysis of *meta*-cleavage pathway operon genes of the TOL plasmid of *Pseudomonas putida* mt-2. *J. Bacteriol.* **160,** 251–255.

Harayama, S., Leppik, R. A., Rekik, M., Mermod, N., Lehrbach, P. R., Reineke, W., and Timmis, K. N. (1986a). Gene order of the TOL catabolic plasmid upper pathway operon and oxidation of both toluene and benzyl alcohol by the *xylA* product. *J. Bacteriol.* **167,** 455–461.

Harayama, S., Mermod, N., Rekik, M., Lehrbach, P. R., and Timmis, K. N. (1987). Roles of the divergent branches of the *meta*-cleavage pathway in the degradation of benzoate and substituted benzoatess. *J. Bacteriol.* **169,** 558–564.

Harayama, S., and Rekik, M. (1990). The *meta* cleavage operon of TOL degradative plasmid pWW0 comprises 13 genes. *Mol. Gen. Genet.* **221,** 113–120.

Harayama, S., Rekik, M., Bairoch, A., Neidle, E. L., and Ornston, L. N. (1991). Potential DNA slippage structure acquired during evolutionary divergence of *Acinetobacter calcoaceticus* chromosomal *benABC* and *Pseudomonas putida* TOL pWWO plasmid *xylXYZ*, genes encoding benzoate dioxygenases. *J. Bacteriol.* **173,** 7540–7548.

Harayama, S., Rekik, M., Ngai, K. L., and Ornston, L. N. (1989a). Physically associated enzymes produce and metabolize 2-hydroxy-2,4-dienoate, a chemically unstable intermediate formed in catechol metabolism via *meta* cleavage in *Pseudomonas putida*. *J. Bacteriol.* **171,** 6251–6258.

Harayama, S., Rekik, M., and Timmis, K. N. (1986b). Genetic analysis of a relaxed substrate specificity aromatic ring dioxygenase, toluate 1,2-dioxygenase, encoded by TOL plasmid pWWO of *Pseudomonas putida*. *Mol. Gen. Genet.* **202,** 226–234.

Harayama, S., Rekik, M., Wubbolts, M., Rose, K., Leppik, A., and Timmis, K. N. (1989b). Characterization of five genes in the upper-pathway operon of TOL plasmid pWWO from *Pseudomonas putida* and identification of the gene products. *J. Bacteriol.* **171,** 5048–5055.

Hartel, U., Eckel, E., Koch, J., Fuchs, G., Linder, D., and Buckel, W. (1993). Purification of glutaryl-CoA dehydrogenase from *Pseudomonas* sp., an enzyme involved in the anaerobic degradation of benzoate. *Arch. Microbiol.* **159,** 174–181.

Harwood, C. S., Burchhardt, G., Herrmann, H., and Fuchs, G. (1998). Anaerobic metabolism of aromatic compounds via the benzoyl-CoA pathway. *FEMS Microbiol. Rev.* **22,** 439–458.

Harwood, C. S., and Parales, R. E. (1996). The β-ketoadipate pathway and the biology of self-identity. *Annu. Rev. Microbiol.* **50,** 533–590.

Heald, S., and Jenkins, R. O. (1994). Trichloroethylene removal and oxidation toxicity mediated by toluene dioxygenase of *Pseudomonas putida*. *Appl. Environ. Microbiol.* **60,** 4634–4637.

Heiden, A. C., Kobel, K., Komenda, M., Koppmann, R., Shao, M., and Wildt, J. (1999). Toluene emissions from plants. *Geophys. Res. Lett.* **26,** 1283–1286.

Heider, J. (2007). Adding handles to unhandy substrates: Anaerobic hydrocarbon activation mechanisms. *Curr. Opin. Chem. Biol* **11,** 188–194.

Hemmi, H., Studts, J. M., Chae, Y. K., Song, J., Markley, J. L., and Fox, B. G. (2001). Solution structure of the toluene 4-monooxygenase effector protein (T4moD). *Biochemistry* **40**, 3512–3524.

Hermuth, K., Leuthner, B., and Heider, J. (2002). Operon structure and expression of the genes for benzylsuccinate synthase in *Thauera aromatica* strain K172. *Arch. Microbiol.* **177**, 132–138.

Holland, H. L., Brown, F. M., Munoz, B., and Ninniss, R. W. (1988). Side chain hydroxylation of aromatic compounds by fungi. Part 1. Isotope effects and mechanism. *J. Chem. Soc. Perkin Trans.* **II**, 1557–1563.

Holtel, A., Goldenberg, D., Giladi, H., Oppenheim, A. B., and Timmis, K. N. (1995). Involvement of IHF protein expression of the Ps promoter of the *Pseudomonas putida* TOL plasmid. *J. Bacteriol.* **177**, 3312–3315.

Holtel, A., Marques, S., Mohler, I., Jakubzik, U., and Timmis, K. N. (1994). Carbon source-dependent inhibition of *xyl* operon expression of the *Pseudomonas putida* TOL plasmid. *J. Bacteriol.* **176**, 1773–1776.

Holzinger, R., Sandoval-Soto, L., Rottenberger, S., Crutzen, P. J., and Kesselmeier, J. (2000). Emissions of volatile organic compounds from *Quercus ilex* L. measured by Proton Transfer Reaction Mass Spectrometry under different environmental conditions. *J. Geophys. Res.* **105**, 20573–20579.

Hopper, D. J. (1976). The hydroxylation of *p*-cresol and its conversion to *p*-hydroxybenzaldehyde in *Pseudomonas putida*. *Biochem. Biophys. Res. Commun.* **69**, 462–468.

Hopper, D. J., and Taylor, D. G. (1977). The purification and properties of *p*-cresol-(acceptor) oxidoreductase (hydroxylating), a flavocytochrome from *Pseudomonas putida*. *Biochem. J.* **167**, 155–162.

Hudlicky, T., Gonzalez, D., and Gibson, D. T. (1999). Enzymatic dihydroxylation of aromatics in enantioselective synthesis: Expanding asymmetric methodology. *Aldrichimica Acta* **32**, 35–62.

Hugouvieux-Cotte-Pattat, N., Kohler, T., Rekik, M., and Harayama, S. (1990). Growth-phase-depenedent expression of the *Pseudomonas putida* TOL plasmid pWW0 catabolic genes. *J. Bacteriol.* **172**, 6651–6660.

Hur, H., Newman, L. M., Wackett, L. P., and Sadowsky, M. J. (1997). Toluene 2-monooxygenase-dependent growth of *Burkholderia cepacia* G4/PR1 on diethyl ether. *Appl. Environ. Microbiol.* **63**, 1606–1609.

Hurtubise, Y., Barriault, D., and Sylvestre, M. (1998). Involvement of the terminal oxygenase β subunit in the biphenyl dioxygenase reactivity pattern toward chlorobiphenyls. *J. Bacteriol.* **180**, 5828–5835.

Inoue, J., Shaw, J. P., Rekik, M., and Harayama, S. (1995). Overlapping substrate specificities of benzaldehyde dehydrogenase (the *xylC* gene product) and 2-hydroxymuconic semialdehyde dehydrogenase (the *xylG* gene product) encoded by TOL plasmid pWW0 of *Pseudomonas putida*. *J. Bacteriol.* **177**, 1196–1201.

Inouye, S., Nakazawa, A., and Nakazawa, T. (1983). Molecular cloning of the regulatory gene *xylR* and operator-promoter regions of the *xylABC* and *xylDEGF* operons of the TOL plasmid. *J. Bacteriol.* **155**, 1192–1199.

Inouye, S., Nakazawa, A., and Nakazawa, T. (1984). Nucleotide sequence of the promoter region of the *xylDEGF* operon on TOL plasmid of *Pseudomonas putida*. *Gene* **29**, 323–330.

Inouye, S., Nakazawa, A., and Nakazawa, T. (1988). Nucleotide sequence of the regulatory gene *xylR* of the TOL plasmid from *Pseudomonas putida*. *Gene* **66**, 301–306.

Inouye, S., Nakazawa, A., and Nakazawa, T. (1981a). Molecular cloning of gene *xylS* of the TOL plasmid: Evidence for positive regulation of the *xylDEGF* operon by *xylS*. *J. Bacteriol.* **148**, 413–418.

Inouye, S., Nakazawa, A., and Nakazawa, T. (1981b). Molecular cloning of TOL genes *xylB* and *xylE* in *Escherichia coli*. *J. Bacteriol.* **145**, 1137–1143.

Inouye, S., Nakazawa, A., and Nakazawa, T. (1985). Determination of the transcription initiation site and identification of the protein product of the regulatory gene *xylR* for *xyl* operons on the TOL plasmid. *J. Bacteriol.* **163**, 863–869.

Inouye, S., Nakazawa, A., and Nakazawa, T. (1987). Overproduction of the *xylS* gene product and activation of the *xylDLEGF* operon on the TOL plasmid. *J. Bacteriol.* **169**, 3587–3592.

Inouye, S., Yamada, M., Nakazawa, A., and Nakazawa, T. (1989). Cloning and sequence analysis of the *ntrA* (*rpoN*) gene of *Pseudomonas putida*. *Gene* **85**, 145–152.

Jiang, H., Parales, R. E., and Gibson, D. T. (1999). The α subunit of toluene dioxygenase from *Pseudomonas putida* F1 can accept electrons from reduced ferredoxin$_{TOL}$ but is catalytically inactive in the absence of the β subunit. *Appl. Environ. Microbiol.* **65**, 315–318.

Jiang, H., Parales, R. E., Lynch, N. A., and Gibson, D. T. (1996). Site-directed mutagenesis of conserved amino acids in the alpha subunit of toluene dioxygenase: Potential mononuclear non-heme iron coordination sites. *J. Bacteriol.* **178**, 3133–3139.

Johnson, G. R., and Olsen, R. H. (1995). Nucleotide sequence analysis of genes encoding a toluene/benzene-2-monooxygenase from *Pseudomonas* sp. strain JS150. *Appl. Environ. Microbiol.* **61**, 3336–3346.

Johnson, G. R., and Olsen, R. H. (1997). Multiple pathways for toluene degradation in *Burkholderia* sp. strain JS150. *Appl. Environ. Microbiol.* **63**, 4047–4052.

Kahng, H.-Y., Byrne, A. M., Olsen, R. H., and Kukor, J. J. (2000). Characterization and role of *tbuX* in utilization of toluene by *Ralstonia pickettii* PKO1. *J. Bacteriol.* **182**, 1232–1242.

Kahng, H.-Y., Malinverni, J. C., Majko, M. M., and Kukor, J. J. (2001). Genetic and functional analysis of the tbc operons for catabolism of alkyl- and chloroaromatic compounds in *Burkholderia* sp. strain JS150. *Appl. Environ. Microbiol.* **67**, 4805–4816.

Kane, S. R., Beller, H. R., Legler, T. C., and Anderson, R. T. (2002). Biochemical and genetic evidence of benzylsuccinate synthase intoluene-degrading, ferric iron-reducing *Geobacter metallireducens*. *Biodegradation* **13**, 149–154.

Kasai, Y., Inoue, J., and Harayama, S. (2001). The TOL palsmid pWWO *xylN* gene product from *Pseudomonas putida* is involved in *m*-xylene uptake. *J. Bacteriol.* **183**, 6662–6666.

Kauppi, B., Lee, K., Carredano, E., Parales, R. E., Gibson, D. T., Eklund, H., and Ramaswamy, S. (1998). Structure of an aromatic ring-hydroxylating dioxygenase- naphthalene 1,2-dioxygenase. *Structure* **6**, 571–586.

Kennes, C., and Veiga, M. C. (2004). Fungal biocatalysts in the biofiltration of VOC-polluted air. *J. Biotechnol.* **113**, 305–319.

Kessler, B., Marques, S., Kohler, T., Ramos, J. L., Timmis, K. N., and de Lorenzo, V. (1994). Cross talk between catabolic pathways in *Pseudomonas putida*: XylS-dependent and -independent activation of the TOL *meta* operon requires the same *cis*-acting sequences within the Pm promoter. *J. Bacteriol.* **176**, 5578–5582.

Kita, A., Kita, S., Fujisawa, I., Inaka, K., Ishida, T., Horiike, K., Nozaki, M., and Miki, K. (1999). An archetypical extradiol-cleaving catecholic dioxygenase: The crystal structure of catechol 2,3-dioxygenase (metapyrocatechase) from *Pseudomonas putida* mt-2. *Structure* **7**, 25–34.

Klecka, G. M., and Gibson, D. T. (1981). Inhibition of catechol 2,3-dioxygenase from *Pseudomonas putida* by 3-chlorocatechol. *Appl. Environ. Microbiol.* **41**, 1159–1165.

Koppmann, R., Khedim, A., Rudolph, J., Poppe, D., Andreae, M. O., Helas, G., Welling, M., and Zenker, T. (1997). Emissions of organic trace gases from savanna fires in southern Africa during the 1992 Southern African Fire Atmosphere Research Initiative and their impact on the formation of tropospheric ozone. *J. Geophys. Res.* **102**, 18879–18888.

Krieger, C. J., Roseboom, W., Albracht, S. P., and Spormann, A. M. (2001). A stable organic free radical in anaerobic benzylsuccinate synthase of *Azoarcus* sp. strain T. *J. Biol. Chem.* **276**, 12924–12927.

Kube, M., Heider, J., Amann, J., Hufnagel, P., Kuhner, S., Beck, A., Reinhardt, R., and Rabus, R. (2004). Genes involved in the anaerobic degradation of toluene in a denitrifying bacterium, strain EbN1. *Arch. Microbiol.* **181,** 182–194.

Kuhner, S., Wohlbrand, L., Fritz, I., Wruck, W., Hultschig, C., Hufnagel, P., Kube, M., Reinhardt, R., and Rabus, R. (2005). Substrate-dependent regulation of anaerobic degradation pathways for toluene and ethylbenzene in a denitrifying bacterium, strain EbN1. *J. Bacteriol.* **187,** 1493–1503.

Kukor, J. J., and Olsen, R. H. (1990a). Diversity of toluene degradation following long term exposure to BTEX *in situ*. In "Biotechnology and Biodegradation" (D. Kamely, A. Chakrabarty, and G. S. Omenn, Eds.), pp. 405–421. Gulf Publishing Co., Houston, TX.

Kukor, J. J., and Olsen, R. H. (1990b). Molecular cloning, characterization, and regulation of a *Pseudomonas pickettii* PKO1 gene encoding phenol hydroxylase and expression of the gene in *Pseudomonas aeruginosa* PAO1c. *J. Bacteriol.* **172,** 4624–4630.

Kukor, J. J., and Olsen, R. H. (1991). Genetic organization and regulation of a *meta* cleavage pathway for catechols produced from catabolism of toluene, benzene, phenol, and cresols by *Pseudomonas pickettii* PKO1. *J. Bacteriol.* **173,** 4587–4594.

Kukor, J. J., and Olsen, R. H. (1992). Complete nucleotide sequence of *tbuD*, the gene encoding phenol/cresol hydroxylase from *Pseudomonas pickettii* PKO1, and functional analysis of the encoded enzymes. *J. Bacteriol.* **174,** 6518–6526.

Kunz, D. A., and Chapman, P. J. (1981). Catabolism of pseudocumene and 3-ethyltoluene by *Pseudomonas putida (arvilla)* mt-2: Evidence for new functions of the TOL (pWWO) plasmid. *J. Bacteriol.* **146,** 179–191.

Lacal, J., Busch, A., Guazzaroni, M. E., Krell, T., and Ramos, J. L. (2006). The TodS-TodT two-component regulatory system recognizes a wide range of effectors and works with DNA-bending proteins. *Proc. Natl. Acad. Sci. USA* **103,** 8191–8196.

Laempe, D., Eisenreich, W., Bacher, A., and Fuchs, G. (1998). Cyclohexa-1,5-diene-1-carboxyl-CoA hydratase, an enzyme involved in anaerobic metabolism of benzoyl-CoA in the denitrifying bacterium *Thauera aromatica*. *Eur. J. Biochem.* **255,** 618–627.

Laempe, D., Jahn, M., and Fuchs, G. (1999). 6-Hydroxycyclohex-1-ene-1-carbonyl-CoA dehydrogenase and 6-oxocyclohex-1-ene-1-carbonyl-CoA hydrolase, enzymes of the benzoyl-CoA pathway of anaerobic aromatic metabolism in the denitrifying bacterium *Thauera aromatica*. *Eur. J. Biochem.* **263,** 420–429.

Lau, P. C. K., Bergeron, H., Labbe, D., Wang, Y., Brousseau, R., and Gibson, D. T. (1994). Sequence and expression of the *todGIH* genes involved in the last three steps of toluene degradation by *Pseudomonas putida* F1. *Gene* **146,** 7–13.

Lau, P. C. K., Wang, Y., Patel, A., Labbé, D., Bergeron, H., Brousseau, R., Konishi, Y., and Rawlings, M. (1997). A bacterial basic region leucine zipper histidine kinase regulating toluene degradation. *Proc. Natl. Acad. Sci. USA* **94,** 1453–1458.

Leahy, J. G., Batchelor, P. J., and Morcomb, S. M. (2003). Evolution of the soluble diiron monooxygenases. *FEMS Microbiol. Rev.* **27,** 449–479.

Leahy, J. G., Byrne, A. M., and Olsen, R. H. (1996). Comparison of factors influencing trichloroethylene degradation by toluene-oxidizing bacteria. *Appl. Environ. Microbiol.* **62,** 825–833.

Leahy, J. G., Johnson, G. R., and Olsen, R. H. (1997). Cross-regulation of toluene monooxygenases by the transcriptional activators TbmR and TbuT. *Appl. Environ. Microbiol.* **63,** 3736–3739.

Lee, J. Y., Park, H. S., and Kim, H. S. (1999). Adenosylcobalamin-mediated methyl transfer by toluate *cis*-dihydrodiol dehydrogenase of the TOL plasmid pWW0. *J. Bacteriol.* **181,** 2953–2957.

Lee, K. (1995). Biochemical studies on toluene and naphthalene dioxygenases. PhD. Dissertation. The University of Iowa, Iowa City, IA.

Lee, K. (1998). Involvement of electrostatic interactions between the components of toluene dioxygenase from *Pseudomonas putida* F1. *J. Microbiol. Biotechnol.* **8**, 416–421.

Lehrbach, P. R., Zeyer, J., Reineke, W., Knackmuss, H.-J., and Timmis, K. N. (1984). Enzyme recruitment *in vitro*: Use of cloned genes to extend the range of haloaromatics degraded by *Pseudomonas* sp. strain B13. *J. Bacteriol.* **158**, 1025–1032.

Lehtio, L., and Goldman, A. (2004). The pyruvate formate lyase family: Sequences, structures and activation. *Protein Eng. Des. Sel.* **17**, 545–552.

Leuthner, B., and Heider, J. (1998). A two-component system involved in regulation of anaerobic toluene metabolism in *Thauera aromatica*. *FEMS Microbiol. Lett.* **166**, 35–41.

Leuthner, B., and Heider, J. (2000). Anaerobic toluene catabolism of *Thauera aromatica*: The *bbs* operon codes for enzymes of beta oxidation of the intermediate benzylsuccinate. *J. Bacteriol.* **182**, 272–277.

Leuthner, B., Leutwein, C., Schulz, H., Hörth, P., Haehnel, W., Schiltz, E., Schägger, H., and Heider, J. (1998). Biochemical and genetic characterization of benzylsuccinate synthase from *Thauera aromatica*: A new glycyl radical enzyme catalysing the first step in anaerobic toluene metabolism. *Mol. Microbiol.* **28**, 615–628.

Leutwein, C., and Heider, J. (1999). Anaerobic toluene-catabolic pathway in denitrifying *Thauera aromatica*: Activation and beta-oxidation of the first intermediate, (*R*)-(+)-benzylsuccinate. *Microbiology* **145**, 3265–3271.

Leutwein, C., and Heider, J. (2001). Succinyl-CoA: (*R*)-benzylsuccinate CoA-transferase: An enzyme of the anaerobic toluene catabolic pathway in denitrifying bacteria. *J. Bacteriol.* **183**, 4288–4295.

Leutwein, C., and Heider, J. (2002). (*R*)-benzylsuccinyl-CoA dehydrogenase of *Thauera aromatica*, an enzyme of the anaerobic toluene catabolic pathway. *Arch. Microbiol.* **178**, 517–524.

Li, S., and Wackett, L. P. (1992). Trichloroethylene oxidation by toluene dioxygenase. *Biochem. Biophys. Res. Comm.* **185**, 443–451.

Lipscomb, J. D. (1994). Biochemistry of the soluble methane monooxygenase. *Annu. Rev. Microbiol.* **48**, 371–399.

Lountos, G. T., Mitchell, K. H., Studts, J. M., Fox, B. G., and Orville, A. M. (2005). Crystal structures and functional studies of T4moD, the toluene 4-monooxygenase catalytic effector protein. *Biochemistry* **44**, 7131–7142.

Luykx, D. M., Prenafeta-Boldu, F. X., and de Bont, J. A. (2003). Toluene monooxygenase from the fungus *Cladosporium sphaerospermum*. *Biochem. Biophys. Res. Commun.* **312**, 373–379.

Lynch, N. A., Jiang, H., and Gibson, D. T. (1996). Rapid purification of the oxygenase component of toluene dioxygenase from a polyol-responsive monoclonal antibody. *Appl. Environ. Microbiol.* **62**, 2133–2137.

Mahendra, S., and Alvarez-Cohen, L. (2006). Kinetics of 1,4-dioxane biodegradation by monooxygenase-expressing bacteria. *Environ. Sci. Technol.* **40**, 5435–5442.

Manjasetty, B. A., Powlowski, J., and Vrielink, A. (2003). Crystal structure of a bifunctional aldolase-dehydrogenase: Sequestering a reactive and volatile intermediate. *Proc. Natl. Acad. Sci. USA* **100**, 6992–6997.

Marques, S., Gallegos, M. T., and Ramos, J. L. (1995). Role of σ^s in transcription from the positively controlled Pm promoter of the TOL plasmid of *Pseudomonas putida*. *Mol. Microbiol.* **18**, 851–857.

Marques, S., Holtel, A., Timmis, K. N., and Ramos, J. L. (1994). Transcriptional induction kinetics from the promoters of the catabolic pathways of TOL plasmid pWWO of *Pseudomonas putida* for metabolism of aromatics. *J. Bacteriol.* **176**, 2517–2524.

McClay, K., Boss, C., Keresztes, I., and Steffan, R. J. (2005). Mutations of toluene-4-monooxygenase that alter regiospecificity of indole oxidation and lead to production of novel indigoid pigments. *Appl. Environ. Microbiol.* **71**, 5476–5483.

McClay, K., Fox, B. G., and Steffan, R. J. (1996). Chloroform mineralization by toluene-oxidizing bacteria. *Appl. Environ. Microbiol.* **62,** 2716–2722.

McClay, K., Fox, B. G., and Steffan, R. J. (2000). Toluene monooxygenase-catalyzed epoxidation of alkenes. *Appl. Environ. Microbiol.* **66,** 1877–1882.

McClay, K., Streger, S. H., and Steffan, R. J. (1995). Induction of toluene oxidatin activity in *Pseudomonas mendocina* KR1 and *Pseudomonas* sp. strain ENVPC5 by chlorinated solvents and alkanes. *Appl. Environ. Microbiol.* **61,** 3479–3481.

McIntire, W., Hopper, D. J., and Singer, T. P. (1985). p-Cresol methylhydroxylase: Assay and general properties. *Biochem. J.* **228,** 325–335.

Menn, F.-M., Zylstra, G. J., and Gibson, D. T. (1991). Location and sequence of the *todF* gene encoding 2-hydroxy-6-oxohepta-2,4-dienoate hydrolase in *Pseudomonas putida* F1. *Gene* **104,** 91–94.

Meulien, P., Downing, R. G., and Broda, P. (1981). Excision of the 40 kb segment of the TOL plasmid from *Pseudomonas putida* mt-2 involves direct repeats. *Mol. Gen. Genet.* **184,** 97–101.

Mitchell, K. H., Rogge, C. E., Gierahn, T., and Fox, B. G. (2003). Insight into the mechanism of aromatic hydroxylation by toluene 4-monooxygenase by use of specifically deuterated toluene and p-xylene. *Proc. Natl. Acad. Sci. USA* **100,** 3784–3789.

Mitchell, K. H., Studts, J. M., and Fox, B. G. (2002). Combined participation of hydroxylase active site residues and effector protein binding in a *para* to *ortho* modulation of toluene 4-monooxygenase regiospecificity. *Biochemistry* **41,** 3176–3188.

Miura, K., Inouye, S., and Nakazawa, A. (1998). The *rpoS* gene regulates OP2, an operon for the lower pathway of xylene catabolism on the TOL plasmid, and the stress response in *Pseudomonas putida* mt-2. *Mol. Gen. Genet.* **259,** 72–78.

Moe, L. A., Bingman, C. A., Wesenberg, G. E., Phillips, G. N. J., and Fox, B. G. (2006a). Structure of T4moC, the Rieske-type ferredoxin component of toluene 4-monooxygenase. *Acta Crystallogr. D Biol. Crystallogr.* **62,** 476–482.

Moe, L. A., and Fox, B. G. (2005). Oxygen-18 tracer studies of enzyme reactions with radical/cation diagnostic probes. *Biochem. Biophys. Res. Commun.* **338,** 240–249.

Moe, L. A., Hu, Z., Deng, D., Austin, R. N., Groves, J. T., and Fox, B. G. (2004). Remarkable aliphatic hydroxylation by the diiron enzyme toluene 4-monooxygenase in reactions with radical or cation diagnostic probes norcarane, 1,1-dimethylcyclopropane, and 1,1-diethylcyclopropane. *Biochemistry* **43,** 15688–15701.

Moe, L. A., McMartin, L. A., and Fox, B. G. (2006b). Component interactions and implications for complex formation in the multicomponent toluene 4-monooxygenase. *Biochemistry* **45,** 5478–5485.

Mondello, F. J., Turcich, M. P., Lobos, J. H., and Erickson, B. D. (1997). Identification and modification of biphenyl dioxygenase sequences that determine the specificity of polychlorinated biphenyl degradation. *Appl. Environ. Microbiol.* **63,** 3096–3103.

Murray, K., Duggleby, C. J., Sala-Trepat, J. M., and Williams, P. A. (1972). The metabolism of benzoate and methylbenzoates via the *meta*-cleavage pathway by *Pseudomonas arvilla* mt-2. *Eur. J. Biochem.* **28,** 301–310.

Nakai, C., Hori, K., Kagamiyama, H., Nakazawa, T., and Nozaki, M. (1983a). Purification, subunit structure, and partial amino acid sequence of metapyrocatechase. *J. Biol. Chem.* **258,** 2916–2922.

Nakai, C., Kagamiyama, H., Nozaki, M., Nakazawa, T., Inouye, S., Ebina, Y., and Nakazawa, A. (1983b). Complete nucleotide sequence of the metapyrocatechase gene on the TOL plasmid of *Pseudomonas putida* mt-2. *J. Biol. Chem.* **258,** 2923–2928.

Nakazawa, T. (2002). Travels of a *Pseudomonas*, from Japan around the world. *Environ. Microbiol.* **4,** 782–786.

Nakazawa, T., Inouye, S., and Nakazawa, A. (1980). Physical and functional mapping of RP4-TOL plasmid recombinants: Analysis of insertion and deletion mutants. *J. Bacteriol.* **144,** 222–231.

Nakazawa, T., and Yokota, T. (1973). Benzoate metabolism in *Pseudomonas putida(arvilla)* mt-2: Demonstration of two benzoate pathways. *J. Bacteriol.* **115**, 262–267.

Neidle, E. L., Hartnett, C., Ornston, L. N., Bairoch, A., Rekik, M., and Harayama, S. (1992). *cis*-Diol dehydrogenases encoded by the TOL pWWO plasmid *xylL* gene and the *Acinetobacter calcoaceticus* chromosomal *benD* gene are members of the short-chain alcohol dehydrogenase superfamily. *Eur. J. Biochem.* **204**, 113–120.

Nelson, M. J. K., Montgomery, S. O., Mahaffey, W. R., and Pritchard, P. H. (1987). Biodegradation of trichloroethylene and involvement of an aromatic biodegradative pathway. *Appl. Environ. Microbiol.* **53**, 949–954.

Nelson, M. J. K., Montgomery, S. O., O'Neill, E. J., and Pritchard, P. H. (1986). Aerobic metabolism of trichloroethylene by a bacterial isolate. *Appl. Environ. Microbiol.* **52**, 383–384.

Newman, L. M., and Wackett, L. P. (1995). Purification and characterization of toluene 2-monooxygenase from *Burkholderia cepacia* G4. *Biochemistry* **34**, 14066–14076.

Nordlund, I., Powlowski, I., and Shingler, V. (1990). Complete nucleotide sequence and polypeptide analysis of multicomponent phenol hydroxylase from *Pseudomonas* sp. stain CF600. *J. Bacteriol.* **172**, 6826–6833.

Notomista, E., Lahm, A., Di Donato, A., and Tramontano, A. (2003). Evolution of bacterial and archaeal multicomponent monooxygenases. *J. Mol. Evol.* **56**, 435–445.

Nozaki, M., Kagamiyama, H., and Hayaishi, O. (1963). Metapyrocatechase I. Purification, crystallization and some properties. *Biochem. Z.* **338**, 582–590.

Nozaki, M., Kotani, S., Ono, K., and Senoh, S. (1970). Metapyrocatechase. III. Substrate specificity and mode of ring fission. *Biochem. Biophys. Acta* **220**, 213–223.

O'Neill, E., Ng, L. C., Sze, C. C., and Shingler, V. (1998). Aromatic ligand binding and intramolecular signalling of the phenol-responsive sigma54-dependent regulator DmpR. *Mol. Microbiol.* **28**, 131–141.

Olsen, R. H., Kukor, J. J., Byrne, A. M., and Johnson, G. R. (1997). Evidence for the evolution of a single component phenol/cresol hydroxylase from a multicomponent toluene monooxygenase. *J. Ind. Microbiol. Biotechnol.* **19**, 360–368.

Olsen, R. H., Kukor, J. J., and Kaphammer, B. (1994). A novel toluene-3-monooxygenase pathway cloned from *Pseudomonas pickettii* PKO1. *J. Bacteriol.* **176**, 3749–3756.

Oppenheim, S. F., Studts, J. M., Fox, B. G., and Dordick, J. S. (2001). Aromatic hydroxylation catalyzed by toluene 4-monooxygenase in organic solvent/aqueous buffer mixtures. *Appl. Biochem. Biotechnol.* **90**, 187–197.

Panke, S., Meyer, A., Huber, C. M., Witholt, B., and Wubbolts, M. G. (1999). An alkane-responsive expression system for the production of fine chemicals. *Appl. Environ. Microbiol.* **65**, 2324–2332.

Parales, J. V., Parales, R. E., Resnick, S. M., and Gibson, D. T. (1998a). Enzyme specificity of 2-nitrotoluene 2,3-dioxygenase from *Pseudomonas* sp. strain JS42 is determined by the C-terminal region of the α subunit of the oxygenase component. *J. Bacteriol.* **180**, 1194–1199.

Parales, R. E., Emig, M. D., Lynch, N. A., and Gibson, D. T. (1998b). Substrate specificities of hybrid naphthalene and 2,4-dinitrotoluene dioxygenase enzyme systems. *J. Bacteriol.* **180**, 2337–2344.

Parales, R. E., Resnick, S. M., Yu, C. L., Boyd, D. R., Sharma, N. D., and Gibson, D. T. (2000). Regioselectivity and enantioselectivity of naphthalene dioxygenase during arene *cis*-dihydroxylation: Control by phenylalanine 352 in the α subunit. *J. Bacteriol.* **182**, 5495–5504.

Phoenix, P., Keane, A., Patel, A., Bergeron, H., Ghoshal, S., and Lau, P. C. K. (2003). Characterization of a new solvent-responsive gene locus in *Pseudomonas putida* F1 and its functionalization as a versatile biosensor. *Environ. Microbiol.* **12**, 1309–1327.

Pikus, J. D., Mitchell, K. H., Studts, J. M., McClay, K., Steffan, R. J., and Fox, B. G. (2000). Threonine 201 in the diiron enzyme toluene 4-monooxygenase is not required for catalysis. *Biochemistry* **39,** 791–799.
Pikus, J. D., Studts, J. M., Achim, C., Kauffmann, K. E., Munck, E., Steffan, R. J., McClay, K., and Fox, B. G. (1996). Recombinant toluene-4-monooxygenase: Catalytic and Mossbauer studies of the purified diiron and Rieske components of a four-protein complex. *Biochemistry* **35,** 9106–9119.
Pikus, J. D., Studts, J. M., McClay, K., Steffan, R. J., and Fox, B. G. (1997). Changes in the regiospecificity of aromatic hydroxylation produced by active site engineering in the diiron enzyme toluene-4-monooxygenase. *Biochemistry* **36,** 9283–9289.
Polissi, A., and Harayama, S. (1993). *In vivo* reactivation of catechol 2,3-dioxygenase mediated by a chloroplast-type ferredoxin: A bacterial strategy to expand the substrate specificity of aromatic degradative pathways. *EMBO J.* **12,** 3339–3347.
Powlowski, J., Sahlman, L., and Shingler, V. (1993). Purification and properties of the physically associated *meta*-cleavage pathway enzymes 4-hydroxy-2-ketovalerate aldolase and aldehyde dehydrogenase (acylating) from *Pseudomonas* sp. strain CF600. *J. Bacteriol.* **175,** 377–385.
Powlowski, J., Sealy, J., Shingler, V., and Cadieux, E. (1997). On the role of DmpK, an auxiliary protein associated with multicomponent phenol hydroxlase from *Pseudomonas* sp. strain CF600. *J. Biol. Chem.* **272,** 945–951.
Prenafeta-Boldú, F. X. (2001). Isolation and characterisation of fungi growing on volatile aromatic hydrocarbons as their sole carbon and energy source. *Mycol. Res.* **105,** 477–484.
Prenafeta-Boldú, F. X., Ballerstedt, H., Gerritse, J., and Grotenhuis, J. T. (2004). Bioremediation of BTEX hydrocarbons: Effect of soil inoculation with the toluene-growing fungus *Cladophialophora* sp. strain T1. *Biodegradation* **15,** 59–65.
Prenafeta-Boldú, F. X., Summerbell, R., and Sybren de Hoog, G. (2006). Fungi growing on aromatic hydrocarbons: Biotechnology's unexpected encounter with biohazard? *FEMS Microbiol. Rev.* **30,** 109–130.
Prenafeta-Boldú, F. X., Vervoort, J., Grotenhuis, J. T. C., and van Groenestijn, J. W. (2002). Substrate interactions during the biodegradation of benzene, toluene, ethylbenzene, and xylene (BTEX) hydrocarbons by the fungus *Cladophialophora* sp. strain T1. *Appl. Environ. Microbiol.* **68,** 2660–2665.
Qi, B., Moe, W. M., and Kinney, K. A. (2002). Biodegradation of volatile organic compounds by five fungal species. *Appl. Microbiol. Biotechnol.* **58,** 684–689.
Rabus, R., and Heider, J. (1998). Initial reactions of anaerobic metabolism of alkylbenzenes in denitrifying and sulfate reducing bacteria. *Arch. Microbiol.* **170,** 377–384.
Rabus, R., Kube, M., Heider, J., Beck, A., Heitmann, K., Widdel, F., and Reinhardt, R. (2005). The genome sequence of an anaerobic aromatic-degrading denitrifying bacterium, strain EbN1. *Arch. Microbiol.* **183,** 27–36.
Rabus, R., and Widdel, F. (1995). Anaerobic degradation of ethylbenzene and other aromatic-hydrocarbons by new denitrifying bacteria. *Arch. Microbiol.* **163,** 96–103.
Ramos, J. L., Duque, E., Huertas, M. J., and Haidour, A. (1995). Isolation and expansion of the catabolic potential of a *Pseudomonas putida* strain able to grow in the presence of high concentrations of aromatic hydrocarbons. *J. Bacteriol.* **177,** 3911–3916.
Ramos, J. L., Marques, S., and Timmis, K. N. (1997). Transcriptional control of the *Pseudomonas* TOL plasmid catabolic operons is achieved through an interplay of host factors and plasmid-encoded regulators. *Annu. Rev. Microbiol.* **51,** 341–373.
Ramos, J. L., Mermod, N., and Timmis, K. N. (1987). Regulatory circuits controlling transcription of TOL plasmid operon encoding *meta*-cleavage pathway for degradation of alkylbenzoates by *Pseudomonas*. *Mol. Microbiol.* **1,** 293–300.

Ramos, J. L., Rojo, F., Zhou, L., and Timmis, K. N. (1990). A family of positive regulators related to the *Pseudomonas putida* plasmid XylS and the *Escherichia coli* AraC activators. *Nucleic Acids Res.* **18,** 2149–2152.

Ramos, J. L., Stolz, A., Reineke, W., and Timmis, K. N. (1986). Altered effector specificities in regulators of gene expression: TOL plasmid *xylS* mutants and their use to engineer expansion of the range of aromatics degraded by bacteria. *Proc. Natl. Acad. Sci.* **83,** 8467–8471.

Ramos-Gonzales, M.-I., Duque, E., and Ramos, J. L. (1991). Conjugational transfer of recombinant DNA in cultures and in soil: Host range of *Pseudomomas putida* TOL plasmid. *Appl. Environ. Microbiol.* **57,** 3020–3027.

Ramos-Gonzalez, M. I., Olson, M., Gatenby, A. A., Mosqueda, G., Manzanera, M., Campos, M. J., Vichez, S., and Ramos, J. L. (2002). Cross-regulation between a novel two-component signal transduction system for catabolism of toluene in *Pseudomonas mendocina* and the TodST system from *Pseudomonas putida*. *J. Bacteriol.* **184,** 7062–7067.

Reddy, B. R., Shaw, L. E., Sayers, J. R., and Williams, P. (1994). Two identical copies of IS*1246*, a 1275 base pair sequence related to other insertion sequences, enclose the *xyl* genes on TOL plasmid pWWO. *Microbiology* **140,** 2305–2307.

Rogers, J. E., and Gibson, D. T. (1977). Purification and properties of *cis*-toluene dihydrodiol dehydrogenase from *Pseudmonas putida*. *J. Bacteriol.* **130,** 1117–1124.

Ryoo, D., Shim, H., Canada, K., Barberi, P., and Wood, T. K. (2000). Aerobic degradation of tetrachloroethylene by toluene-*o*-monooxygenase of *Pseudomonas stutzeri* OX1. *Nat. Biotechnol.* **18,** 775–778.

Sazinsky, M. H., Bard, J., Di Donato, A., and Lippard, S. J. (2004). Crystal structure of the toluene/*o*-xylene monooxygenase hydroxylase from *Pseudomonas stutzeri* OX1. *J. Biol. Chem.* **279,** 30600–30610.

Scognamiglio, R., Notomista, E., Barbieri, P., Pucci, P., Dal Piaz, F., Tramontano, A., and Di Donato, A. (2001). Conformational analysis of putative regulatory subunit D of the toluene/*o*-xylene-monooxygenase complex from *Pseudomonas stutzeri* OX1. *Protein Sci.* **10,** 482–490.

Selmer, T., Pierik, A. J., and Heider, J. (2005). New glycyl radical enzymes catalysing key metabolic steps in anaerobic bacteria. *Biol Chem.* **386,** 981–988.

Shanklin, J., Achim, C., Schmidt, H., Fox, B. G., and Münck, E. (1997). Mössbauer studies of alkane omega-hydroxylase: Evidence for a diiron cluster in an integral-membrane enzyme. *Proc. Natl. Acad. Sci. USA* **94,** 2981–2986.

Shanklin, J., Whittle, E., and Fox, B. G. (1994). Eight histidine residues are catalytically essential in a membrane-associated iron enzyme, stearoyl-CoA desaturase, and are conserved in alkane hydroxylase and xylene monooxygenase. *Biochemistry* **33,** 12787–12794.

Sharp, J. O., Wood, T. K., and Alvarez-Cohen, L. (2005). Aerobic biodegradation of N-nitrosodimethylamine (NDMA) by axenic bacterial strains. *Biotechnol. Bioeng.* **89,** 608–618.

Shaw, J. P., and Harayama, S. (1990). Purification and characterisation of TOL plasmid-encoded benzyl alcohol dehydrogenase and benzaldehyde dehydrogenase of *Pseudomonas putida*. *Eur. J. Biochem.* **191,** 705–714.

Shaw, J. P., and Harayama, S. (1992). Purification and characterisation of the NADH: Acceptor reductase component of xylene monooxygenase encoded by the TOL plasmid pWW0 of *Pseudomonas putida* mt-2. *Eur. J. Biochem.* **209,** 51–61.

Shaw, J. P., and Harayama, S. (1995). Characterization *in vitro* of the hydroxylase component of xylene monooxygenase, the first enzyme of the TOL-plasmid-encoded pathway for the mineralization of toluene and xylenes. *J. Ferment. Bioeng.* **79,** 195–199.

Shaw, J. P., Rekik, M., Schwager, F., and Harayama, S. (1993). Kinetic studies on benzyl alcohol dehydrogenase encoded by TOL plasmid pWWO. A member of the zinc-containing long chain alcohol dehydrogenase family. *J. Biol. Chem.* **268,** 10842–10850.

Shaw, J. P., Schwager, F., and Harayama, S. (1992). Substrate-specificity of benzyl alcohol dehydrogenase and benzaldehyde dehydrogenase encoded by TOL plasmid pWW0. Metabolic and mechanistic implications. *Biochem. J.* **283,** 789–794.

Sheldrake, G. N. (1992). In "Chirality in industry: The commercial manufacture and application of optically active compounds" (A. N. Collins, G. N. Sheldrake, and J. Crosby, Eds.), pp. 127–166. Wiley, Chichester, UK.

Shields, M. S., and Francesconi, S. C. (1996). Microbial degradation of trichloroethylene dichloroethylenes and aromatic pollutants. US Patent 5,543317.

Shields, M. S., Montgomery, S. O., Chapman, P. J., Cuskey, S. M., and Pritchard, P. H. (1989). Novel pathway of toluene catabolism in the trichloroethylene-degrading bacterium G4. *Appl. Environ. Microbiol.* **55,** 1624–1629.

Shields, M. S., Montgomery, S. O., Cuskey, S. M., Chapman, P. J., and Pritchard, P. H. (1991). Mutants of *Pseudomonas cepacia* G4 defective in catabolism of aromatic compounds and trichloroethylene. *Appl. Environ. Microbiol.* **57,** 1935–1941.

Shields, M. S., and Reagin, M. J. (1992). Selection of a *Pseudomonas cepacia* strain constitutive for the degradation of trichloroethylene. *Appl. Environ. Microbiol.* **58,** 3977–3983.

Shields, M. S., Reagin, M. J., Gerger, R. R., Campbell, R., and Somerville, C. (1995). TOM, a new aromatic degradative plasmid from *Burkholderia* (*Pseudomonas*) *cepacia* G4. *Appl. Environ. Microbiol.* **61,** 1352–1356.

Shim, H., and Wood, T. K. (2000). Aerobic degradation of mixtures of chlorinated aliphatics by cloned toluene-*o*-xylene monooxygenase and toluene *o*-monooxygenase in resting cells. *Biotechnol. Bioeng.* **70,** 693–698.

Shingler, V., Powlowski, J., and Marklund, U. (1992). Nucleotide sequence and functional analysis of the complete phenol/3,4-dimethylphenol catabolic pathway of *Pseudomonas* sp. strain CF600. *J. Bacteriol.* **174,** 711–724.

Shingleton, J. T., Applegate, B. M., Nagel, A. C., Bienkowski, P. R., and Sayler, G. S. (1998). Induction of the *tod* operon by trichloroethylene in *Pseudomonas putida* TVA8. *Appl. Environ. Microbiol.* **64,** 5049–5052.

Shinoda, Y., Sakai, Y., Uenishi, H., Uchihashi, Y., Hiraishi, A., Yukawa, H., Yurimoto, H., and Kato, N. (2004). Aerobic and anaerobic toluene degradation by a newly isolated denitrifying bacterium, *Thauera* sp. strain DNT-1. *Appl. Environ. Microbiol.* **70,** 1385–1392.

Sinninghe Damste, J. S., de las Heras, F. X. C., and de Leeuw, J. W. (1992). Molecular analysis of sulphur-rich brown coals by flash pyrolysis-gas chromatography-mass spectrometry: The type III-S kerogen. *J. Chromatogr.* **607,** 361–376.

Skjeldal, L., Peterson, F. C., Doreleijers, J. F., Moe, L. A., Pikus, J. D., Westler, W. M., Markley, J. L., Volkman, B. F., and Fox, B. G. (2004). Solution structure of T4 moC, the Rieske ferredoxin component of the toluene 4-monooxygenase complex. *J. Biol. Inorg. Chem.* **9,** 945–953.

Snyder, R. (2002). Benzene and leukemia. *Crit. Rev. Toxicol.* **32,** 155–210.

Solera, D., Arenghi, F. L. G., Woelk, T., Galli, E., and Barbieri, P. (2004). TouR-mediated effector-independent growth phase-dependent activation of the σ^{54} P*tou* promoter of *Pseudomonas stutzeri* OX1. *J. Bacteriol.* **186,** 7353–7363.

Spain, J. C., and Nishino, S. F. (1987). Degradation of 1,4-dichlorobenzene by a *Pseudomonas* sp. *Appl. Environ. Microbiol.* **53,** 1010–1019.

Spooner, R. A., Bagdasarian, M., and Franklin, F. C. H. (1987). Activation of the *xylDLEGF* promoter of the TOL toluene-xylene degradation pathway by overproduction of the *xylS* regulatory gene product. *J. Bacteriol.* **169,** 3581–3586.

Spooner, R. A., Lindsay, K., and Franklin, F. C. H. (1986). Genetic, functional and sequence analysis of the *xylR* and *xylS* regulatory genes of the TOL plasmid pWW0. *J. Gen. Microbiol.* **132,** 1347–1358.

Spormann, A. M., and Widdel, F. (2000). Metabolism of alkylbenzenes, alkanes, and other hydrocarbons in anaerobic bacteria. *Biodegradation* **11,** 85–105.

Studts, J. M., and Fox, B. G. (1999). Application of a fed-batch fermentaion to the preparation of isotopically labeled or selenomethionine-labeled proteins. *Protein Expr. Purif.* **16,** 109–119.

Studts, J. M., Mitchell, K. H., Pikus, J. D., McClay, K., Steffan, R. J., and Fox, B. G. (2000). Optimized expression and purification of toluene 4-monooxygenase hydroxylase. *Protein Expr. Purif.* **20,** 58–65.

Subramanian, V., Liu, T.-N., Yeh, W.-K., and Gibson, D. T. (1979). Toluene dioxygenase: Purification of an iron-sulfur protein by affinity chromatography. *Biochem. Biophys. Res. Commun.* **91,** 1131–1139.

Subramanian, V., Liu, T.-N., Yeh, W.-K., Narro, M., and Gibson, D. T. (1981). Purification and properties of NADH-ferredoxin$_{TOL}$ reductase: A component of toluene dioxygenase from *Pseudomonas putida*. *J. Biol. Chem.* **256,** 2723–2730.

Subramanian, V., Liu, T.-N., Yeh, W.-K., Serdar, C. M., Wackett, L. P., and Gibson, D. T. (1985). Purification and properties of ferredoxin$_{TOL}$: A component of toluene dioxygenase from *Pseudomonas putida* F1. *J. Biol. Chem.* **260,** 2355–2363.

Subramanya, H. S., Roper, D. I., Dauter, Z., Dodson, E. J., Davies, G. J., Wilson, K. S., and Wigley, D. B. (1996). Enzymatic ketonization of 2-hydroxymuconate: Specificity and mechanism investigated by the crystal structures of two isomerases. *Biochemistry* **35,** 792–802.

Suzuki, M., Hayakawa, T., Shaw, J. P., Rekik, M., and Harayama, S. (1991). Primary structure of xylene monooxygenase: Similarities to and differences from the alkane hydroxylation system. *J. Bacteriol.* **173,** 1690–1695.

Tao, Y., Bentley, W. E., and Wood, T. K. (2005). Regiospecific oxidation of naphthalene and fluorene by toluene monooxygenases and engineered toluene 4-monooxygenases of *Pseudomonas mendocina* KR1. *Biotechnol. Bioeng.* **90,** 85–94.

Tao, Y., Fishman, A., Bentley, W. E., and Wood, T. K. (2004). Altering toluene 4-monooxygenase by active-site engineering for the synthesis of 3-methoxycatechol, methoxyhydroquinone, and methylhydroquinone. *J. Bacteriol.* **186,** 4705–4713.

Taylor, A. B., Czerwinski, R. M., Johnson, W. H. J., Whitman, C. P., and Hackert, M. L. (1998). Crystal structure of 4-oxalocrotonate tautomerase inactivated by 2-oxo-3-pentynoate at 2.4-Å resolution: Analysis and implications for the mechanism of inactivation and catalysis. *Biochemistry* **37,** 14692–14700.

Tschech, A., and Fuchs, G. (1987). Anaerobic degradation of phenol by pure cultures of newly isolated denitrifying pseudomonads. *Arch. Microbiol.* **148,** 213–217.

Tsuda, M., and Iino, T. (1987). Genetic analysis of a transposon carrying toluene degrading genes on a TOL plasmid pWW0. *Mol. Gen. Genet.* **210,** 270–276.

Tsuda, M., and Iino, T. (1988). Identification and characterization of Tn*4653*, a transposon covering the toluene transposon Tn*4651* on TOL plasmid pWW0. *Mol. Gen. Genet.* **213,** 72–77.

Vercellone-Smith, P., and Herson, D. S. (1997). Toluene elicits a carbon starvation response in *Pseudomonas putida* mt-2 containing the TOL plasmid pWWO. *Appl. Environ. Microbiol.* **63,** 1925–1932.

Vrkocova, P., Valterova, I., Vrkoc, J., and Koutek, B. (2000). Volatiles released from oak, a host tree for the bark beetle *Scolytus intricatus*. *Biochem. Syst. Ecol.* **28,** 933–947.

Wackett, L. P. (1984). Aromatic hydrocarbon metabolism by fungal and bacterial enzymes. Ph. D. Dissertation, The University of Texas, Austin.

Wackett, L. P. (2002). Mechanism and applications of Rieske non-heme iron dioxygenases. *Enz. Microbial Technol.* **31,** 577–587.

Wang, Y., Rawlings, M., Gibson, D. T., Labbé, D., Bergeron, H., Brousseau, R., and Lau, P. C. K. (1995). Identification of a membrane protein and a truncated LysR-type regulator associated with the toluene degradation pathway in *Pseudomonas putida* F1. *Mol. Gen. Genet.* **246,** 570–579.

Washer, C. E., and Edwards, E. A. (2007). Identification and expression of benzylsuccinate synthase genes in a toluene-degrading methanogenic consortium. *Appl. Environ. Microbiol.* **73,** 1367–1369.

Weber, F. J., Hage, K. C., and de Bont, J. A. (1995). Growth of the fungus *Cladosporium sphaerospermum* with toluene as the sole carbon and energy source. *Appl. Environ. Microbiol.* **61,** 3562–3566.

White, G. P., and Dunn, N. W. (1978). Compatibility and sex-specific phage plating characteristics of the TOL and NAH catabolic plasmids. *Genet. Res.* **32,** 207–213.

Whited, G. M., and Gibson, D. T. (1991a). Separation and partial characterization of the enzymes of the toluene-4-monooxygenase catabolic pathway in *Pseudomonas mendocina* KR1. *J. Bacteriol.* **173,** 3017–3020.

Whited, G. M., and Gibson, D. T. (1991b). Toluene-4-monooxygenase, a three-component enzyme system that catalyzes the oxidation of toluene to *p*-cresol in *Pseudomonas mendocina* KR1. *J. Bacteriol.* **173,** 3010–3016.

Widdel, F., and Rabus, R. (2001). Anaerobic biodegradation of saturated and aromatic hydrocarbons. *Curr. Opin. Biotechnol.* **12,** 259–276.

Wilkins, P. C., Dalton, H., Samuel, C. J., and Green, J. (1994). Further evidence for multiple pathways in soluble methane-monooxygenase-catalyzed oxidations from the measurement of deuterium kinetic isotope effects. *Eur. J. Biochem.* **226,** 555–560.

Williams, P., and Murray, K. (1974). Metabolism of benzoate and the methyl benzoates by *Pseudomonas putida* (arvilla) mt-2: Evidence for the existence of a TOL plasmid. *J. Bacteriol.* **120,** 416–423.

Williams, P. A., and Sayers, J. R. (1994). The evolution of pathways for aromatic hydrocarbon oxidation in *Pseudomonas*. *Biodegradation* **5,** 195–217.

Williams, P. A., Shaw, L. M., Pitt, C. W., and Vrecl, M. (1996). *xylUW*, two genes at the start of the upper pathway operon of TOL plasmid pWW0, appear to play no essential part in determining its catabolic phenotype. *Microbiology* **143,** 101–107.

Winter, R. B., Yen, K.-M., and Ensley, B. D. (1989). Efficient degradation of trichloroethylene by a recombinant *Escherichia coli*. *Bio/Technology* **7,** 282–285.

Woertz, J. R., Kinney, K. A., McIntosh, N. D., and Szaniszlo, P. J. (2001). Removal of toluene in a vapor-phase bioreactor containing a strain of the dimorphic black yeast *Exophiala lecanii-corni*. *Biotechnol. Bioeng.* **75,** 550–558.

Wong, C. L., and Dunn, N. W. (1974). Transmissible plasmid coding for the degradation of benzoate and *m*-toluate in *Pseudomonas arvilla* mt-2. *Genet. Res.* **23,** 227–232.

Worsey, M. J., Franklin, F. C. H., and Williams, P. A. (1978). Regulation of the degradative pathway enzymes coded for by the TOL plasmid (pWWO) from *Pseudomonas putida* mt-2. *J. Bacteriol.* **134,** 757–764.

Worsey, M. J., and Williams, P. A. (1975). Metabolism of toluene and xylenes by *Pseudomonas putida* (arvilla) mt-2: Evidence for a new function of the TOL plasmid. *J. Bacteriol.* **124,** 7–13.

Worsey, M. J., and Williams, P. A. (1977). Characterization of a spontaneously occurring mutant of the TOL20 plasmid in *Pseudomonas putida* MT20: Possible regulatory implications. *J. Bacteriol.* **130,** 1149–1158.

Wright, A., and Olsen, R. H. (1994). Self-mobilization and organization of the genes encoding the toluene metabolic pathway of *Pseudomonas mendocina* KR1. *Appl. Environ. Microbiol.* **60,** 235–242.

Wubbolts, M. G., Reuvekamp, P., and Witholt, B. (1994). TOL plasmid-specified xylene oxygenase is a wide substrate range monooxygenase capable of olefin epoxidation. *Enz. Microbial Technol.* **16,** 608–615.

Xia, B., Pikus, J. D., Xia, W., McClay, K., Steffan, R. J., Chae, Y. K., Westler, W. M., Markley, J. L., and Fox, B. G. (1999). Detection and classification of hyperfine-shifted

^{1}H, ^{2}H, and ^{15}N resonances of the Rieske ferredoxin component of toluene 4-monooxygenase. *Biochemistry* **38,** 727–739.

Yadav, S., and Reddy, C. A. (1993). Degradation of benzene, toluene, ethylbenzene, and xylenes (BTEX) by the lignin-degrading basidiomycete *Phanerochaete chrysosporium*. *Appl. Environ. Microbiol.* **59,** 756–762.

Yen, K.-M., and Karl, M. R. (1992). Identification of a new gene, *tmoF*, in the *Pseudomonas mendocina* KR1 gene cluster encoding toluene-4-monooxygenase. *J. Bacteriol.* **174,** 7253–7261.

Yen, K.-M., Karl, M. R., Blatt, L. M., Simon, M. J., Winter, R. B., Fausset, P. R., Lu, H. S., Harcourt, A. A., and Chen, K. K. (1991). Cloning and characterization of a *Pseudomonas mendocina* KR1 gene cluster encoding toluene-4-monooxygenase. *J. Bacteriol.* **173,** 5315–5327.

Zengler, K., Heider, J., Rossello-Mora, R., and Widdel, F. (1999). Phototrophic utilization of toluene under anoxic conditions by a new strain of *Blastochloris sulfoviridis*. *Arch. Microbiol.* **172,** 204–212.

Zielinski, M., Backhaus, S., and Hofer, B. (2002). The principle determinants for the structure of the substrate-binding pocket are located within a central core of a biphenyl dioxygenase α subunit. *Microbiology* **148,** 2439–2448.

Zylstra, G. J., and Gibson, D. T. (1989). Toluene degradation by *Pseudomonas putida* F1: Nucleotide sequence of the *todC1C2BADE* genes and their expression in *E. coli*. *J. Biol. Chem.* **264,** 14940–14946.

Zylstra, G. J., and Gibson, D. T. (1991). Aromatic hydrocarbon degradation: A molecular approach. *Genet. Eng.* **13,** 183–203.

Zylstra, G. J., McCombie, W. R., Gibson, D. T., and Finette, B. A. (1988). Toluene degradation by *Pseudomonas putida* F1: Genetic organization of the *tod* operon. *Appl. Environ. Microbiol.* **54,** 1498–1503.

Zylstra, G. J., Wackett, L. P., and Gibson, D. T. (1989). Trichloroethylene degradation by *Escherichia coli* containing the cloned *Pseudomonas putida* F1 toluene dioxygenase genes. *Appl. Environ. Microbiol.* **55,** 3162–3166.

CHAPTER 2

Microbial Endocrinology: Experimental Design Issues in the Study of Interkingdom Signalling in Infectious Disease

Primrose P. E. Freestone* and **Mark Lyte**[†,1]

Contents			
	I.	Microbial Endocrinology as a New Scientific Discipline	76
		A. Overview and goal of review	76
		B. The stress response and its influence on the immune system and the infectious agent	77
		C. Mechanisms by which stress-related neuroendocrine hormones can enhance bacterial growth and virulence	79
		D. The spectrum of catecholamine responsive bacteria	84
		E. Catecholamine specificity in enteric bacteria	85
		F. Molecular analyses of bacterial catecholamine responsiveness	88
	II.	Experimental Design Issues in the Study of Microbial Endocrinology	89
		A. Medium comparability	90
		B. Bacterial inoculum size and its influence on stress neurohormone responsiveness	92

* Department of Infection, Immunity and Inflammation, University of Leicester School of Medicine, Leicester, United Kingdom
† Department of Pharmacy Practice, School of Pharmacy, Texas Tech University Health Sciences Center, Lubbock, Texas 79430-8162
[1] Corresponding author: Department of Pharmacy Practice, 3601 4th Street, STOP 8162, Texas Tech University Health Sciences Center, Lubbock, Texas 79430-8162

Advances in Applied Microbiology, Volume 64
ISSN 0065-2164, DOI: 10.1016/S0065-2164(08)00402-4

© 2008 Elsevier Inc.
All rights reserved.

       C. The importance of neuroendocrine hormone
          concentration   94
       D. Neuroendocrine stress neurohormone
          assay methodologies   97
   III. Conclusions and Future Directions   101
   References   102

I. MICROBIAL ENDOCRINOLOGY AS A NEW SCIENTIFIC DISCIPLINE

A. Overview and goal of review

Microbial Endocrinology is an interdisciplinary research field that represents the intersection of microbiology, endocrinology and neurophysiology. It is directed at providing a new paradigm to examine and understand the ability of microorganisms to interact with a host under various physical and behavioral conditions present in both health (i.e. normal homeostasis) and disease. For example, Microbial Endocrinology has been used to get a new understanding of how stress could increase the ability of small inocula of bacteria to cause disease. At its foundation is the tenet that microorganisms have evolved systems for sensing human neurohormones, which they use as environmental cues to determine that they are inside a host and that it is time to initiate growth and pathogenic processes. The field was founded by one of us (ML) in 1992 (Lyte and Ernst, 1992) when it was first observed that stress-related neuroendocrine hormones, specifically the catecholamines, could directly influence bacterial growth through neuroendocrine–bacterial interactions, as described in this chapter. Microbial endocrinology is constantly evolving, as the scientific and medical communities are increasingly becoming aware that neuroendocrine hormones produced by the body can influence bacterial growth and virulence, and that the more classic nutrient-centered view may need modification.

Although this review could cover the whole spectrum of neuroendocrine effector molecules, because of the clear role that stress has been found to play in the pathophysiology of infection, bacterial interactions with the catecholamine family of stress-related molecules have by far received the greatest attention. For this reason, this article will focus primarily upon the analysis of bacterial responses to the catecholamine neurohormones norepinephrine (or as known outside of the United States as noradrenaline), epinephrine (also known as adrenaline), dopamine, and related molecules. However, the reader should keep in mind that other neurohormones can also markedly influence infectious agents. For instance, several reports have described the ability of *Burkholderia pseudomallei* to respond to human insulin, which makes it unsurprising that this species is a major pathogen of Type 1 diabetics (Woods *et al.*, 1993). Alverdy and co-workers have recently shown that the endogenous opioid peptide dynorphin produced by intestinal neurons can be used by the quorum-sensing apparatus in *P. aeruginosa* to increase virulence (Zaborina *et al.*, 2007). Although bacteria have so far been the principal microbial species investigated for their responses to eukaryotic neurohormones, pathogenic yeast have also been shown to recognize mammalian neurohormones. For instance, *Candida albicans* has been shown to possess receptors for estrogen and progesterone, which may explain why pregnant women are so susceptible to thrush infections (Powell *et al.*, 1983; Sonnex, 1998).

It is also important to recognize that a careful examination of the past literature will reveal that the association of neuroendocrine hormones and microorganisms was well recognized to lead to often disastrous medical complications, although the mechanism of direct neuroendocrine–bacterial interactions was not envisioned. The reader is directed to a recent review (Lyte, 2004) which highlights many, though certainly not all, of these pre-1992 studies extending back to the 1930s when the administration of epinephrine to patients experiencing urticaria (itching) often led to fulminating sepsis and death.

The goal of this review, therefore, is to provide the reader not only with an understanding of the theoretical underpinnings of Microbial Endocrinology, but also to provide examples of its application to the clinical arena as well as to provide specific methodological advice as to its performance in the laboratory setting.

B. The stress response and its influence on the immune system and the infectious agent

The catecholamine family of stress neurohormones play an integral role in the acute "fight and flight" stress response in metazoa (Reiche *et al.*, 2004). Chemically, catecholamines are a group of tyrosine-derived effector compounds comprising a benzene ring with two adjacent hydroxyl groups and an opposing amine side chain. In mammals, the catecholamine synthesis pathway is L-dopa (usually obtained from dietary sources) → Dopamine → Norepinephrine → Epinephrine (Fig. 2.1).

The ability of stress to modulate susceptibility to infection is most often perceived in terms of adversely affecting the immune system function (Peterson *et al.*, 1991; Yang and Glaser, 2000). The immune and central nervous systems (CNS) are well recognized to be involved in functionally relevant cross talk to maintain homeostasis under normal conditions. Further, it is becoming increasingly clear that the immune system and CNS are also communicative especially during infection, as varied immune cell populations receive signals from the CNS through

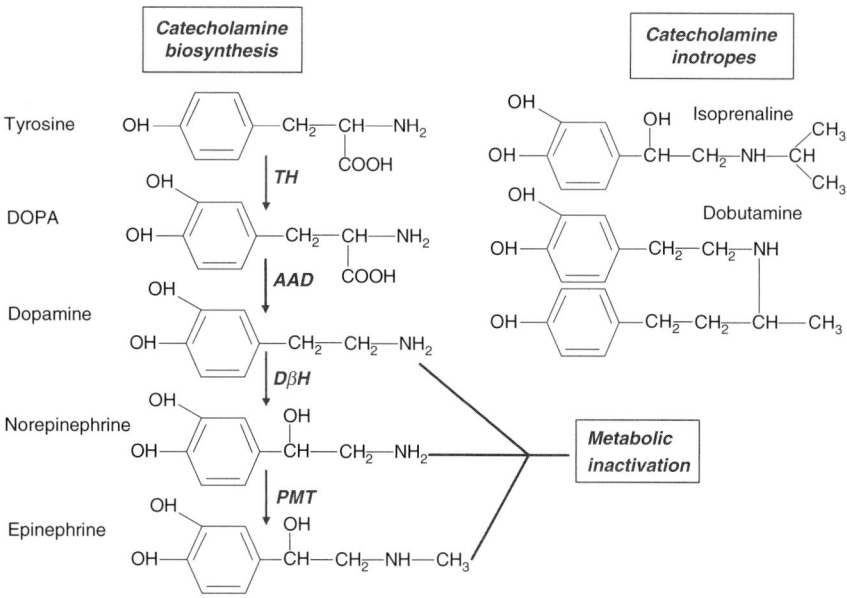

FIGURE 2.1 Catecholamine and inotrope structures. Catecholamines are synthesized from L-DOPA, obtained from dietary sources (tyrosine and phenylalanine). Key: Catecholamine biosynthesis: TH – tyrosine hydroxylase; AAD – aromatic L-amino decarboxylase; DβH – dopamine β-hydroxylase; PMT – phenylethanolamine-N-methyltransferase.

possession of receptors for the catecholamines (principally norepinephrine and epinephrine), as well as stress elaborated glucocorticoids and neuropeptides and even thyroid and sex hormones (reviewed by Reiche et al., 2004). But is this the complete "story" to explain the ability of stress to affect pathogenesis of infectious disease? While the role of stress neurohormone modulation of host immunity must be considered in any holistic analysis of how stress can modulate mammalian susceptibility to infection, in the last decade it has become equally clear that infectious agents can also perceive and use to their advantage the neurohormonal products of the host's physiological response to stress.

That the bacteria can use stress-related human neuroendocrine hormones as environmental cues to recognize that they are inside a host should not be surprising. The evolutionary development of microorganisms preceded that of vertebrates, and many reports have shown a wide distribution throughout nature of what were once thought to be almost exclusively vertebrate neuroendocrine hormones (e.g., norepinephrine has been found in insects (Pitman, 1971) as well as protozoa (Janakidevi et al., 1966), and dopamine has been isolated from broad beans (Apaydin, 2000; Rabey et al., 1992) and bananas (Lyte et al., 1997a; Waalkes et al., 1958).

This ubiquitous distribution of catecholamines suggests that microorganisms would have had ample time during evolution to come into contact with, and to develop mechanisms to recognize catecholamines and initiate changes in both growth and elaboration of virulence-associated factors. Evidence for such direct neuroendocrine–bacterial interactions resulting in physiological changes in bacteria can be seen in Table 2.1, which shows that the enhancement of bacterial growth and virulence production has now been documented for over 50 bacterial species.

C. Mechanisms by which stress-related neuroendocrine hormones can enhance bacterial growth and virulence

1. Catecholamines mediate access to host sequestered iron

The iron restriction imposed by a majority of host tissues is a major obstacle to the growth of most microbial pathogens. A strategy that infectious bacteria often employ to scavenge nutritionally essential iron is the production and utilization of siderophores, which are, secreted low molecular weight catecholate or hydroxamate molecules that possess high affinity for ferric iron (Ratledge and Dover, 2000). However, siderophores are often less than effective at obtaining iron from high affinity ferric iron binding proteins such as transferrin in blood and lactoferrin in mucosal secretions. The same catechol moiety found in many siderophores is also present in the catecholamine family of stress-related neurohormones, and Freestone *et al.* have shown that norepinephrine, epinephrine and dopamine, various inotropes (isoprenaline, dobutamine), and certain of their metabolites (dihydroxymandelic acid and dihydroxyphenylglycol) all share the ability to facilitate bacterial acquisition of normally inaccessible transferrin and lactoferrin bound iron (Freestone *et al.*, 2000, 2002, 2003, 2007a,b). Catecholamine-facilitated iron provision from host iron binding proteins has been demonstrated for a range of Gram-positive and Gram-negative bacteria (Freestone *et al.*, 2000, 2002, 2003, 2007a,b; Lyte *et al.*, 2003; Neal *et al.*, 2001).

Mechanistically, the catechol moiety has been shown to form a complex with the ferric iron bound by transferrin and lactoferrin, reducing the iron binding affinity of these proteins (Freestone *et al.*, 2000, 2002, 2003, 2007a,b), and thereby rendering transferrin and lactoferrin much more susceptible to "theft" by bacterial ferric iron binding siderophores such as enterobactin (Freestone *et al.*, 2003). Although the precise molecular details of how catecholamines liberate transferrin and lactoferrin-complexed iron remains to be elucidated, for bacteria such as *E. coli* (Burton *et al.*, 2002; Freestone *et al.*, 2003) and *Salmonella* (Williams *et al.*, 2006) both siderophore synthesis and uptake systems appear to be integral elements in the mechanism by which stress-related neuroendocrine hormones induce growth of these Gram-negative species. The role of the

TABLE 2.1 Stress-related neuroendocrine hormone responsive bacteria

Species	References
Aeromonas hydrophila	Kinney et al. (1999)
Acinetobacter lwoffii	Freestone et al. (1999)
Bordetella bronchiseptica, B. pertussis	Anderson and Armstrong (2006)
Borrelia burgdorferi	Scheckelhoff et al. (2007)
Campylobacter jejuni	Roberts et al. (2002); Cogan et al. (2007)
Citrobacter freundii	Freestone et al. (1999)
Enterobacter agglomerans, E. sakazaki	Freestone et al. (1999)
Enterococcus faecalis, E. cloacae	Lyte and Ernst (1993); Freestone et al. (1999)
Escherichia coli (commensal and pathogenic)	Lyte and Ernst (1992, 1993); Freestone et al. (1999, 2000, 2002, 2003, 2007a,b,c); Chen et al. (2003); Green et al. (2003, 2004); Vlisidou et al. (2004); Lyte et al. (1996a,b, 1997a,b,c,d); Schreiber and Brown (2005); Sperandio et al. (2003); Walters and Sperandio (2006); Bansal et al. (2007); Dowd (2007)
Hafnia alvei	Freestone et al. (1999)
Klebsiella oxytoca, K. pneumoniae	Freestone et al. (1999); Belay et al. (2003)
Listeria monocytogenes	Freestone et al. (1999); Coulanges et al. (1998)
L. innocua, L. ivanovii, L. welshimeri, L. seeligeri, L. grayi	Coulanges et al. (1998)
Morganella morgani	Freestone et al. (1999)
Proteus mirabilis	Freestone et al. (1999)
Pseudomonas aeruginosa	Freestone et al. (1999); Alverdy et al. (2000); O'Donnell et al. (2006).
Salmonella enterica	Lyte and Ernst (1992, 1993), Freestone et al. (1999, 2007a,b); Green et al. (2003, 2004); Williams et al. (2006)
Shigella sonnei, S. flexneri	O'Donnell et al. (2006)
Staphylococcus aureus	Freestone et al., 1999; Neal et al., 2001; O'Donnell et al., 2006
Staphylococcus epidermidis, S. capitis, S. saprophyticus, S. haemolyticus, S. hominis	Freestone et al. (1999); Neal et al. (2001); Lyte et al. (2003).

(continued)

TABLE 2.1 (*continued*)

Species	References
Streptococcus dysgalactica	Freestone *et al.* (1999).
Vibrio parahaemolyticus, V. mimicus, V. vulnificus	Lacoste *et al.* (2001); Nakano *et al.* (2007a,b)
Xanthomonas maltophila	Freestone *et al.* (1999)
Yersinia enterocolitica	Lyte and Ernst (1992); Freestone *et al.* (1999, 2007a,b).
Oral/Periodontal pathogens	Roberts *et al.* (2002, 2005)
Actinomyces gerenscseriae, A. naeslundii, A. odontolyticus	
Campylobacter gracilis	
Capnocytophaga sputigena, C. gingivalis	
Eikenella corrodens	
Eubacterium saburreum	
Fusobacterium periodonticum, F. nucleatum subsp. *Vincentii*	
Leptotrichia buccalis	
Neisseria mucosa	
Peptostreptococcus anaerobius, P. micros	
Prevotella denticola, P. melaninogenica	
Staphylococcus intermedius	
Streptococcus gordonii, S. constellatus, S. mitis, S. mutans, S. sanguis	

The references cited in Table 2.1 represent reports that have shown direct effects of neuroendocrine hormones on bacteria following the first report to propose and test direct neuroendocrine–bacterial interaction as a mechanism to account for the ability of stress-related neurohormones to alter bacterial growth (Lyte and Ernst, 1992). The Table does not represent an exhaustive examination of the microbiological literature that has reported the association of neuroendocrine hormones with altered bacterial growth since the 1930s (for review see Lyte, 2004). For example, the ability of a number of dihydroxyphenyl compounds, including norepinephrine to stimulate the growth of *C. fetus* by alteration of aerotolerance was reported over 30 years ago (Bowdre *et al.*, 1976).

siderophore, therefore, is to facilitate internalization into the bacterial cell of the transferrin or lactoferrin iron liberated by the catecholamine (Freestone *et al.*, 2003).

2. Catecholamines and autoinducer activity

Enabling access to transferrin and lactoferrin-complexed iron is not the only mechanism by which catecholamines may affect bacterial growth. Using a serum-based culture medium, two studies have shown that catecholamines also induce production of a novel autoinducer of growth (Freestone *et al.*, 1999; Lyte *et al.*, 1996a). This novel autoinducer termed the norepinephrine-induced autoinducer (NE-AI) is heat stable, highly cross-species acting, and is able to enhance the growth of bacterial cell numbers to a level similar to that achievable with the catecholamines (Freestone *et al.*, 1999). Significantly, the NE-AI activity works independent of transferrin or lactoferrin, and is also able to stimulate recovery of viable but nonculturable *E. coli* O157:H7 and *Salmonella* (Freestone *et al.*, 1999; Reissbrodt *et al.*, 2002; Voigt *et al.*, 2006) as well as increase the rates of germination and subsequent growth of *Bacillus anthracis* spores (Reissbrodt *et al.*, 2004). The NE-AI is also recognized by many periodontal pathogens (Roberts *et al.*, 2005). Induction of the NE-AI in Gram-negative bacteria requires only a 4–6 hour exposure to norepinephrine (Freestone *et al.*, 1999; Lyte *et al.*, 1996a), which obviates the need for any additional catecholamine exposure to increase further bacterial proliferation. Translation of these observations to the host environment provides an explanation for why catecholamine release during acute stress could have lasting and widely acting effects on bacteria long after catecholamine levels have returned to normal.

As discussed earlier, the NE-AI is produced during the initial stages of catecholamine-growth induction, and because the growth stimulator induces its own synthesis, its production is, for Gram-negatives, as important as catecholamine-mediated iron delivery. The work of Sperandio *et al.* (2003) suggests that another autoinducer-like activity, AI-3, may also be involved in catecholamine-mediated effects on enteric bacteria. AI-3 is a novel signaling molecule produced by *E. coli* O157:H7 whose production requires the involvement of LuxS. The structure of AI-3, although not fully characterized, has been shown by mass spectroscopy to be a low molecular weight, and chemically unlike AI-2 (a furanosyl borate diester), or the homoserine lactone family of signaling molecules, both of which are synthesized by LuxS (Sperandio *et al.*, 2003). AI-3 is involved in quorum sensing (high cell-density dependent) regulation of the transcription of genes encoding the *E. coli* O157:H7 LEE type III secretion system. Based on the observation that epinephrine can restore expression of virulence gene expression in an *E. coli* O157:H7 *luxS* mutant defective in AI-3 synthesis, it has been proposed that there is a crosscommunication between the luxS/AI-3 quorum sensing system and the epinephrine host signaling system (Sperandio *et al.*, 2003; Walters and Sperandio, 2006). Additional studies have shown that norepinephrine also interacts with the sensor for AI-3, QseC (Clarke *et al.*, 2006 – see Section I.E). However, several questions still remain about the ligand specificity of QseC, since the complete structure of AI-3 is still unknown, and whether QseC generally recognizes low molecular weight aromatic molecules (which are environmentally common) (Freestone *et al.*, 2007c). The role of AI-3 relative to the NE-AI in the mechanism(s) by which catecholamine stress neurohormones modulate bacterial growth and virulence also remains to be determined, as *in vitro* work has shown that AI-3 is involved in virulence gene regulation at high cell density, while the NE-AI exerts its effects during the very first stages of bacterial interactions with catecholamines when cell numbers are very low as reflective of the initial infectious event.

As shown in Table 2.1, the ability of catecholamines to induce growth either by provision of host sequestered iron or by induction of the novel growth stimulating NE-AI has been shown to be important factors for the growth of over 50 bacterial species ranging from commensals such as *E. coli* (Freestone *et al.*, 2002) and the coagulase-negative staphylococci (Freestone *et al.*, 1999; Lyte *et al.*, 2003; Neal *et al.*, 2001) where it has been shown to be a critical factor in inotrope-mediated induction of staphylococcal biofilm formation in intravenous lines (Lyte *et al.*, 2003), to pathogenic bacteria such as *Bordetella* (Anderson and Armstrong, 2006), *Campylobacter jejuni* (Cogan *et al.*, 2007), enteropathogenic and enterohemorrhagic *E. coli* (Freestone *et al.*, 1999, 2000, 2003, 2007a,b) *S. enterica* (Freestone *et al.*, 1999, 2007a,b), *Vibrio* species (Nakano *et al.*, 2007a), and *Y. enterocolitica* (Freestone *et al.*, 1999, 2007a,b). In terms of access to host iron, these species lack specific systems for acquisition of iron from transferrin or lactoferrin, and the ability to use host neuroendocrine hormones to access such sequestered iron allows rapid growth in normally hostile environments such as blood or serum. *In vitro* work using serum or blood-supplemented media has further shown that the growth stimulation possible with catecholamines is up to 5 logs over unsupplemented controls, in a timeframe of 14–24 hours (Freestone *et al.*, 1999, 2002, 2003, 2007a,b; Neal *et al.*, 2001).

3. Catecholamines can enhance bacterial attachment to host tissues

In addition to stimulating the growth of bacteria, stress-related neuroendocrine hormones has been reported to significantly enhance both commensal and pathogen attachment to gut tissues. Norepinephrine has been shown to increase the expression of K99 pilus adhesin of enterotoxigenic *E. coli* (Lyte *et al.*, 1997c) as well as type 1 fimbriae of commensal *E. coli* (Hendrickson *et al.*, 1999). Vlisidou and coworkers examined the effect of norepinephrine on the adherence and enteropathogenicity of *E. coli* O157:H7 in a bovine ligated ileal loop model of infection (Vlisidou *et al.*, 2004). Norepinephrine was found to increase *E. coli* O157:H7-induced intestinal inflammatory and secretory responses as well as bacterial adherence to intestinal mucosa. In this study, norepinephrine-induced modulation of enteritis and adherence was found to be dependent on the ability of *E. coli* O157:H7 to form attaching and effacing lesions. Interestingly, later work by Chen *et al.* (2006) revealed that norepinephrine also promoted cecal adherence of a non-O157 *E. coli* strain as well as *E. coli* O157:H7 *eae* or *espA* mutants incapable of intimate mucosal attachment. Using tissue culture assays and a rat ileal loop model of virulence, Nakano *et al.* (2007b) found that norepinephrine enhanced the cytotoxicity and enterotoxicity of *V. parahaemolyticus*, as well as increased transcription of several Type III secretion system TSS1-related genes. Recently, Cogan *et al.* (2007) also found that norepinephrine increased the host cell invasiveness of *Campylobacter jejuni*.

In addition to its direct effects on enteric bacteria, catecholamines and other neuroendocrine hormones released in response to stress have also been shown to act upon the intestinal mucosa to indirectly increase attachment of intestinal pathogens to gut epithelial tissues. Chamber analyses of porcine gut explants showed that contraluminal application of norepinephrine (Chen *et al.*, 2006) or the noncatecholamine adrenocorticotrophic hormone (Schreiber and Brown, 2005) resulted in significant enhancement in the attachment of *E. coli* O157:H7 to the luminal mucosal surface.

D. The spectrum of catecholamine responsive bacteria

Table 2.1 lists the catecholamine-responsive bacterial species. Although this list is heavily weighted toward enteric pathogens, it should be pointed out that since catecholamines have a ubiquitous distribution in the mammalian body it might be expected that bacteria occupying extra-intestinal niches could similarly respond to changes in the stress levels of their host's environment. For instance, in humans stress is a well-recognized risk factor in initiation and progression of periodontal disease and stress neurohormones such as cortisol, epinephrine and norepinephrine have been isolated from saliva and gingival crevicular fluid (Johannsen *et al.*, 2006; Kennedy *et al.*, 2001). Using a serum-based medium and anaerobic culture conditions, Roberts and coworkers showed that both norepinephrine and epinephrine were recognized by oral bacteria implicated in causing periodontal disease (Roberts *et al.*, 2002, 2005). Of 43 species tested, more than half evidenced significant growth enhancement in the presence of the catecholamines (Table 2.1), thereby providing additional insight into why stress can trigger or exacerbate gum disease. Catecholamines have also recently been shown to enhance the growth of *Bordetella* species, including respiratory pathogens such as *B. bronchiseptica* and *B. pertussis*. Anderson and Armstrong (2006) showed that under iron-limited conditions, or in serum-based media, norepinephrine stimulated the growth of *B. bronchiseptica* as well as inducing expression of BfeA, a siderophore receptor important for growth *in vivo*. This suggests that respiratory pathogens may also be able to perceive catecholamines as a host environmental cue within the lung. Stress-neurohormone-enhanced virulence is not restricted to pathogens of mammalian hosts, as norepinephrine is shown to be recognized by oyster pathogens (Lacoste *et al.*, 2001) and by frog-infecting bacteria (Kinney *et al.*, 1999).

E. Catecholamine specificity in enteric bacteria

The enteric nervous system (ENS) consists of nearly 500 million neurons that innervate the entire length of the gastrointestinal tract and effectively constitute the third nervous system in the body in addition to the more generally recognizable central and peripheral nervous systems (for a review of the ENS, see Furness, 2006). Noradrenergic and dopaminergic neurons are abundant in the ENS, although neurons expressing phenylethanolamine *N*-methyltransferase, the enzyme needed for the synthesis of epinephrine from norepinephrine, have yet to be found (Costa *et al.*, 2000). This may explain why epinephrine, in contrast to norepinephrine and dopamine, has not been isolated from the mammalian intestinal tract (Costa *et al.*, 2000; Eisenhofer *et al.*, 1996). Studies that have investigated the responses of bacterial pathogens to a range of catecholamines have shown that enteric bacteria appear to possess specificity for the catecholamine that they are most likely to encounter in their particular intestinal niche. Sperandio and coworkers have reported that several *E. coli* O157:H7 virulence genes are regulated by epinephrine (Sperandio *et al.*, 2003; Walters and Sperandio, 2006) and have proposed that epinephrine is a specific host neurohormonal cue for enteropathogenic bacteria. However, analyses of the specificity of catecholamine responsiveness of three enteric pathogens noted for either their propensity to primarily colonize the gut (*Y. enterocolitica*), or to additionally inhabit extraintestinal sites (*E. coli* O157:H7 and *S. enterica*) revealed a clear preference for norepinephrine and dopamine over epinephrine (Freestone *et al.*, 2007a). Analysis of the catecholamine responsiveness of 14 *Y. enterocolitica* isolates revealed no significant growth responsiveness to epinephrine. Indeed, epinephrine even antagonized *Y. enterocolitica* responses to norepinephrine and dopamine (Fig. 2.2A and B), although the same combinations of catecholamine were stimulatory for *E. coli* (Fig. 2.2B and C). Similarly Nakano and colleagues found that the growth responses of several *Vibrio* species to epinephrine were at least a log lower in magnitude than those to either norepinephrine or dopamine (Nakano *et al.*, 2007a). In a separate

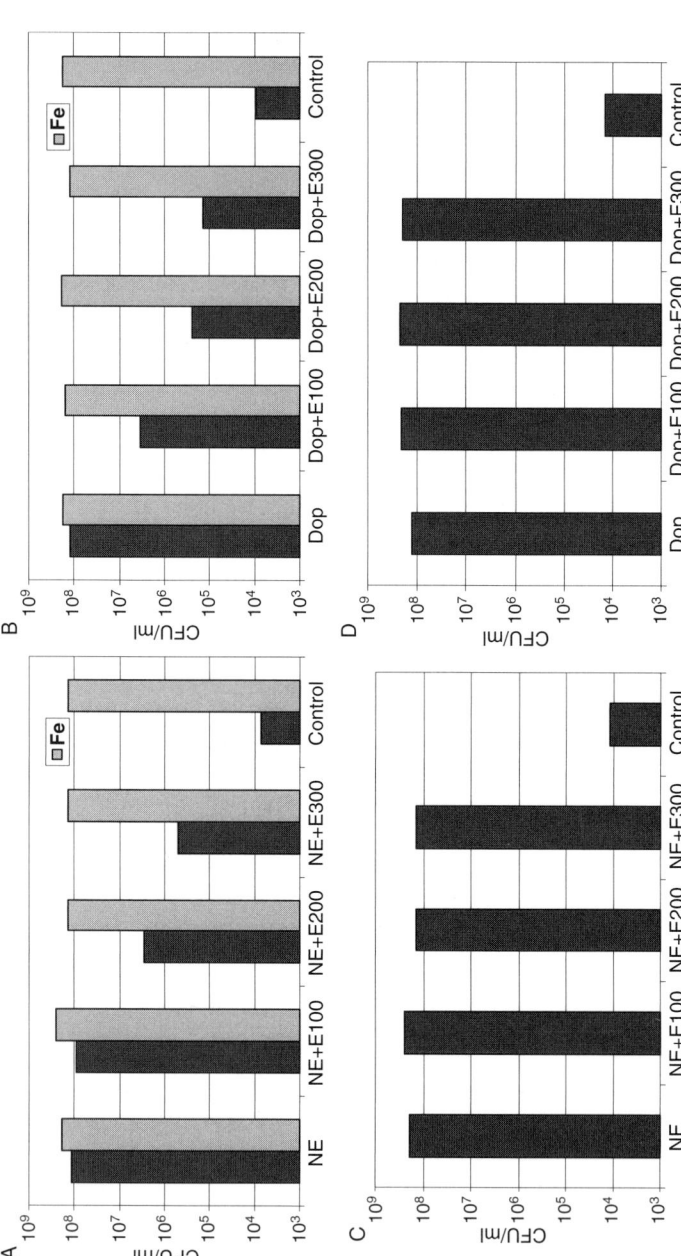

FIGURE 2.2 The ability of epinephrine to inhibit *Y. enterocolitica* growth induction by norepinephrine and dopamine. *Y. enterocolitica* (A, B) and *E. coli* O157:H7 (C, D) were inoculated at approximately 10^2 CFU/ml into duplicate 1 ml aliquots of serum-SAPI containing the combination of catecholamines shown, and incubated for either 40 hours (*Y. enterocolitica*) or 18 hours (*E. coli* O157:H7), and enumerated for growth (CFU/ml) as per standard technique (Freestone, *et al.* 2000). The results shown are representative data from four separate experiments; data points typically showed variation of no more than 3%. Black bar: catecholamine/catecholamine combination only; light gray bar: catecholamine/catecholamine combinations plus 100 mM Fe(NO$_3$)$_3$. Key: Norepinephrine (NE), Dopamine (Dop), and Epinephrine

study, Bansal *et al.* (2007) reported that epinephrine was markedly less effective than norepinephrine at enhancing the attachment of *E. coli* O157:H7 to gut tissue culture cells (Bansal *et al.*, 2007).

The lack of *Y. enterocolitica* responsiveness to epinephrine (Freestone *et al.*, 2007a) indicates that catecholamines can no longer be considered as agents that only mediate access to host sequestered iron for bacterial growth. The epinephrine antagonism of norepinephrine and dopamine responsiveness also provides some insight into the catecholamine environment of the gut. If epinephrine were present in concentrations equivalent to those of norepinephrine and dopamine, *Y. enterocolitica* should have significant problems in growing within the intestinal environment, but clearly it does not. Based upon the lack of biochemical evidence for epinephrine synthesis in the gut, it is questionable whether an infecting enteric bacterium is likely to be exposed to significant levels of epinephrine while in the intestinal tract (Freestone *et al.*, 2007a; Lyte, 2004). It may therefore be speculated that norepinephrine, rather than epinephrine, is likely to be the crosscommunicating adrenergic neurohormonal signal between host and enteric pathogens, and that the bacterial response to epinephrine observed in *E. coli* O157:H7 (Sperandio *et al.*, 2003; Walters and Sperandio, 2006) may be more related to its structural similarity to norepinephrine. Given the demonstration that specificity in catecholamine recognition exists for *E. coli* O157:H7, analyses of the effects of the gut catecholamines norepinephrine and dopamine on *E. coli* O157:H7 virulence gene expression would be informative (see following Section I.F).

In mammalian systems catecholamine antagonists have been used therapeutically to block catecholamine effects (e.g., α- and β-adrenergic antagonists regulate cardiovascular function) and pharmacologically, to locate and characterize catecholamine receptors. In bacterial systems, recent work has shown that catecholamine antagonists can block catecholamine effects in bacteria, even though there is no genomic evidence for either adrenergic or dopaminergic receptors in prokaryotic organisms. A pharmacological investigation of the effects of a wide range of α- and β-adrenergic and dopaminergic receptor antagonists on norepinephrine, epinephrine and dopamine dependent growth induction in *Y. enterocolitica*, *E. coli* O157:H7 and *S. enterica* showed that each species possessed separate

---

(Epi) (black bar). NE, 50 μM NE; NE + E100, 50 μM NE plus 100 μM Epi; NE + E200, 50 μM NE plus 200 μM Epi; NE + E300, 50 μM NE plus 300 μM Epi; Dop, 50 μM Dop; Dop + E100, 50 μM Dop plus 100 μM Epi; D + E200, 50 μM Dop plus 200 μM Epi; Dop + E300, 50 μM Dop plus 300 μM Epi. The results shown are representative data from three separate experiments; data points typically showed variation of no more than 3%. This figure was taken with permission from Freestone, P. P., Haigh, R. D., and Lyte, M. (2007a). Specificity of catecholamine-induced growth in *Escherichia coli* O157:H7, *salmonella enterica* and *yersinia enterocolitica*. *FEMS Microbiol. Lett.* **269**, 221–228.

and distinct response systems for each catecholamine (Freestone *et al.*, 2007b). α-Adrenergic antagonists (phentolamine, prazosin, and phenoxybenzamine), but not β-adrenergic antagonists (propranolol, labetalol or apomorphine), were able to block bacterial responses to norepinephrine and epinephrine, but did not inhibit dopamine responsiveness. Conversely, dopaminergic antagonists (chlorpromazine) blocked responses to dopamine, but did not affect response to either norepinephrine or epinephrine. Considered collectively, the specificity and catecholamine antagonist studies (Freestone *et al.*, 2007a,b) suggest that enteric bacteria have evolved specific response systems for the catecholamine(s) they are likely to encounter in the anatomical region they are most likely to inhabit within the mammalian body. It will therefore be interesting in future investigations to observe the catecholamine specificity of species more exclusively occupying extraintestinal locations.

In animals, specificity in catecholamine recognition usually indicates the existence of a specific adrenergic or dopaminergic signaling pathway and receptor. Although the ability of catecholamines to stimulate bacterial growth has been known for over a decade (Lyte, 2004), little is still known concerning the nature of the putative bacterial receptor(s) to which norepinephrine, epinephrine and dopamine may bind and exert their effects, or even whether the binding properties of such a receptor are similar between different bacterial species. As already mentioned, adrenergic and dopaminergic receptors do not exist in prokaryotic species. However, a recent report (Clarke *et al.*, 2006) used *in vitro* constructs to show that norepinephrine and epinephrine were recognized by the *E. coli* O157:H7 two-component regulator sensor kinase QseC, leading to the proposal that this is the bacterial receptor for these catecholamines. QseC also recognizes the LuxS-dependentAI-3, which has led to the suggestion of a link between the pathways for interkingdom (Microbial Endocrinology) signalling and intercellular (quorum sensing) signalling. It is interesting that although *Y. enterocolitica* shows responsiveness to norepinephrine, but not to epinephrine, it does not contain homologues for QseC (or the related receptor QseE). This suggests that a different response system for norepinephrine is likely to exist in *Y. enterocolitica*, and possibly in other bacterial species (Freestone *et al.*, 2007a,b).

F. Molecular analyses of bacterial catecholamine responsiveness

Although there have been reports of catecholamine effects on expression of several bacterial virulence genes (Anderson and Armstrong, 2006; Nakano *et al.*, 2007a,b; Sperandio *et al.*, 2003; Walters and Sperandio, 2006) as of yet there have been only two investigations which have examined how stress-related catecholamine neurohormones can modulate bacterial gene expression in a global context. Bansal *et al.* (2007) used DNA microarrays to obtain gene expression profiles of *E. coli* O157:H7 grown in a biofilm in the presence of either norepinephrine or epinephrine. Although the culture medium was Luria broth-based and therefore not analogous to the host environment, plus no evidence presented as to the absolute equivalence of test and control culture growth levels, this study reported that both norepinephrine and epinephrine upregulated expression of genes involved in surface colonization and virulence. Importantly, this study demonstrated that the profiles of genes regulated by the two catecholamines were different, which agreed well with observations of differential catecholamine responsiveness observed by Freestone and coworkers (Freestone *et al.*, 2007a,b). Over 900 genes were differentially expressed following treatment with epinephrine and norepinephrine, with around half of the genes showing differential expression common between the two catecholamines. The Bansal study also found that norepinephrine and epinephrine modulated expression of genes involved in a number of cellular processes, but interestingly none of the *eae* (intimin adherence protein) genes showed any significant changes in expression following exposure to either of the catecholamines.

The observation by Bansal *et al.* (2007) that epinephrine did not activate expression of *eae* or associated genes, is in marked contrast to several earlier studies (Sperandio *et al.*, 2003; Walters and Sperandio, 2006). A second *E. coli* O157:H7 transcriptional profiling study has used a smaller array of 610 *E. coli* O157:H7 virulence genes to the Basal investigation. Dowd (2007) found that *E. coli* O157:H7 cultured in serum-SAPI medium containing norepinephrine displayed differential regulation of 101 genes compared to nonsupplemented controls similarly grown in serum-SAPI. Genes that showed substantially higher levels of expression included *eae, espB* and *espA*, and the shiga toxin *stx1, stx2* genes (Dowd, 2007). These results agree well with that of earlier studies that norepinephrine increased Shiga-like toxin production of *E. coli* O157:H7 cultured in serum-based media (Lyte *et al.*, 1996b), and supports the evidence that acute stress resulting in elevated norepinephrine levels in the gut could result in enhanced virulence of *E. coli* O157:H7 *in vivo* (Lyte *et al.*, 1996b, 1997a,b,c,d; Sperandio *et al.*, 2003; Vlisidou *et al.*, 2004).

II. EXPERIMENTAL DESIGN ISSUES IN THE STUDY OF MICROBIAL ENDOCRINOLOGY

Central to the *in vitro* demonstration that any neuroendocrine hormone is able to influence bacterial growth or production of virulence-associated factors is the use of a medium that mimics the *in vivo* milieu as closely as possible. The following sections therefore consider the influence that test media can have on analysis of bacterial catecholamine stress neurohormone responsiveness. The following section is also intended to provide a guide to the study of other neurohormones as well, in particular the need for relevant bacterial inocula that more closely reflects that present at the actual time of infection when the bacterium first encounters the neuroendocrine hormone.

A. Medium comparability

In any comparative analysis the equivalence of the test and control experimental environments can, without argument, determine the entire validity of the results. Microbial Endocrinology-related studies require that the investigation of neuroendocrine stress hormones on bacterial growth or virulence be conducted in media that are as host-like as possible. The use of traditional rich microbiological media does not provide the type of stressful environmental conditions that a microorganism would encounter upon entrance into a host, and use of such media may not show whether a bacterium is responding to the stress neuroendocrine hormone under evaluation. In their analysis of the role of the sympathetic nervous system in bacteremia, Straub *et al.* (2006) used Luria broth as a test medium for analyzing stress neurohormone effects on bacteria. Because they saw no effects of norepinephrine on bacterial growth, they prematurely concluded that the catecholamine was not directly interacting with the bacterial species investigated. However, without additionally testing the catecholamine-bacterial combination in a medium that more accurately reflected the *in vivo* milieu, they may have erroneously concluded that bacterial-neuroendocrine interactions did not play a role in their model of bacteremia, especially given the number of reports in Table 2.1 that show that neuroendocrine-bacterial interactions are widespread.

To create a suitably host-like medium for the investigation of bacterial-neuroendocrine stress hormone effects, Lyte and Ernst (1992) developed serum-SAPI, a minimal salts medium supplemented with 30% (v/v) adult bovine serum (for the composition of serum-SAPI see section 2.4.1). Serum-SAPI is a highly stressful bacteriostatic medium that exposes bacteria to conditions similar to that inside a host, such as iron limitation, limited nutrient availability, and immune defence proteins such as antibodies and complement. Most bacteria usually grow poorly in serum-SAPI, thus making it environmentally similar to most mammalian tissue fluids. And as will be discussed in Section II.D.4, there are important differences between serum and freshly obtained plasma that can influence bacterial responses to catecholamines, such as the ability to form biofilms (Lyte *et al.*, 2003).

Serum-SAPI medium, or variants of it (Roberts et al., 2002, 2005), has been used extensively in the analysis of stress neuroendocrine hormone effects on bacteria, particularly growth induction (Belay et al., 2003; Freestone et al., 1999, 2000, 2002, 2003, 2007a,b; Kinney et al., 1999; Lyte and Ernst, 1992, 1993; Neal et al., 2001; Roberts et al., 2002, 2005). Most bacteria grow poorly in serum-SAPI, at least in part because of the iron restriction imposed by the serum iron binding protein transferrin, and generally achieve only a few log increases in growth over starting inocula (which are usually low – typically around 10^1–10^2 CFU/ml, to reflect the numbers of bacteria likely to be present at the start of an infection (Lyte and Ernst, 1992, 1993). However, catecholamine supplementation allows bacteria to access the iron sequestered by transferrin, and so enables growth [addition of exogenous iron can also support growth (Freestone et al., 2000)]. The stimulation of growth in the presence of catecholamines can be considerable – up 5 log orders over non-supplemented control cultures (Freestone et al., 1999, 2002; Lyte and Ernst, 1992, 1993; Lyte et al., 1996a,b; Neal et al., 2001).

The bacteriostatic nature of serum or plasma-SAPI, and the poor growth of control cultures, can complicate creation of equivalent conditions between test and control cultures. This challenge is well exemplified in a recent study by Toscano et al. (2007) who showed that *in vitro* pretreatment of *S. typhimurium* with norepinephrine prior to infection of the pathogen into young pigs markedly increased its virulence. Examination of the tissue distribution of *Salmonella* 24 hours postchallenge revealed that norepinephrine treated bacteria were present in greater numbers and more widely distributed in gut tissues than control bacteria. Although this is an interesting study that has implications for food safety, given that widespread nature of catecholamines and related compounds in human foods (Freestone et al., 2007c) there are, however, some methodological issues highlighted by the Toscano et al. (2007) investigation that typify the experimental design challenges central to the study of Microbial Endocrinology. The stress neuroendocrine hormone preincubation test medium for the Toscano study was serum-SAPI supplemented with 2 mM norepinephrine, but because of poor growth in noncatecholamine supplemented serum-SAPI, the control cultures were grown in Luria broth, a traditional, rich microbiological media. This extreme difference between the test and control preincubation media composition makes it difficult to assign the origin of the observed differences between test and control cultures to the particular stress neuroendocrine hormone tested, or possibly to host factors in the serum component of the test culture media. It would have been additionally informative to have investigated the effects of 2 mM norepinephrine in Luria broth on dissemination of *Salmonella,* since Vlisidou et al. (2004) showed that 5 mM norepinephrine in Luria broth markedly enhanced both the attachment and enterotoxicity of *E. coli* O157:H7 in a calf ileal loop model of virulence.

B. Bacterial inoculum size and its influence on stress neurohormone responsiveness

Also important to the proper conduct of Microbial Endocrinology-related analyses is the use of a bacterial inoculum that accurately reflects the infectious dose encountered *in vivo*. Starting inocula are therefore usually low – typically around $10^1–10^2$ CFU/ml, which is intended to reflect the numbers of bacteria likely to be present at the start of an infection (Freestone *et al.*, 1999, 2002, 2003, 2007a,b,c; Lyte and Ernst, 1992, 1993). This is in direct contrast to the majority of studies in the Microbial Endocrinology-related microbiological literature which have utilized log orders higher inocula to start test cultures. Yet a number of studies have analyzed, possibly for technical reasons, responses to catecholamines in high cell density cultures (around 10^8 CFU/ml), which have then been extrapolated back to *in vivo* scenarios where the infecting bacteria are likely to be in much lower numbers (Sperandio *et al.*, 2003; Vlisidou *et al.*, 2004; Walters and Sperandio, 2006). The lack of growth responsiveness of *Y. enterocolitica* isolates at low cell densities (10^2 CFU/ml) to epinephrine, and the ability of epinephrine to antagonize *Yersinia* responses to other catecholamines (Freestone *et al.*, 2007a) has already been discussed earlier in this review. What is intriguing is that when analyzed at high cell density (10^8 CFU/ml), this species could be able to respond to epinephrine specifically because of the population density-related ability to now acquire iron from transferrin (Freestone *et al.*, 2007a).

Further experimentation has shown that the population density of a bacterial culture can markedly influence the specificity of catecholamine responsiveness. Freestone *et al.* (2007a) used serum-SAPI medium to examine the growth responses to catecholamines of *E. coli* O157:H7, *S. enterica* and *Y. enterocolitica* over an 8-log dilution curve (Fig. 2.3). It can be seen that the ability of the catecholamines to affect bacterial growth in a serum-based medium is initially evident at low ($<10^4$ CFU/ml) cell densities with the greatest differences observed at very low ($<10^2$ CFU/ml) cell densities. Although Fig. 2.3 shows that both *E. coli* O157:H7 and *S. enterica* are able to respond to epinephrine at very low population densities, which are reflective of the bacterial numbers that are likely to be present at the initial stages of an infection, there exists an order of catecholamine preference with growth responses to norepinephrine and dopamine being at least a log-fold greater than those to epinephrine. These observations and the *Y. enterocolitica* epinephrine response data (Freestone *et al.*, 2007a) suggests that in the context of analysis of microbe–host catecholamine interactions caution should be exercised before assuming that conclusions reached from *in vitro* observations of responses of high density cultures can be directly extrapolated to the low

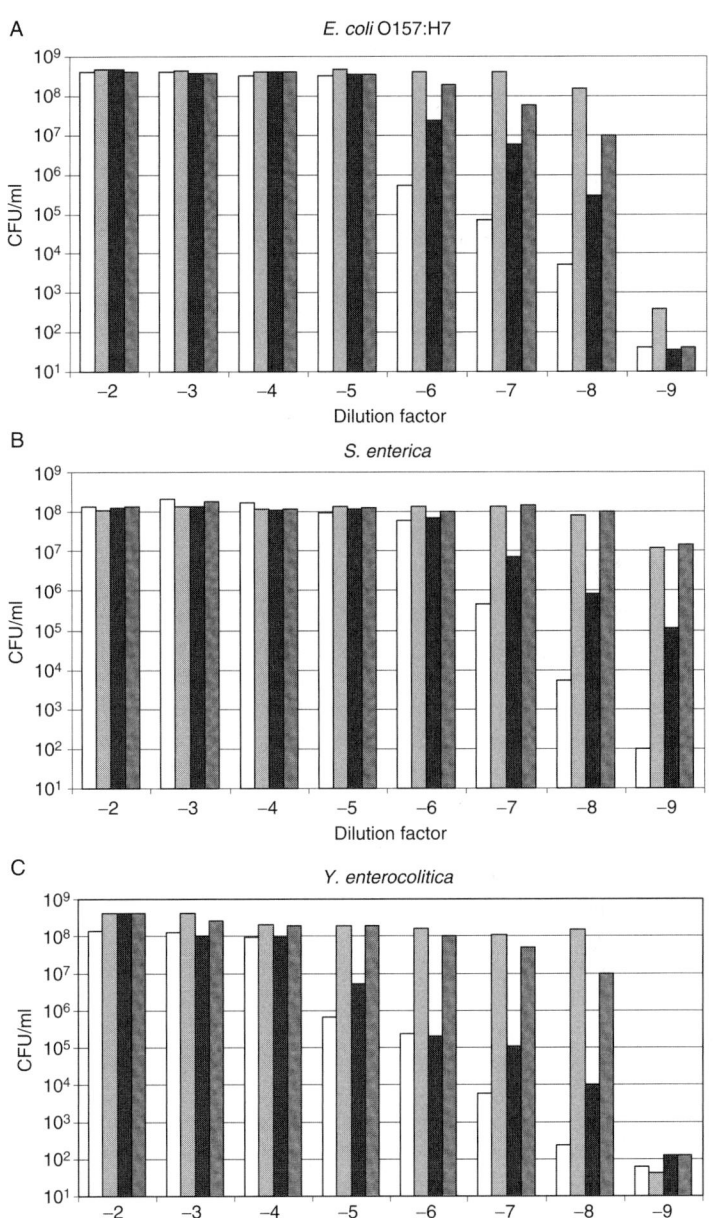

FIGURE 2.3 Bacterial population density influences catecholamine specificity. Histograms A–C shows the growth response of varying inoculum size on the specificity of growth response to catecholamines norepinephrine (NE), epinephrine (Epi), and dopamine (Dop). Cultures were diluted in 10-fold steps into serum-SAPI medium, incubated for either 18 hours (*E. coli* O157:H7 and *S. enterica*) or 40 hours (*Y. enterocolitica*),

population numbers typically present during the initial stage of an infection (Tarr and Neill, 2001).

It cannot be overemphasized that even using host-like media such as serum-SAPI, failure to adhere to the use of sufficiently low levels of inocula can lead to such rapid growth that the evaluation of response to a specific neuroendocrine hormone can be problematic. As an example, differential effects of catecholamines on the growth of several Gram-negative bacterial species have been reported (Belay et al., 2003). In these experiments the inoculum size was so great that no lag phase occurred in the serum-based medium and maximal growth was achieved by both control and catecholamine-supplemented cultures in less than 12 hours (Belay et al., 2003). When aspects of this experiment were later repeated using lower inocula (O'Donnell et al., 2006), a lag phase that is more characteristic of a bacteriostatic medium (Freestone et al., 1999, 2002, 2003, 2007a,b,c; Lyte and Ernst, 1992, 1993) was reported as was increased responsiveness to norepinephrine not observed in the previous report. It should also be pointed out that given the inherent variability in any optical density measurement that is typically used to estimate CFU/ml for culture initiation, the actual plate counts of the inocula also needs to be performed to confirm the numbers of bacteria that actually were used to initiate the culture.

C. The importance of neuroendocrine hormone concentration

Optimally, the concentration of any one particular neuroendocrine hormone used in bacterial response analyses should be reflective of the concentration likely to be encountered in the host. It is important to realize that the concentrations of most neuroendocrine hormones are determined from fluid-based samples, such as plasma or urine, which are easy to obtain from a subject. Since the site of action of the vast majority of neurohormones is in the target tissue and not the circulation (or urine), concentrations derived from fluid specimens may underestimate the true tissue level by log orders (Leinhardt et al., 1993). This situation will, of

---

and enumerated for growth (CFU/ml) as described in Freestone et al. (2007a). The inoculum sizes of the E. coli, S. enterica and Y. enterocolitica cultures were 5.29, 2.84 and 4.44 × 10^8 CFU/ml, respectively. The results shown are representative data from four separate experiments; individual data points showed variation of no more than 5%. White bar: no additions (control); light grey bar: 50 µM NE; black bar: 100 µM Epi; diagonal hatch: 50 µM Dop. This figure was taken with permission from Freestone, P. P., Haigh, R. D., and Lyte, M. (2007a). Specificity of catecholamine-induced growth in *Escherichia coli* O157:H7, *salmonella enterica* and *yersinia enterocolitica*. FEMS Microbiol. Lett. **269**, 221–228.

course, affect design elements of the experiment, most notably the concentration(s) at which the specific neuroendocrine hormone under study should be tested (for further discussion, see Lyte, 2004). If the concentration of the neurohormone at the site of action in the body is not known, or is very variable, then dose responses analyses should be performed over a reasonably wide range of neurohormone concentrations. A typical dose response profile of catecholamine growth responsiveness for *E. coli* O157:H7 and *Y. enterocolitica* inoculated into serum-SAPI medium is shown in Fig. 2.4. As already mentioned, epinephrine had little effect on growth of *Y. enterocolitica*, and on a concentration-dependent basis was less potent at stimulating growth of *E. coli* O157:H7 than either norepinephrine or dopamine (see also Fig. 2.3). Previous catecholamine dose response analyses for bacterial species such as *E. coli, Salmonella, Yersinia,* and the coagulase-negative staphylococci suggest that initial test concentrations of 50 µM norepinephrine or dopamine, and 100 µM epinephrine (Fig. 2.4) would be suitable. Higher concentrations of catecholamine (>200 µM) start to become inhibitory (Fig. 2.4).

Catecholamines can induce growth by chelating host sequestered iron, which in iron limited media such as serum or blood can lead to a microbiologically important growth stimulation (Cogan *et al.,* 2007; Coulanges *et al.,* 1998; Freestone *et al.,* 2000, 2002, 2003, 2007a,b,c; Lyte *et al.,* 2003; Neal *et al.,* 2001). However, given the role that iron, and in particular iron restriction, plays in the regulation of virulence factor expression, the question arises as to the physiological basis of neurohormone effects determined in noniron limited non-host-like media. Some investigations of the effects of catecholamines on bacteria virulence have used catecholamine concentrations in the mM range. For instance, Vlisidou *et al.* (2004) used 5 mM norepinephrine in Luria broth to investigate the effects of catecholamines on *E. coli* O157:H7 adherence to gut tissues. Luria broth is an iron-replete rich culture medium, and it is possible that a 5 mM addition of an iron chelator may have made the norepinephrine-supplemented test medium Fe limited. The possibility of a role of iron limitation in the observed results could be investigated by incubating cultures in iron-limited Luria broth (made, for example by Chelex pretreatment) or by the direct addition to test cultures of a ferric iron chelator such as dipyridyl. The chelation of iron by catecholamines in iron replete media, can also lead to the generation of oxygen-derived free radicals, a cell damaging process that has been implicated in development of neurodegenerative diseases such as Parkinson's (Borisenko *et al.,* 2000); therefore, particularly at high catecholamine concentrations, the role of oxidative stress influencing bacterial gene expression also needs consideration.

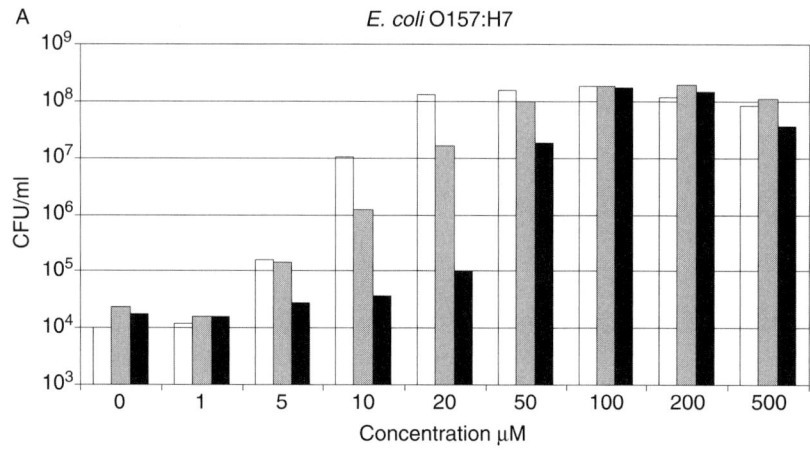
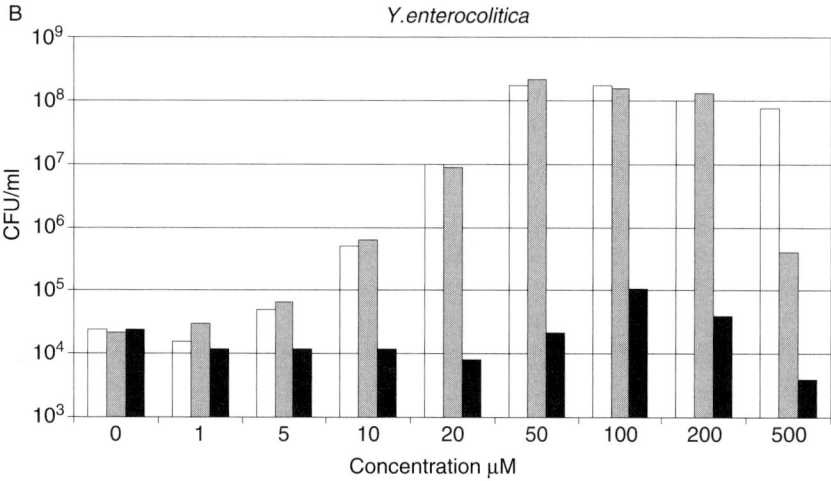

FIGURE 2.4 Dose–response effect of catecholamines on *E. coli* O157:H7 and *Y. enterocolitica*. *E. coli* O157:H7 and *Y. enterocolitica* were inoculated at approximately 10^2 CFU/ml into duplicate 1 ml aliquots of serum-SAPI containing the concentrations of the catecholamines shown and incubated for either 40 hours (*Y. enterocolitica*) or 18 hours (*E. coli* O157:H7), and enumerated for growth (CFU/ml) as described in Freestone et al. (2007). The results shown for norepinephrine (NE) (grey bar), dopamine (Dop) (white bar), and epinephrine (Epi) (black bar) are representative data from two separate experiments; data points showed variation of less than 3%. This figure was taken with permission from Freestone, P. P., Haigh, R. D., and Lyte, M. (2007a). Specificity of catecholamine-induced growth in *Escherichia coli* O157:H7, *salmonella enterica* and *yersinia enterocolitica*. FEMS Microbiol. Lett. **269**, 221–228.

D. Neuroendocrine stress neurohormone assay methodologies

The following methodologies describe assay methodologies widely used in the analysis of bacterial interactions with neuroendocrine hormones.

1. Medium preparation and culture conditions

As discussed earlier, a large number of studies that have examined bacterial growth responses to catecholamines have been performed in serum-supplemented SAPI medium (Freestone *et al.*, 1999, 2000, 2002, 2003, 2007a,b,c; Lyte and Ernst, 1992, 1993; Lyte *et al.*, 1996a,b, 1997a,c,d). SAPI minimal medium consists of 6.25 mM NH_4NO_3, 1.84 mM KH_2PO_4, 3.35 mM KCl, 1.01 mM $MgSO_4$ and 2.77 mM glucose, pH 7.5, supplemented with 30% (v/v) adult bovine serum. Fetal calf serum is to be avoided in design of any host-like media as it is naturally rich in a wide range of neuroendocrine hormones and therefore is nearly as close to being as supportive of bacterial growth as traditional rich laboratory culture media. SAPI is autoclaved without the glucose for 20 min at 121 °C; the glucose and serum are added aseptically afterwards. Serum-SAPI medium is stable up to a year when stored at −20 °C, although it is not recommended to refreeze frozen media once thawed, as the serum component of the medium will become less bacteriostatic, which results in higher growth levels in unsupplemented control assays. Filter sterilization of serum-SAPI is also not recommended, as this again reduces the bacteriostatic nature of the serum and may produce anomalously high cell numbers in control assays.

Bacteria are usually grown overnight in Luria broth or in other laboratory culture media such as Tryptone Soya Broth (TSB), harvested by centrifugation at 5000–10,000g, washed once in serum-SAPI, resuspended in serum-SAPI, and diluted into 1 ml aliquots of serum-SAPI medium to give an initial inoculum of approximately 10^2 CFU/ml (for an overnight Luria broth or TSB culture, a dilution of around 1:10^7 of the original culture usually yields an inoculum of around 10^2 CFU/ml; precise inocula sizes should be determined for each experiment by plating out a suitable dilution of the inoculum). Controls for serum-SAPI growth assays consist of an equivalent volume of the solvent used to dissolve the test compound (double-distilled water in the case of catecholamine bitartrate or hydrochloride salt solutions). For bacteria such as *E. coli* or *Salmonella*, assays are incubated statically for 16 hours in a humidified 37 °C incubator in air supplemented with 5% (v/v) CO_2; other species may need longer periods to reach maximal growth levels (1–5 × 10^8 CFU/ml), for example *Y. enterocolitica* requires 36–40 hours (Freestone *et al.*, 1999, 2007a,b; Lyte and Ernst, 1992).

The presence of CO_2 is crucial to the experimental design of microbial–neuroendocrine hormone analyses as it provides necessary buffering of the medium components, as well as a more host-like atmosphere. Failure to grow cultures in the presence of CO_2 may result in a greatly diminished bacterial response to neurohormones. After incubation, cultures are mixed and aliquots withdrawn for estimation of growth by serial dilution into PBS and spot or pour-plate analysis using media such as Luria agar. Individual growth assays and plate counts are carried out in duplicate or triplicate and all experiments should be performed on at least two separate occasions. It is also important to note that the use of racemic mixtures of neuroendocrine hormones, such as the +/− orientation of norepinephrine should be avoided. The preferred composition should be the L-form as this is the physiological form present in the host. Catecholamine stock solutions should be freshly prepared to avoid the possibility of oxidation that may be encountered during periods of storage or light exposure.

2. Transferrin-iron removal analyses

Iron-saturated (diferric, holo-) transferrin (0.50 mg/ml in 100 mM Tris-HCl, pH 7.5 containing 10% (v/v) glycerol (Freestone et al., 2000) is incubated at 37 °C with a range of catecholamine concentrations (50 μg in a final volume of 50 μl is a convenient volume for electrophoretic analysis). The precise time of incubation needs to be determined empirically, but our experience shows that 15 hours is sufficient. The catecholamine is omitted from negative controls, while a ferric chelator such as 10 mM dihydroxybenzoic acid, which is known to remove all the iron from saturated transferrin in the conditions of this assay, is used as a positive control. After incubation, transferrin samples are analyzed for iron loss by electrophoresis in 6% polyacrylamide gels containing 6 M urea (Freestone et al., 2000), using a vertical minigel system. Electrophoresis is at a constant 70 V for approximately 6–7 hours at room temperature, using 100 mM Tris, 20 mM boric acid, 1.6 mM EDTA (pH 8.4) as the running buffer. A partially iron-saturated human transferrin preparation, containing a total of 10 μg of iron-depleted apo, iron-saturated holo- and both N-terminal and C-terminal domain monoferric isoforms of transferrin, are used as a marker standard. After electrophoresis, gels are fixed and stained for 10 min in 0.05% (w/v) Coomassie blue in 40% methanol, 10% acetic acid, and destained in 7.5% (v/v) methanol, 5% (v/v) acetic acid.

3. Bacterial transferrin iron uptake assays

To prepare ^{55}Fe-labelled transferrin, apo-transferrin is incubated at 37 °C for a minimum of 5 hours with 25 μCi of ^{55}FeCl$_3$ in a reaction mixture containing 100 mM Tris-HCl, pH 8.0 plus a total of 1.5 μg of Fe per mg of apo-transferrin protein using sodium citrate (2 mM) as the iron donor (Freestone et al., 2000); this labelling method gives a transferrin preparation of approximately 30% iron saturation, which is similar to the iron saturation status of transferrin in mammalian blood. To prepare ^{55}Fe-lactoferrin, if apo-lactoferrin is not available commercially, iron-saturated lactoferrin must first be depleted of iron by sequential dialysis against for 24 hours against 0.2 M citric acid (pH 2.3), distilled water and reneutralization using 100 mM Tris-HCl (pH 7.5) (Freestone et al., 2000). For both apo-lactoferrin and citric acid iron-depleted lactoferrin, labelling conditions are as for transferrin, except that the incubation with ^{55}Fe needs to be for up to 15 hours. When the iron labelling is complete, nonincorporated ^{55}Fe needs to be removed from the ^{55}Fe-transferrin and lactoferrin using one or two rounds of spin-column chromatography (BioRad Micro Bio-spin 6 columns equilibrated in 100 mM Tris-HCl pH 7.5). Note that because of the very high affinity of lactoferrin for ferric iron it is difficult to obtain completely iron-depleted preparations of the protein using citric acid as an iron depletion agent Consequently, the specific activity of the ^{55}Fe-labeled lactoferrin used may be less than that of ^{55}Fe-labeled transferrin (Freestone et al., 2000).

To test the ability of bacterial strains to acquire Fe from transferrin, exponentially grown bacteria (to ensure they are metabolically active) are washed once and then inoculated at a cell density of around 10^8 CFU/ml into a suitable sterile medium (SAPI, serum-SAPI or a neutral buffered medium containing a readily metabolized carbon source such as glucose) that is supplemented 2×10^5 CPM/ml of filter-sterilized ^{55}Fe-transferrin or lactoferrin, with and without the desired concentration of catecholamine or other compound (it is best to start with a concentration which can induce growth in serum-based medium). Bacteria are then incubated statically for approximately 6 hours (this is the optimum incubation time for enteric bacteria such as *E. coli* or *Salmonella*) (Freestone et al., 2003, 2007a,b,c). However, other bacterial species may require longer incubation, and so performance of a time course of ^{55}Fe-uptake may be necessary. When the ^{55}Fe-uptake assay is complete, bacteria are harvested by low speed centrifugation (5000–10,000g) and washed once in PBS (repeat washes are generally not necessary, and for some species can result in lysis if the bacteria were incubated in serum-based assay media). The bacterial pellet is then resuspended in 50 µl of PBS, added to 2 ml of a scintillation fluid suitable for aqueous samples, and counted in the tritium channel of a scintillation counter. All assays should carried out in triplicate, and performed on at least two separate occasions.

4. Catecholamine-mediated biofilm production
We have previously shown that catecholamine-based inotropic drugs (isoprenaline and dobutamine), which are used clinically to maintain heart and kidney function in acutely ill patients, were as effective as catecholamines

norepinephrine, epinephrine and dopamine in their ability to induce growth of both pathogenic and commensal bacteria (Freestone *et al.*, 2002; Neal *et al.*, 2001). Approximately 50% of patients in intensive care units (ICU) receive catecholamine inotropic support during their hospital stay (Smythe *et al.*, 1993). It is also significant that ICU patients are recognized as being more likely to develop a serious hospital-acquired infection than patients in the general hospital population. Intravenous catheter-associated infections are a particular problem in the ICU which prompted us to investigate whether the widely prescribed inotropic agents might play a role in the increased susceptibility of such patients to colonization of catheters with coagulase-negative staphylococci, in particular the skin commensal *Staphylococcus epidermidis* (Lyte *et al.*, 2003).

To examine whether inotropic agents influenced the formation of staphylococcal biofilm, a plasma-based biofilm assay was conducted on medical grade plastic substrates. The results of this investigation demonstrated that inotropes were able to markedly increase both staphylococcal growth and biofilm formation (Lyte *et al.*, 2003). Fig. 2.5 shows

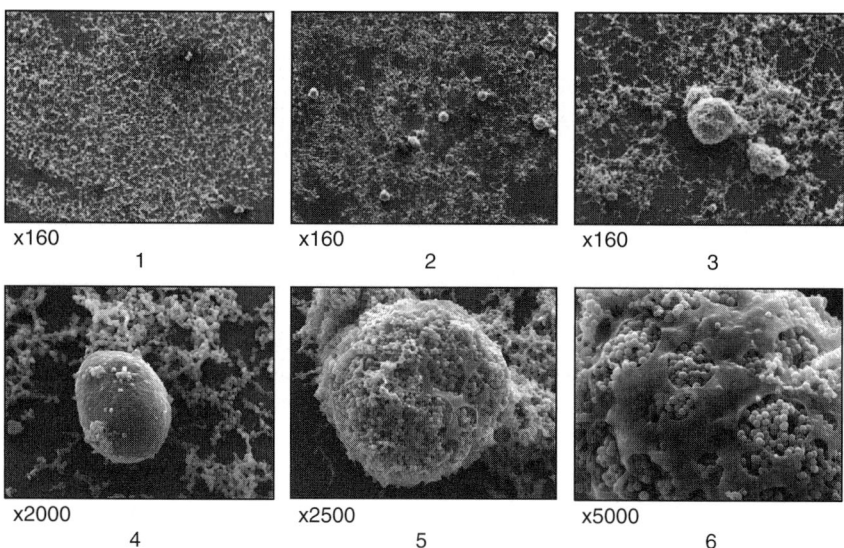

FIGURE 2.5 Catecholamine-induced biofilm formation in *S. epidermidis*. The image panels show scanning electron micrographs (SEM) of biofilms of *Staphylococcus epidermidis* adhering to polystyrene after overnight growth in freshly prepared plasma-SAPI medium in the absence (SEM 1) or presence (SEM 2–6) of 0.1 mM norepinephrine as described previously (Lyte *et al.*, 2003). Initial inocula for both *S. epidermidis* cultures were approximately 10^2 CFU/ml. Biofilms were prepared and analyzed as described in Section II.D.5 and reference (Lyte *et al.*, 2003). Higher magnification showing details of the bacterial-exopolysaccharide clusters are shown in SEM panels 3–6.

additional biofilm images obtained from the Lancet report that were not included in the original publication. As detailed in the report, biofilms were only formed in the presence of freshly obtained, not previously frozen, plasma, although both plasma and serum supported increased planktonic growth of *S. epidermidis* in the presence of norepinephrine (Lyte *et al.*, 2003). The clinical relevance of this *in vitro* study was emphasized by the use of inotropes at concentrations at or below those at which they are used clinically, in conjunction with low bacterial inocula that mimic bacterial numbers likely to be present at the initiation of an opportunistic infection.

III. CONCLUSIONS AND FUTURE DIRECTIONS

The realization that microorganisms can use the neuroendocrine environment of the host as part of the infectious process as embodied by Microbial Endocrinology opens up new avenues with which to examine the pathogenesis of infectious disease. The patient, whether in the community or hospital setting, is a biologically complex organism in which the microorganisms it coexists with, species both beneficial and potentially harmful, have had a long evolutionary timeframe in which to interact with their host, and learn to "sense" when it is stressed. In addition to the body's own neurophysiological output during stress, current therapies also employ neuroendocrine-based drugs, such as the catecholamine inotropes, which may interact with microorganisms in unintended ways as illustrated by inotrope-induced *S. epidermidis* biofilm formation (Lyte *et al.*, 2003).

Ongoing and future research endeavors in Microbial Endocrinology will undoubtedly reveal additional mechanisms by which neuroendocrine–bacterial interactions can influence the infectious disease process. For example, a very recent study from our laboratories has shown that catecholamine inotropes can actually restore the growth of antimicrobial-damaged bacteria that otherwise do not grow in standard clinical laboratory assays (Freestone *et al.*, 2008). Not only does this provide a possible explanation of the well-recognized clinical observation that antimicrobials determined to be effective in the clinical laboratory may not show the same efficacy when administered to the patient, but also the exciting possibility that the efficiency of recovery of antibiotic-damaged bacteria in the clinical laboratory can be enhanced by the design of recovery media that incorporates the same neuroendocrine hormones that are most prominently elaborated in the patient. Studies are currently underway to examine such possibilities.

REFERENCES

Alverdy, J., Holbrook, C., Rocha, F., Seiden, L., Wu, R. L., Musch, M., Chang, E., Ohman, D., and Suh, S. (2000). Gut-derived sepsis occurs when the right pathogen with the right virulence genes meets the right host: Evidence for *in vivo* virulence expression in *Pseudomonas aeruginosa*. *Ann. Surg.* **232**, 480–489.

Anderson, M. T., and Armstrong, S. K. (2006). The *Bordetella* Bfe system: Growth and transcriptional response to siderophores, catechols, and neuroendocrine catecholamines. *J. Bacteriol.* **188**, 5731.

Apaydin, H. (2000). Broad bean (*Vicia faba*)—a natural source of L-DOPA—prolongs 'on' periods in patients with Parkinson's disease who have 'on–off' fluctuations. *Mov. Disord.* **15**, 164–166.

Bansal, T., Englert, D., Lee, J., Hegde, M., Wood, T. K., and Jayaraman, A. (2007). Differential effects of epinephrine, norepinephrine, and indole on *Escherichia coli* O157:H7 chemotaxis, colonization and gene expression. *Infect. Immun.* **75**, 4597–4607.

Belay, T., Aviles, H., Vance, M., Fountain, K., and Sonnenfeld, G. (2003). Catecholamines and *in vitro* growth of pathogenic bacteria: Enhancement of growth varies greatly among bacterial species. *Life Sci.* **73**, 1527–1535.

Borisenko, G., Kagan, A., Hsia, C., and Schor, N. F. (2000). Interaction between 6-Hydroxydopamine and Transferrin: "Let My Iron Go". *Biochemistry* **39**, 3392–3400.

Bowdre, J. H., Krieg, N. R., Hoffman, P. S., and Smibert, R. M. (1976). Stimulatory effect of dihydroxyphenyl compounds on the aerotolerance of *Spirillum volutans* and *Campylobacter fetus* subspecies *jejuni*. *Appl. Environ. Microbiol.* **31**, 127–133.

Burton, C. L., Chhabra, S. R., Swift, S., Baldwin, T. J., Withers, H., Hill, S. J., and Williams, P. (2002). The growth response of *Escherichia coli* to neurotransmitters and related catecholamine drugs requires a functional enterobactin biosynthesis and uptake system. *Infect. Immun.* **70**, 5913–5923.

Chen, C., Brown, D. R., Xie, Y., Green, B. T., and Lyte, M. (2003). Catecholamines modulate *Escherichia coli* O157:H7 adherence to murine cecal mucosa. *Shock* **20**, 183–188.

Chen, C., Lyte, M., Stevens, M. P., Vulchanova, L., and Brown, D. R. (2006). Mucosally directed adrenergic nerves and sympathomimetric drugs enhance non-intimate adherence of *Escherichia coli* O157:H7 to porcine cecum and colon. *Eur. J. Pharmacol.* **539**, 116–124.

Clarke, M. B., Hughes, D. T., Zhu, C., Boedeker, E. C., and Sperandio, V. (2006). The QseC sensor kinase: A bacterial adrenergic receptor. *Proc Natl. Acad. Sci. USA.* **103**, 10420–10425.

Cogan, T. A., Thomas, A. O., Rees, L. E. N., Taylor, A. H., Jepson, M. A., Williams, P. H., Ketley, J., and Humphrey, T. J. (2007). Norepinephrine increases the pathogenic potential of *Campylobacter jejuni*. *Gut* **56**, 1060–1065.

Costa, M., Brookes, S. J., and Hennig, G. W. (2000). Anatomy and physiology of the enteric nervous system. *Gut* **47**(Suppl 4), 15–19.

Coulanges, V., Andre, P., and Vidon, D. J. (1998). Effect of siderophores, catecholamines, and catechol compounds on *Listeria* spp. Growth in iron-complexed medium. *Biochem. Biophys. Res. Commun.* **249**, 526–530.

Dowd, S. E. (2007). *Escherichia coli* O157:H7 gene expression in the presence of catecholamine norepinephrine. *FEMS Micro. Lett.* **273**, 214–223.

Eisenhofer, G., Åneman, A., Hooper, D., Rundqvist, B., and Friberg, P. (1996). Mesenteric organ production, hepatic metabolism, and renal elimination of norepinephrine and its metabolites in humans. *J. Neurochem.* **66**, 1565–1573.

Freestone, P. P., Haigh, R. D., Williams, P. H., and Lyte, M. (1999). Stimulation of bacterial growth by heat-stable, norepinephrine-induced autoinducers. *FEMS Microbiol. Lett.* **172**, 53–60.

Freestone, P. P., Lyte, M., Neal, C. P., Maggs, A. F., Haigh, R. D., and Williams, P. H. (2000). The mammalian neuroendocrine hormone norepinephrine supplies iron for bacterial growth in the presence of transferrin or lactoferrin. *J. Bacteriol.* **182**, 6091–6098.

Freestone, P. P., Williams, P. H., Haigh, R. D., Maggs, A. F., Neal, C. P., and Lyte, M. (2002). Growth stimulation of intestinal commensal *Escherichia coli* by catecholamines: A possible contributory factor in trauma-induced sepsis. *Shock* **18**, 465–470.

Freestone, P. P., Haigh, R. D., Williams, P. H., and Lyte, M. (2003). Involvement of enterobactin in norepinephrine-mediated iron supply from transferrin to enterohaemorrhagic *Escherichia coli*. *FEMS Microbiol. Lett.* **222**, 39–43.

Freestone, P. P., Haigh, R. D., and Lyte, M. (2007a). Specificity of catecholamine-induced growth in *Escherichia coli* O157:H7, *Salmonella enterica* and *Yersinia enterocolitica*. *FEMS Microbiol. Lett.* **269**, 221–228.

Freestone, P., Haigh, R. D., and Lyte, M. (2007b). Blockade of catecholamine-induced growth by adrenergic and dopaminergic receptor antagonists in *Escherichia coli* O157:H7, *Salmonella enterica* and *Yersinia enterocolitica*. *BMC Microbiol.* **7**, 8.

Freestone, P. P., Walton, N., Haigh, R. D., and Lyte, M. (2007c). Influence of Dietary Catechols on the Growth of Enteropathogenic Bacteria. *Int. J. Food Micro.* **119**, 159–169.

Freestone, P. P., Haigh, R. D., and Lyte, M. (2008). Catecholamine inotrope resuscitation of antibiotic-damaged *Staphylococci* and its blockade by specific receptor antagonists. *J. Infect. Dis.* **197**, 1044–1052.

Furness, J. B. (2006). The Enteric Nervous System. Blackwell Pub, Malden, MA.

Green, B. T., Lyte, M., Kulkarni-Narla, A., and Brown, D. R. (2003). Neuromodulation of enteropathogen internalization in Peyer's patches from porcine jejunum. *J. Neuroimmunol.* **141**, 74–82.

Green, B. T., Lyte, M., Chen, C., Xie, Y., Casey, M. A., Kulkarni-Narla, A., Vulchanova, L., and Brown, D. R. (2004). Adrenergic modulation of *Escherichia coli* O157:H7 adherence to the colonic mucosa. *Am. J. Physiol. Gastrointest. Liver Physiol.* **287**, G1238–G1246.

Hendrickson, B. A., Guo, J., Laughlin, R., Chen, Y., and Alverdy, J. C. (1999). Increased type 1 fimbrial expression among commensal *Escherichia coli* isolates in the murine cecum following catabolic stress. *Infect. Immun.* **67**, 745–753.

Janakidevi, K., Dewey, V. C., and Kidder, G. W. (1966). The Biosynthesis of catecholamines in two genera of Protozoa. *J. Biol. Chem.* **241**, 2576–2578.

Johannsen, A., Rylander, G., Söder, B., and Asberg, M. (2006). Dental plaque, gingival inflammation, and elevated levels of interleukin-6 and cortisol in gingival crevicular fluid from women with stress-related depression and exhaustion. *J. Periodontol.* **77**, 1403–1409.

Kennedy, B., Dillon, E., Mills, P. J., and Ziegler, M. G. (2001). Catecholamines in human saliva. *Life Sciences* **69**, 87–99.

Kinney, K. S., Austin, C. E., Morton, D. S., and Sonnenfeld, G. (1999). Catecholamine enhancement of *Aeromonas hydrophila* growth. *Microb. Pathog.* **26**, 85–91.

Lacoste, A., Jalabert, F., Malham, S. K., Cueff, A., and Poulet, S. A. (2001). Stress and stress-induced neuroendocrine changes increase the susceptibility of juvenile oysters (*Crassostrea gigas*) to *Vibrio splendidus*. *Appl. Environ. Microbiol.* **67**, 2304–2309.

Leinhardt, D. J., Arnold, J., Shipley, K. A., Mughal, M. M., Little, R. A., and Irving, M. H. (1993). Plasma NE concentrations do not accurately reflect sympathetic nervous system activity in human sepsis. *Am. J. Physiol.* **265**, E284–E288.

Lyte, M., and Ernst, S. (1992). Catecholamine induced growth of gram negative bacteria. *Life Sci.* **50**, 203–212.

Lyte, M., and Ernst, S. (1993). Alpha and beta adrenergic receptor involvement in catecholamine-induced growth of gram-negative bacteria. *Biochem. Biophys. Res. Commun.* **190**, 447–452.

Lyte, M., Frank, C. D., and Green, B. T. (1996a). Production of an autoinducer of growth by norepinephrine cultured *Escherichia coli* O157:H7. *FEMS Microbiol. Lett.* **139**, 155–159.

Lyte, M., Arulanandam, B. P., and Frank, C. D. (1996b). Production of Shiga-like toxins by *Escherichia coli* O157:H7 can be influenced by the neuroendocrine hormone norepinephrine. *J. Lab. Clin. Med.* **128**, 392–398.

Lyte, M. (1997a). Induction of gram-negative bacterial growth by neurochemical containing banana (Musa x paradisiaca) extracts. *FEMS Microbiol. Lett.* **154**, 245–250.

Lyte, M., and Bailey, M. T. (1997b). Neuroendocrine-bacterial interactions in a neurotoxin-induced model of trauma. *J. Surg. Res.* **70**, 195–201.

Lyte, M., Arulanandam, B., Nguyen, K., Frank, C., Erickson, A., and Francis, D. (1997c). Norepinephrine induced growth and expression of virulence associated factors in enterotoxigenic and enterohemorrhagic strains of *Escherichia coli*. *Adv. Exp. Med. Biol.* **412**, 331–339.

Lyte, M., Erickson, A. K., Arulanandam, B. P., Frank, C. D., Crawford, M. A., and Francis, D. H. (1997d). Norepinephrine-induced expression of the K99 pilus adhesin of enterotoxigenic *Escherichia coli*. *Biochem. Biophys. Res. Commun.* **232**, 682–686.

Lyte, M., Freestone, P. P., Neal, C. P., Olson, B. A., Haigh, R. D., Bayston, R., and Williams, P. H. (2003). Stimulation of *Staphylococcus epidermidis* growth and biofilm formation by catecholamine inotropes. *Lancet* **361**, 130–135.

Lyte, M. (2004). Microbial endocrinology and infectious disease in the 21st century. *Trends Microbiol.* **12**, 14–20.

Nakano, M., Takahashi, A., Sakai, Y., Kawano, M., Harada, N., Mawatari, K., and Nakaya, Y. (2007a). Catecholamine-induced stimulation of growth in *Vibrio* species. *Lett. Appl. Microbiol.* **44**, 649–653.

Nakano, M., Takahashi, A., Sakai, Y., and Nakaya, Y. (2007b). Modulation of pathogenicity with norepinephrine related to the type III secretion system of *Vibrio parahaemolyticus*. *J. Infect. Dis.* **195**, 1353–1360.

Neal, C. P., Freestone, P. P., Maggs, A. F., Haigh, R. D., Williams, P. H., and Lyte, M. (2001). Catecholamine inotropes as growth factors for *Staphylococcus epidermidis* and other coagulase-negative staphylococci. *FEMS Microbiol. Lett.* **194**, 163–169.

O'Donnell, P. M., Aviles, H., Lyte, M., and Sonnenfeld, G. (2006). Enhancement of *in vitro* growth of pathogenic bacteria by norepinephrine: Importance of inoculum density and role of transferrin. *Appl. Environ. Microbiol.* **72**, 5097–5099.

Peterson, P. K., Chao, C. C., Molitor, T., Murtaugh, M., Strgar, F., and Sharp, B. M. (1991). Stress and pathogenesis of infectious disease. *Rev. Infect. Dis.* **13**, 710–720.

Pitman, R. M. (1971). Transmitter substances in insects: A review. *Comp. Gen. Pharmacol.* **2**, 347–371.

Powell, B. L., Drutz, D. J., Hulpert, M., and Hun, S. H. (1983). Relationship of progesterone and oestradiol-binding proteins in *Coccidiodes immitis* to coccidioidal dissemination in pregnancy. *Infect. Immun.* **40**, 478–485.

Rabey, J. M., Vered, Y., Shabtai, H., Graff, E., and Korczyn, A. D. (1992). Improvement of Parkinsonian features correlate with high plasma levodopa values after broad bean (*Vicia faba*) consumption. *J. Neurol. Neurosurg. Psychiatry* **55**, 725–727.

Ratledge, C., and Dover, L. G. (2000). Iron metabolism in pathogenic bacteria. *Ann. Rev. Microbiol.* **54**, 881–941.

Reiche, E. M. V., Nunes, S. O. V., and Morimoto, H. K (2004). Stress, depression, the immune system, and cancer. *Lancet Oncol.* **5**, 617–625.

Reissbrodt, R., Rienaecker, I., Romanova, J. M., Freestone, P. P. E., Haigh, R. D., Lyte, M., Tschäpe, H., and Williams, P. H. (2002). Resuscitation of *Salmonella* enterica serovar typhimurium and enterohemorrhagic *Escherichia coli* from the viable but nonculturable state by heat-stable enterobacterial autoinducer. *Appl. Environ. Microbiol.* **68**, 4788–4794.

Reissbrodt, R., Raßbach, A., Burghardt, B., Rienäcker, I., Mietke, H., Schleif, J., Tschäpe, H., and Williams, P. H. (2004). Assessment of a new selective chromogenic *Bacillus cereus* group plating medium and use of enterobacterial autoinducer of growth for cultural identification of *Bacillus* species. *J. Clin. Microbiol.* **42**, 3795–3798.

Roberts, A., Matthews, J., Socransky, S., Freestone, P., Williams, P., and Chapple, I. (2002). Stress and the periodontal diseases: Effects of catecholamines on the growth of periodontal bacteria *in vitro*. *Oral Microbiol. Immunol.* **17**, 96–303.

Roberts, A., Matthews, J., Socransky, S., Freestone, P., Williams, P., and Chapple, I. (2005). Stress and the periodontal diseases: Growth responses of periodontal bacteria to *Escherichia coli* stress-associated autoinducer and exogenous Fe. *Oral Microbiol. Immunol.* **20**, 147–153.

Scheckelhoff, M. R., Telford, S. R., Wesley, M., and Hu, L. T. (2007). Borrelia burgdorferi intercepts host hormonal signals to regulate expression of outer surface protein A. *Proc. Natl. Acad. Sci. USA* **104**, 7247–7252.

Schreiber, K. L., and Brown, D. R. (2005). Adrenocorticotrophic hormone modulates *Escherichia coli* O157:H7 adherence to porcine colonic mucosa. *Stress* **8**, 185–190.

Sonnex, C. (1998). Influence of ovarian hormones on urogenital infection. *Sex. Transm. Infect.* **74**, 11–19.

Smythe, M. A., Melendy, S., Jahns, B., and Dmuchowski, C. (1993). An exploratory analysis of medication utilization in a medical intensive care unit. *Crit. Care Med.* **21**, 1319–1323.

Sperandio, V., Torres, A. G., Jarvis, B., Nataro, J. P., and Kaper, J. (2003). Bacteria-host communication: The language of hormones. *Proc. Natl. Acad. Sci. USA.* **100**, 8951–8956.

Straub, R. H., Wiest, R., Strauch, U. G., Härle, P., and Schölmerich, J. (2006). The role of the sympathetic nervous system in intestinal inflammation. *Gut* **55**, 1640–1649.

Tarr, P. I., and Neill, M. A. (2001). *Escherichia coli* O157:H7. *Gastroenterol. Clin. North Am.* **30**, 735–751.

Toscano, M. J., Stabel, T. J., Bearson, S. M. D., Bearson, B. L., and Lay, D. C., Jr. (2007). Cultivation of *Salmonella enterica* serovar Typhimurium in a norepinephrine-containing medium alters *in vivo* tissue prevalence in swine. *J. Exp. Anim. Sci.* **43**, 329–338.

Voigt, W., Fruth, A., Tschäpe, H., Reissbrodt, R., and Williams, P. H. (2006). Enterobacterial autoinducer of growth enhances Shiga toxin production by enterohemorrhagic *Escherichia coli*. *J. Clin. Microbiol.* **44**, 2247–2249.

Vlisidou, I., Lyte, M., Van Diemen, P. M., Hawes, P., Monaghan, P., Wallis, T. S., and Stevens, M. P. (2004). The neuroendocrine stress hormone norepinephrine augments *Escherichia coli* O157:H7-induced enteritis and adherence in a bovine ligated ileal loop model of infection. *Infect. Immun.* **72**, 5446–5451.

Waalkes, T. P. A., Sjoerdsma, C. R., Creveling, H., Weissbach, H., and Udenfriend, S. (1958). Serotonin, norepinephrine, and related compounds in bananas. *Science* **127**, 648–650.

Walters, M., and Sperandio, V. (2006). Autoinducer-3 and epinephrine signaling in the kinetics of locus of enterocyte effacement gene expression in enterohemorrhagic *Escherichia coli*. *Infect. Immun.* **74**, 5445–5455.

Williams, P. H., Rabsch, W., Methner, U., Voigt, W., Tschäpe, H., and Reissbrodt, R. (2006). Catecholate receptor proteins in Salmonella enterica: Role in virulence and implications for vaccine development. *Vaccine* **24**, 3840–3844.

Woods, D. E., Jones, A. L., and Hill, P. J. (1993). Interaction of insulin with *Pseudomonas pseudomallei*. *Infect. Immun.* **61**, 4045–4050.

Yang, E. V., and Glaser, R. (2000). Stress-induced immunomodulation: Impact on immune defenses against infectious disease. *Biomed. Pharmacother.* **54**, 245–250.

Zaborina, O., Lepine, F., Xiao, G., Valuckaite, V., Chen, Y., Li, T., Ciancio, M., Zaborin, A., Petrof, E. O., Turner, J. R., Rahme, L. G., Chang, E., *et al.* (2007). Dynorphin activates quorum sensing quinolone signaling in *Pseudomonas aeruginosa*. *PLoS Pathogens* **3**, 3.

CHAPTER 3

Molecular Genetics of Selenate Reduction by *Enterobacter cloacae* SLD1a-1

Nathan Yee[*,1] and Donald Y. Kobayashi[†]

Contents

	I. Introduction	108
	II. Physiology and Biochemistry of *E. cloacae* SLD1a-1	108
	III. Basic Systems of Molecular Genetics in Facultative Se-reducing Bacteria	110
	A. Growth of *E. cloacae* SLD1a-1 on selenate-containing agar plates	110
	B. Direct cloning experiments	110
	C. Gene specific mutagenesis	112
	D. Random mutagenesis using transposons	113
	IV. Genetic Analysis of Selenate Reduction in *E. cloacae* SLD1a-1	115
	V. A Molecular Model for Selenate Reduction in *E. cloacae* SLD1a-1	118
	VI. Conclusions and Future Prospects	120
	Acknowledgments	120
	References	120

---

[*] Department of Environmental Sciences, Rutgers, The State University of New Jersey, New Brunswick, New Jersey 08901
[†] Department of Plant Biology and Pathology, Rutgers, The State University of New Jersey, New Brunswick, New Jersey 08901
[1] Corresponding author: Department of Environmental Sciences, Rutgers, The State University of New Jersey, New Brunswick, New Jersey 08901

Advances in Applied Microbiology, Volume 64
ISSN 0065-2164, DOI: 10.1016/S0065-2164(08)00403-6

© 2008 Elsevier Inc.
All rights reserved.

I. INTRODUCTION

Selenium contamination of natural waters is a global environmental problem. A variety of anthropogenic activities, including coal combustion, mining, petroleum refining, and agricultural irrigation, are responsible for releasing toxic levels of selenium into the environment, (see review by Lemly, 2004). Elevated concentrations of selenium in surface waters have been shown to cause serious ecological problem, most notably reproductive failure and teratogenic deformities in aquatic wildlife (Hoffman and Heinz, 1988; Hoffman et al., 1988; Lemly, 2002; Ohlendorf et al., 1986). In these polluted environments, the toxicity of selenium results from the accumulation of dissolved selenate [Se(VI), SeO_4^{2-}] and selenite [Se(IV), SeO_3^{2-}] oxyanions. The reduction of selenate and selenite to sparingly soluble elemental selenium [Se(0)] is one of the primary processes by which selenium can be detoxified.

Microorganisms inhabiting selenium-contaminated sediments can reduce dissolved Se(VI) to Se(0) and precipitate elemental selenium particles (Dungan et al., 2003; Losi and Frankenberger, 1997; Macy et al., 1993; Oremland et al., 1989). The reduction and precipitation of selenium by Se-reducing bacteria has been proposed as a potential bioremediation strategy (Cantafio et al., 1996; Frankenberger and Arshad, 2001; Macy, et al., 1993). However, to employ Se-reducing bacteria for practical bioremediation, reliable models must be developed to predict microbial behavior, including when they are active, what mechanisms are involved, and under what conditions the mechanisms function. Identification of the genes that control microbial selenate reduction is central to understanding the environmental factors that regulate Se-reducing activity. In this review, we describe the molecular biology of microbial selenate reduction, focusing specifically on a genetically tractable Se-reducing bacterium, *Enterobacter cloacae* SLD1a-1.

II. PHYSIOLOGY AND BIOCHEMISTRY OF *E. CLOACAE* SLD1a-1

E. cloacae SLD1a-1 is a gram-negative facultative bacterium isolated from seleniferous waters in San Joaquin, California (Losi and Frankenberger, 1997). This organism can catalyze the reduction of Se(VI) to Se(0), and precipitate extracellular elemental selenium particles. The optimum pH and temperature for selenate reduction activity are 7.0 °C and 36 °C, respectively. Advantageously, *E. cloacae* SLD1a-1 can be handled under ambient air conditions, and readily cultivated aerobically to high cell densities. This heterotrophic bacterium can use a variety of electron donors, including glucose, mannose, and galactose, to reduce Se(VI) (Losi and Frankenberger, 1998). Other carbohydrate sources that promote selenate reduction by *E. cloacae* SLD1a-1 include disaccharides (lactose and maltose) and ketose (e.g., fructose), and to a lesser extent monosaccharides (arabinose and xylose). In the absence of oxygen, *E. cloacae* SLD1a-1 can grow anaerobically by respiring nitrate or by fermenting on sugars. Interestingly, this bacterium is capable of reducing Se(VI) during fermentative growth on glucose, but anaerobic respiration of nitrate inhibits selenate reduction activity (Losi and Frankenberger, 1997; Ridley *et al.*, 2006).

Several researchers have noted that *E. cloacae* SLD1a-1 catalyzes the reduction of Se(VI) only after oxygen is depleted from the nutrient medium (Losi and Frankenberger, 1998; Yee *et al.*, 2007). When inoculated in selenate-containing LB broth, *E. cloacae* SLD1a-1 forms visible red elemental selenium precipitates approximately after 1 day of incubation. Growth curve and dissolved oxygen measurements indicate that selenate reduction occurs once bacterial cultures reach stationary growth phase, and after oxygen has been completely consumed. Selenate reduction experiments conducted at varying oxygen levels have demonstrated that decreasing oxygen concentrations results in increasing selenate reduction activity (Losi and Frankenberger, 1998). Therefore, suboxic conditions are necessary to promote the reduction of Se(VI) and precipitation of Se(0) by *E. cloacae* SLD1a-1.

Biochemical studies have shown that the selenate reductase in *E. cloacae* is a membrane-bound heterotrimeric complex with the active site facing the periplasmic side of the cytoplasmic membrane (Watts *et al.*, 2003). The enzyme complex has apparent molecular mass of $\sim$600 kDa, and an overall subunit composition of $\alpha_3\beta_3\gamma_3$ with corresponding subunit molecular masses of $\sim$100, $\sim$55, and $\sim$36 kDa, respectively (Ridley *et al.*, 2006). The selenate reductase contains molybdenum, heme, and nonheme iron as prosthetic constituents, and a *b*-type cytochrome in the active complex. The estimated cytochrome content of the enzyme is 0.9 ± 0.14 mol of heme per mol of protein. Cultures grown in the presence of tungstate inhibit selenate reduction activity, suggesting the selenate reductase is a molybdo-enzyme, resembling other membrane-bound proteins involved in the electron transport chain. *In vitro* assays have shown that the selenate reductase can reduce selenate but cannot reduce nitrate, nitrite, sulfate, arsenate, dimethylsulfoxide (DMSO), or trimethylamine *N*-oxide (TMAO) (Watts *et al.*, 2003). Recently, it has been demonstrated that the selenate reductase can also reduce chlorate and bromate (Ridley *et al.*, 2006), suggesting that specificity of reductase activity does not tightly adhere to selenate alone.

III. BASIC SYSTEMS OF MOLECULAR GENETICS IN FACULTATIVE Se-REDUCING BACTERIA

A. Growth of *E. cloacae* SLD1a-1 on selenate-containing agar plates

E. cloacae SLD1a-1 readily grows on nutrient rich media such as Luria-Broth (LB) agar and makes single colonies overnight on the bench top. The ability of this bacterium to rapidly grow on LB agar is a major benefit in establishing a genetic system, as it greatly facilitates mutant isolation. When grown on selenate-containing LB agar, *E. cloacae* SLD1a-1 precipitates red elemental selenium particles. The precipitate consists of discrete spherical particles composed of pure Se(0) (Losi and Frankenberger, 1998; Yee *et al.*, 2007). These particles are remarkably monodispersed and approximately 200 nm in diameter. Macroscopically, the microbe-Se(0) particle aggregates appear as bright red colonies on LB plates. In contrast, bacterial strains unable to reduce Se(VI) do not precipitate element selenium and form white colonies on selenate-containing agar. Therefore, the formation of red elemental selenium as a product of selenate reduction can be used to easily screen for bacterial strains that display selenate reduction activity.

B. Direct cloning experiments

Identification of the genes that confer selenate reduction activity can be accomplished by constructing a genomic library and cloning *E. cloacae* SLD1a-1 genes in a non Se-reducing host strain, such as *Escherichia coli* S17-1. Yee *et al.* (2007) demonstrated that heterologous expression of *E. cloacae* SLD1a-1 genes involved in the selenate reduction pathway confers Se(VI) reduction activity in *E. coli*. A genomic library can be constructed using the following procedure:

1. Isolate total genomic DNA of *E. cloacae* SLD1a-1 by sodium dodecyl sulfate (SDS) lysis and purify on a cesium chloride gradient.
2. Partially digest the DNA with a restriction enzyme and then size-fractionate the DNA on a 10–40% glycerol gradient centrifuged at 194,000g for 33 hours.
3. Collect DNA fragments greater than 15 kb and ligate into the cosmid cloning vector, for example pLAFR3 (Staskawicz *et al.*, 1986), and package the vector into lambda phage by using Gigapak Gold extracts (Strategene).
4. Transduce the *E. cloacae* genomic library into an *E. coli* strain, such as S17-1, that does not display selenate reduction activity and plate onto LB agar containing tetracycline. Because the vector pLAFR3 confers

resistance to the antibiotic tetracycline, growth on LB agar containing tetracycling can be used to select for colonies containing the cosmid vector.
5. Pick individual colonies and transfer to LB plates containing 10 mM sodium selenate.

The precipitation of red elemental selenium as a product of Se(VI) reduction provides a simple screen to identify clones within the genomic library that confer selenate reduction activity to the recipient *E. coli* strain. Clone-containing strains that are unable to reduce selenate form white colonies on selenate-containing agar, while clones that are able to catalyze the reduction of Se(VI) to Se(0) form red colonies. Fig. 3.1 shows a clone, designated pECL29, that forms red colonies when streaked on selenate-containing agar. *E. coli* S17-1 carrying the vector pLAFR3 is shown as an experimental control. When streaked on selenate-containing agar, *E. coli* S17-1 colonies form white colonies, indicating that the strain cannot reduce selenate to elemental selenium. Insertion of the 9.75 kb DNA fragment pECL29 confers selenate reduction activity in *E. coli* S17-1. Subcloning experiments and DNA sequencing has shown that cells harboring a DNA fragment containing the *fnr* (Fumurate Nitrate Reduction regulator) gene is responsible for activating selenate reduction activity in *E. coli* (Yee *et al.*, 2007).

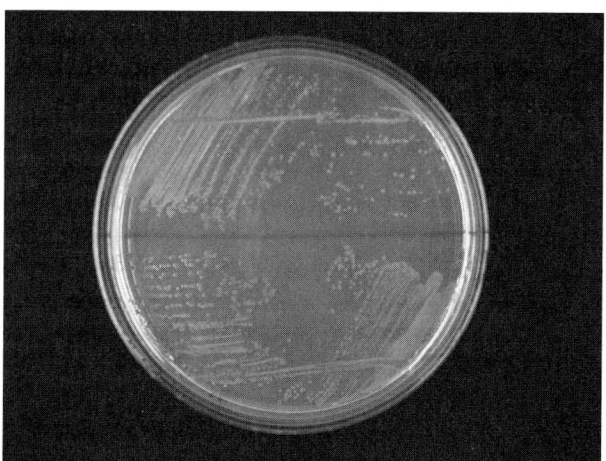

FIGURE 3.1 *E. coli* S17-1 grown on selenate containing agar. (*Top*) *E. coli* S17-1 carrying the cosmid vector pLAFR3. (*Bottom*) *E. coli* S17-1 carrying the cosmid clone pECL29 (9.75 kb insert). (See Color Plate Section in the back of the book.)

C. Gene specific mutagenesis

If direct cloning experiments reveal that a particular gene is involved in selenate reduction, a gene specific knock-out mutation can be introduced into the wild-type strain to verify its involvement. Loss of selenate reduction activity due to the gene knockout can be determined by growing the derivative mutant strains on selenate-containing agar and screening for the formation of white colonies. Below is an example of a targeted gene knock-out procedure:

1. Construct a plasmid vector containing the upstream and downstream flanking sequences of a gene of interest. A suitable vector is the *sacB*-based plasmid pJQ200SK, which confers gentamycin resistance. To ensure homologous recombination, the flanking sequences should at least be of 500 base pairs.
2. Insert an antibiotic (e.g., kanamycin) resistance cassette between the cloned fragments within the plasmid vector. Mate the mutant construct into the wild type strain using *E. coli* and select transconjugants by plating the mutants on agar plates containing the appropriate antibiotics. For example, if pJQ200SK is used as the plasmid vector, then the agar plates should contain kanamycin (the cassette confers kanamycin resistance) and gentamycin (pJQ200SK confers gentamycin resistance). Only strains that are resistant to both antibiotics can grow on the selective media. Colonies that carry the antibiotic resistance genes in the plasmid have undergone a single crossover event, resulting in the integration of the plasmid into the genome.
3. Isolate single colonies that display kanamycin and gentamycin resistance, then inoculate the isolates in 3 ml LB and grow overnight at 30 °C. Dilute overnight cultures (1:1000) in fresh LB broth medium containing 5% sucrose. Expression of the *sacB* gene in the presence of the sucrose is lethal to the bacterium, thereby permitting selection for loss of the vector pJQ200SK.
4. After 6 hours of growth, plate cultures onto LB agar containing 5% sucrose and kanamycin. Colonies that grow on LB agar supplemented with kanamycin, but are unable to grow on LB agar supplemented with kanamycin and gentamycin have undergone a second recombination event. This results in the loss of plasmid DNA from the genome and the replacement of the targeted gene with the kanamycin cassette. These colonies are putative knock-out mutants.
5. To determine if the mutants are defective in selenate reduction, plate the derivative strains on LB agar containing selenate. If the derivative strains form white colonies on the LB agar supplemented with selenate, the mutants are unable to reduce Se(VI) to Se(0).

6. To complement the mutation, conjugally mate a plasmid containing the wild-type sequence of the particular gene into the mutants using *E. coli*. Transconjugants can be selected using the appropriate antibiotics and screened for selenate reduction activity using the LB plate assay. If the complemented mutants form red colonies on the LB agar supplemented with selenate, then selenate reduction activity has been restored.

Yee *et al.* (2007) demonstrated that knock-out mutagenesis of the *fnr* gene in *E. cloacae* SLD1a-1 result in derivative mutant strains that are deficient in selenate reduction activity. These mutants are unable to precipitate elemental selenium and form white colonies when grown on selenate containing agar. Complementation of mutations using the wild-type *fnr* gene rescues the ability of mutant strains to reduce Se(VI).

D. Random mutagenesis using transposons

Transposon mutagenesis can be used to generate a large number of different mutants deficient in selenate reduction activity. This approach involves the random insertion of a genetic element into the chromosomal DNA of a Se-reducing bacterium to disrupt genes that are either directly or indirectly required for selenate reduction activity. Tranposon mutants can be screened and isolated on agar plates containing selenate. Mutant strains that have lost the ability to reduce Se(VI) to Se(0) can be identified based on the color of the colony. A suitable transposon is the mini-Tn5-*lacZ1* element (De Lorenzo *et al.*, 1990), which is a genetically engineered transposon that contains a kanamycin resistance gene and a promoterless β-galactosidase gene placed within a suicide plasmid containing ampicillin resistance that is used as a carrier for delivery into the recipient strain. The donor strain is an *E. coli* host that encodes the functions needed for the transposon carrier plasmid to replicate and mate into other gram-negative bacteria. If the recipient strain does not encode the function required for plasmid replication, then all derivative kanamycin resistant colonies are expected to be the result of transposon incorporation into the genome. Below is a procedure for generating *E. cloacae* mutants using the mini-Tn5 transposon system:

1. Spread plate an overnight culture of *E. cloacae* SLD1a-1 cells on LB agar containing the antibiotic rifampicin, and incubate until single colonies appear on the plate. Colonies that can grow on this media are naturally resistant to rifampicin. These colonies are selected for the mutagenesis experiment.

2. To generate transposon mutants of *E. cloacae*, grow cultures of *E. coli* S17-λpir(miniTn5*lacZ*1) and *E. cloacae* in LB broth, and harvest cells at late log phase (O.D.$_{600}$ of 0.7–0.9). Mix 10 µl of *E. coli* S17-λpir(miniTn5*lacZ*1) with 10 µl of *E. cloacae* SLD1a-1 in 3 ml of sterilized distilled deionized water. Filter the cell suspension through a 13 mm 0.45 µm filter. Transfer the filter containing the cells to LB plates and incubate at 30 °C.
3. After 4 hours, remove the cells from the filter and spread plate onto LB plates containing rifampicin and kanamycin. Incubate plates at 30 °C until single colonies appear on the agar. Only *E. cloacae* cells containing the transposon insertion can grow on this selective media.
4. Replica-plate single colonies onto LB agar containing selenate and kanamycin. Incubate plates at 30 °C overnight. If a mutant strain forms red colonies, then the mutation did not affect the bacterium's ability to reduce Se(VI) to Se(0). However, if a mutant strain forms white colonies, then mutation disrupted a gene required for selenate reduction. Mutants deficient in selenate reduction activity are selected for further characterization.
5. To determine the insertion site of the miniTn5*lacZ*1 element in the mutant strain, construct a genomic library of the mutant. An appropriate vector is the cosmid pLAFR3 (Staskawicz *et al.*, 1986). Because pLAFR3 confers tetracycline resistance, and the transposon confers kanamycin antibiotic resistance, cosmid clones containing the mini-Tn5*lacZ*1 insertion can be directly selected by plating cells on LB agar containing tetracycline and kanamycin.
6. For mutant complementation, design primers to amplify the wild-type sequence of the gene from *E. cloacae* SLD1a-1 chromosomal DNA. Clone the PCR product and transform the DNA into *E. coli* S17-1, then conjugally mate the wild-type gene into the mutant strain. If the complemented mutants form red colonies on the LB agar supplemented with selenate, then selenate reduction activity has been restored.

Figure 3.2 shows seven transposon mutants of *E. cloacae* grown on LB agar supplemented with 10 mM selenate. Also shown is the wild type strain and *E. coli* S17-1 as controls. The wild type strain *E. cloacae* SLD1-1 forms a red colony, indicating the formation of red elemental selenium. The *E. coli* strain forms a white colony, confirming that it cannot reduce Se(VI) to Se(0). Three of the mutants shown in Fig. 3.2 form red colonies, indicating that the mutation did not affect their ability to reduce selenate. However four of the mutants produce white colonies, signifying that the transposon insertion disrupted the genes required for selenate reduction. Interestingly, all four mutants that grow as white colonies on selenate-containing agar, form red colonies on agar supplemented with selenite (Fig. 3.3). Mutants deficient in selenate reduction activity are unaffected in their ability to reduce selenite into elemental selenium. This observation suggests that Se(VI) and Se(IV) reduction are mediated by different genetic systems.

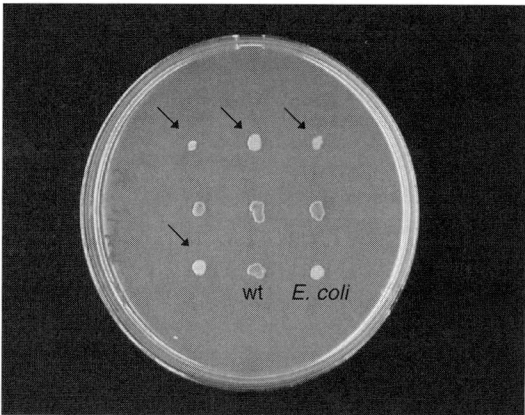

FIGURE 3.2 Transposon mutants of *E. cloacae* SLD1a-1 on selenate-containing agar. Seven mutants are shown. Arrows indicate that the four mutants have lost the ability to reduce selenate to elemental selenium. Two controls are also shown: the wild-type strain (wt), and *E. coli* S17-1. (See Color Plate Section in the back of the book.)

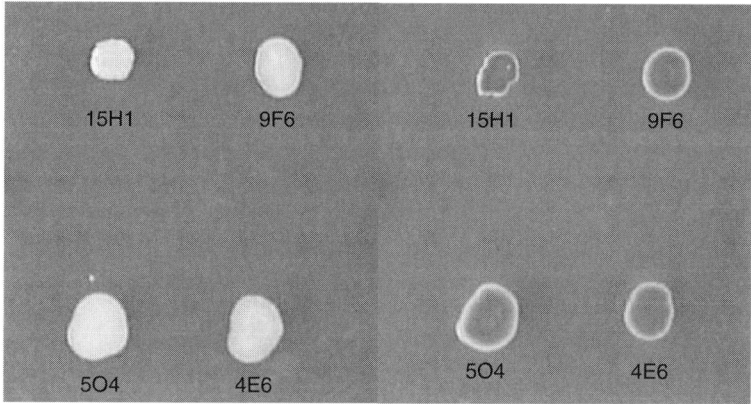

FIGURE 3.3 Four transposon mutants of *E. cloacae* SLD1a-1 grown on selenate-containing agar (*left*) and selenate containing agar (*right*). (See Color Plate Section in the back of the book.)

IV. GENETIC ANALYSIS OF SELENATE REDUCTION IN *E. CLOACAE* SLD1a-1

Mutagenesis experiments have identified three genes that are essential for selenate reduction activity in *E. cloacae*. Mutations in *fnr*, *tatC*, or *menD* gene result in derivative strains that are unable to reduce selenate (Table 3.1). For each mutated gene, complementation by the wild-type sequences rescues their abolished phenotype.

TABLE 3.1 Genes in *E. cloacae* SLD1a-1 required for selenate reduction

Gene	Function	Reference
fnr	Encodes for the Fumurate Nitrate Reduction regulator. FNR regulates the expression of anaerobic genes.	Yee *et al.*, 2007
tatC	Encodes for a central protein in the twin arginine translocation (TAT) system. This system is responsible for exporting folded proteins across the inner membrane.	Ma *et al.*, 2007
menD	Encodes for a principle protein in the menaquinone biosynthesis pathway. Menaquinones are anaerobic electron carriers.	Ma *et al.*, in preparation

The *fnr* gene encodes for the transcription regulator FNR. The FNR protein is located in the cytoplasm and that functions as an oxygen sensing transcription factor that monitors the availability of oxygen (Guest, 1995; Kiley and Beinert, 1999). The functional form of FNR contains a $[4Fe-4S]^{2+}$ cluster ligated by four essential cysteine residues, which is involved in site-specific DNA binding activity (Khoroshilova *et al.*, 1997; Kiley and Beinert, 1999). FNR is a global regulator that controls the expression of genes required for growth under oxygen limiting conditions. Transcriptomic analysis of *E. coli* have demonstrated that FNR plays a critical role in modulating gene expression to mediate adaptation between aerobic and anaerobic modes of energy generation (Constantinidou *et al.*, 2006; Kang *et al.*, 2005; Salmon *et al.*, 2003). Depletion of oxygen from the environment activates FNR regulation, which induces the expression of at least 103 different operons (Constantinidou *et al.*, 2006). As the bacterium is challenged with oxygen deprivation, FNR induces the synthesis of essential anaerobic proteins, including enzymes for the anaerobic oxidation of carbon sources such as glycerol and formate dehydrogenases and anaerobic respiratory enzymes for the reduction of alternate electron acceptors such as nitrate, fumarate, and dimethyl sulfoxide (DMSO) reductases (Spiro, 1994; Unden *et al.*, 1995). Mutagenesis experiments have shown that the *fnr* gene in *E. cloacae* SLD1a-1 is required for selenate reduction activity. Although, FNR regulation in *E. cloacae* is not fully understood, the present data suggest that the FNR transcription factor regulates the expression of the selenate reductase as the bacterium transitions from oxic to suboxic growth. This is consistent with growth studies that have shown that decreasing oxygen levels promotes increasing selenate reduction activity.

The *tatC* gene encodes for a central protein in the twin arginine translocation (Tat) system. The bacterial Tat system is comprised of three principle membrane components (TatA, TatB, and TatC) and is responsible for transporting folded proteins from the cytoplasm across the inner membrane. A diverse range of proteins are exported by this pathway, including anaerobic redox enzymes such as the nitrate reductase, TMAO reductase, DMSO reductase, hydrogenase, and formate dehydrogenase (Lee *et al.*, 2006). Characteristic of all proteins transported by the Tat system is a twin-arginine dipeptide signal sequence that is required to initiate transport (Berks, 1996; Cristobal *et al.*, 1999). TatC serves as the initial docking site for the signal peptides. Mutagenesis experiments have shown that mutation of the *tatC* gene in *E. cloacae* SLD1a-1 results in complete loss of selenate reduction activity. This observation suggests that the enzyme(s) conferring selenate reductase activity in *E. cloacae* is a folded protein that is exported across the inner membrane through the Tat pathway. Once exterior to the cytoplasm, the reductase enzyme is embedded in the cytoplasmic membrane with the active site facing the periplasmic space (Watts *et al.*, 2003).

The *menD* genes encodes for an essential protein in menaquinone biosynthesis. Menaquinones are small lipid-soluble molecules that act as electron shuttles in the bacterial electron transport chain. Diverse species of facultative anaerobes are known to synthesize menaquinone under oxygen-limiting conditions to mediate anaerobic respiration of alternate electron acceptors. For example, menaquinone is essential for anaerobic respiration in *E. coli* K-12 and *Shewanella oneidensis* MR-1. *E. coli* mutants deficient in menaquinone biosynthesis are unable to employ fumarate or TMAO as a terminal electron acceptors for anaerobic growth (Guest, 1977, 1979; Lambden and Guest, 1976; Meganathan, 1984). In the case of facultative metal-reducing bacterium *S. oneidensis* MR-1, menaquinone is required for nitrate, iron(III), and fumarate reductase activity (Myers and Myers, 1993; Saffarini *et al.*, 2002). Menaquinone also contributes to, but is not absolutely required for, anaerobic manganese(IV) reduction (Myers and Myers, 1993). In *E. cloacae* SLD1a-1, mutation of the *menD* gene results in complete loss of selenate reduction activity. Supplementation of the growth medium with menaquinone analogues and/or intermediates such as 1,1,4-dihydroxy-2-napthoic acid, restores selenate reduction activity in *menD* mutants (Ma *et al.*, in preparation). This observation indicates that anaerobic electron carriers are required for selenate reduction activity in *E. cloacae* SLD1a-1.

V. A MOLECULAR MODEL FOR SELENATE REDUCTION IN *E. CLOACAE* SLD1a-1

The genes required for selenate reduction indicate that *E. cloacae* SLD1a-1 catalyzes the reduction of selenate via an electron transport chain that is regulated by the global transcription regulator FNR. Fig. 3.4 depicts a working model for the reduction mechanism. Under suboxic conditions, the FNR protein becomes active and binds onto the chromosomal DNA of *E. cloacae* SLD1a-1. Site-specific DNA binding by C-terminal domain of FNR induces the transcription of mRNA carrying the genetic information for the selenate reductase enzyme. After translation, the selenate reductase enzyme folds into the proper conformation for transport through the Tat pathway. Like many other redox enzymes that are secreted by the Tat system, the selenate reductase may incorporate a co-factor concomitant with folding in the cytoplasm (Palmer *et al.*, 2005). The folded enzyme then targets the TatBC receptor complex for export across the inner membrane. The twin-arginine dipeptide signal sequence of the selenate reductase binds onto the docking site of TatC to initiate transport. The proton motive force drives the formation of an active translocase and the selenate reductase is exported through a TatA pore. As the enzyme is released on the periplasmic side of membrane, the signal peptide is removed.

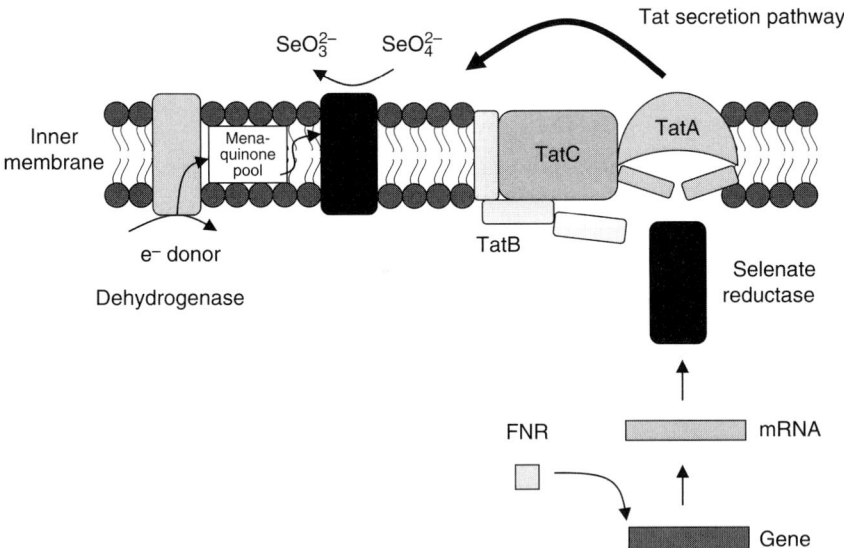

FIGURE 3.4 A molecular model of selenate [Se(VI), SeO_4^{2-}] reduction by *E. cloacae* SLD1a-1. The model depicts the role of FNR, TatABC, and menaquinone in the selenate reduction process.

Concurrent to the expression of the selenate reductase, anaerobic conditions trigger the cellular production of menaquinone. Biosynthesis of menaquinone involves seven genes designated as *menA, B, C, D, E, F,* and *ubiE* that encode various steps of the pathway. The *menF* gene encodes for isochorismate synthase, which catalyzes the first step in the menaquinone biosynthesis pathway by forming isochorismate from the precursor chorismate (Dahm *et al.*, 1998). The *menD* gene product then catalyzes the decarboxylation of 2-oxoglutarate, and the synthesis of 2-succinyl-6-hydroxy-2,4-cyclohexadiene-1-carboxylate from isochorismate and succinic semialdehyde (Bhasin *et al.*, 2003). The menaquinone intermediates O-succinylbenzoate, O-succinylbenzoyl-CoA, and 1,4-dihydroxy-2-naphthoate are subsequently formed by the enzymes encoded by the *menC, menE,* and *menB* genes respectively (Sharma *et al.*, 1992, 1993, 1996). The *menA* gene, which encodes for the inner membrane protein 1,4-dihydroxy-2-naphthoate octaprenyltransferase, is required for the transfer of an octaprenyl side chain to 1,4-dihydroxy-2-naphthoate (Shineberg *et al.*, 1976; Suvarna *et al.*, 1998). In this step of the reaction, the menaquinone biosynthesis pathway becomes associated with the membrane, and the product of the reaction, demethylmenaquinone, is the first major quinone produced for anaerobic respiration. Finally, a C-methyltransferase encoded by the *ubiE* gene catalyzes the conversion of demethylmenaquinone to menaquinone (Lee *et al.*, 1997). Both demethylmenaquinone and menaquinone operate as electron carriers within the inner membrane, and are putatively required for reduction of selenate to selenite.

Once menaquinone is synthesized, electron transfer through the electron transport chain is initiated by the oxidation of the electron donor. Hydrogenase or flavin-containing dehydrogenase catalyzes electron transfer from the electron donor to the menaquinone pool. In the cytoplasmic membrane, menaquinone diffuses to the oxidation site of the iron–sulfur and heme containing holocomplex of selenate reductase. Because of the large size of the selenate reductase, electrons are likely to be transported within the reductase molecule by inter-heme electron transfer (according to decreasing heme redox potential). The γ and β subunits of the enzyme form a conventional electron transfer chain mediating the transfer of electrons from the menaquinone pool to the active site. The active site in selenate reductase is located in the molybdenum-containing α subunit and faces the periplasmic side of the cytoplasmic membrane. Once the electrons reach the catalytic site, they are transferred to the selenate oxyanions bound to the α subunit of the enzyme. This final electron transfer step catalyzes the reduction of selenate to selenite. The reduction product is then released into the periplasmic space. A separate and distinct electron transfer system subsequently mediates the reduction of selenite to elemental selenium.

VI. CONCLUSIONS AND FUTURE PROSPECTS

This brief review summarizes the recent advances in our understanding of selenate reduction in *E. cloacae* SLD1a-1. Molecular genetic studies have revealed that anaerobic respiratory operons play an essential role in the selenate reduction pathway. The proposed molecular model for selenate reduction in *E. cloacae* SLD1a-1 successfully explains the effect of oxygen levels in the environment on Se(VI) reduction activity. Further genetic characterization will ultimately allow for the development of a comprehensive molecular model that accurately predicts selenium reduction activity over a broad range of environmental conditions.

From a molecular prospective, our understanding of microbial selenium reduction is in its naissance. Phylogenetically diverse species of *Bacteria* and *Archaea* in soils, sediments, and wastewater could reduce selenium, but the genes required for the activity in these organisms are virtually unknown. Future genome-enabled investigations, such as whole-genome expression studies, will allow further elucidation of selenate reduction pathways in diverse Se-reducing prokaryotes. Resolution of the genetic basis of microbial selenium reduction will undoubtedly aid in the practical implementation of bioreactor systems and *in situ* bioremediation designs.

An additional key area in need of further investigation is the identification of the molecular determinants that catalyze selenite reduction. Microbial reduction of selenate generates selenite, which is then rapidly reduced to elemental selenium by a highly efficient electron transfer pathway. Unfortunately the genetic characterization of selenite reduction in *E. cloacae* SLD1a-1 and other microbes remain elusive. Transposon mutagenesis studies of *E. cloacae* SLD1a-1 have failed to mention the isolation of a mutant deficient in selenite reduction activity, suggesting that selenite reduction may be catalyzed by inherently redundant enzymes, or mutation of the selenite reductase gene is lethal to the organism. Alternatively, the reduction of selenite is not enzymatic at all, but instead electron transfer is mediated by auxiliary detoxification compounds and reducing agents. Future work should include both genetic and biochemical studies to illuminate the basic molecular reactions that control selenite reduction activity.

ACKNOWLEDGMENTS

This work was supported by the National Research Initiative of the USDA Cooperative State Research, Education, and Extension Service, grant no. 2005-35107-16230.

REFERENCES

Berks, B. C. (1996). A common export pathway for proteins binding complex redox cofactors? *Mol. Microbiol.* **22**, 393–404.

Bhasin, M., Billinsky, J. L., and Palmer, D. R. (2003). Steady-state kinetics and molecular evolution of *Escherichia coli* MenD [(1R,6R)-2-succinyl-6-hydroxy-2,4-cyclohexadiene-1-carboxylate synthase], an anomalous thiamin diphosphate-dependent decarboxylase-carboligase. *Biochemistry* **42**, 13496–13504.
Cantafio, A. W., Hagen, K. D., Lewis, G. E., Bledsoe, T. L., Nunan, K. M., and Macy, J. M. (1996). Pilot-scale selenium bioremediation of San Joaquin drainage water with Thauera selenatis. *Appl. Environ. Microbiol.* **62**, 3298–3303.
Constantinidou, C., Hobman, J. L., Griffiths, L., Patel, M. D., Penn, C. W., Cole, J. A., and Overton, T. W. (2006). A reassessment of the FNR regulon and transcriptomic analysis of the effects of nitrate, nitrite, NarXL, and NarQP as *Escherichia coli* K12 adapts from aerobic. *J. Biol. Chem.* **281**, 4802–4815.
Cristobal, S., de Gier, J. W., Nielsen, H., and von Heijne, G. (1999). Competition between Sec- and TAT-dependent protein translocation in *Escherichia coli*. *EMBO J.* **18**, 2982–2990.
Dahm, C., Muller, R., Schulte, G., Schmidt, K., and Leistner, E. (1998). The role of isochorismate hydroxymutase genes entC and menF in enterobactin and menaquinone biosynthesis in *Escherichia coli*. *Biochim. Biophys. Acta* **1425**, 377–386.
De Lorenzo, V., Herrero, M., Jakubzik, U., and Timmis, K. N. (1990). Mini-Tn5 transposon derivatives for insertion mutagenesis, promoter probing and chromosomal insertion of cloned DNA in Gram-negative eubacteria. *J. Bacteriol.* **172**, 6568–6572.
Dungan, R. S., Yates, S. R., and Frankenberger, W. T., Jr. (2003). Transformation of Se(VI) and Se(IV) by *Stenophomonas maltophilia* isolated from a seleniferous agricultural drainage pond sediment. *Environ. Microbiol.* **5**, 287–295.
Frankenberger, W. T., and Arshad, M. (2001). Bioremediation of selenium-contaminated sediments and water. *Biofactors* **14**, 241–254.
Guest, J. R. (1977). Menaquinone biosynthesis: Mutants of *Escherichia coli* K-12 requiring 2-succinylbenzoate. *J. Bacteriol.* **130**, 1038–1046.
Guest, J. R. (1979). Anaerobic growth of *Escherichia coli* K12 with fumarate as terminal electron acceptor. Genetic studies with menaquinone and fluoroacetate-resistant mutants. *J. Gen. Microbiol.* **115**, 259–271.
Guest, J. R. (1995). Adaptation to life without oxygen. *Philos. Trans. R. Soc. Lond. B. Biol. Sci.* **350**, 189–202.
Hoffman, D. J., and Heinz, G. H. (1988). Embryotoxic and teratogenic effects of selenium in the diet of mallards. *J. Toxicol. Environ. Health* **24**, 477–490.
Hoffman, D. J., Ohlendorf, H. M., and Aldrich, T. W. (1988). Selenium teratogenesis in natural populations of aquatic birds in central California. *Arch. Environ. Contam. Toxicol.* **17**, 519–525.
Kang, Y. S., Weber, K. D., Yu, Q., Kiley, P. J., and Blattner, F. R. (2005). Genome-wide expression analysis indicates that FNR of *Escherichia coli* K-12 regulates a large number of genes of unknown function. *J. Bacteriol.* **18**, 1135–1160.
Khoroshilova, N., Popescu, C., Munck, E., Beinert, H., and Kiley, P. (1997). Iron–sulfur cluster diassembly in the FNR protein of *Escherichia coli* by O_2: [4Fe-4S] to [2Fe-2S] conversion with loss of biological activity. *Proc. Natl. Acad. Sci. USA* **94**, 6087–6092.
Kiley, P. J., and Beinert, H. (1999). Oxygen sensing by the global regulator, FNR: The role of the iron–sulfur cluster. *FEMS Microbiol. Rev* **22**, 341–352.
Lambden, P. R., and Guest, J. R. (1976). Mutants of *Escherichia coli* K12 unable to use fumarate as an anaerobic electron acceptor. *J. Gen. Microbiol.* **97**, 145–160.
Lee, P. A., Tullman-Ercek, D., and Georgiou, G. (2006). The bacterial twin-arginine translocation pathway. *Annu. Rev. Microbiol.* **60**, 373–395.
Lee, P. T., Hsu, A. Y., Ha, H. T., and Clarke, C. F. (1997). A C-methyltransferase involved in both ubiquinone and menaquinone biosynthesis: Isolation and identification of the *Escherichia coli ubiE* gene. *J. Bacteriol.* **179**, 1748–1754.
Lemly, A. D. (2002). Symptoms and implications of selenium toxicity in fish: The Belews Lake case example. *Aquat. Toxicol.* **57**, 39–49.

Lemly, A. D. (2004). Aquatic selenium pollution is a global environmental safety issue. *Ecotoxicol. Environ. Saf.* **59**, 44–56.
Losi, M. E., and Frankenberger, W. T. (1997). Reduction of selenium oxyanions by *Enterobacter cloacae* S LD1a-1: Isolation and growth of the bacterium and its expulsion of selenium particles. *Appl. Environ. Microbiol.* **63**, 3079–3084.
Losi, M. E., and Frankenberger, W. T. (1998). Reduction of selenium oxyanions by *Enterobacter cloacae* strain SLD1a-1. *In* "Environmental Chemistry of Selenium" (W. T. Frankenberger and R. A. Engberg, eds.), pp. 515–544. Marcel Dekker, New York.
Ma, J., Kobayashi, D. Y., and Yee, N. (2007). A chemical kinetic and molecular genetic study of selenium oxyanion reduction by *Enterobacter cloacae* SLD1a-1. *Environ. Sci. Technol.* **41**, 7795–7801.
Macy, J. M., Lawson, S., and DeMoll-Decker, H. (1993). Bioremediation of selenium oxyanions in San Joaquin drainage water using *Thauera selenatis* in a biological reactor system. *Appl. Microbiol. Biotechnol.* **40**, 588–594.
Macy, J. M., Rech, S., Auling, G., Dorsch, M., Stackenbrandt, E., and Sly, L. I. (1993). *Thauera selenatis* gen-nov, sp-nov, a member of the beta-subclass of proteobacteria with a novel type of anaerobic respiration. *Int. J. Syst. Bacteriol.* **43**, 135–142.
Meganathan, R. (1984). Inability of *men* mutants of *Escherichia coli* to use trimethylamine-N-oxide as an electron acceptor. *FEMS Microbiol. Lett.* **24**, 57–62.
Myers, C. R., and Myers, J. M. (1993). Role of menaquinone in the reduction of fumarate, nitrate, iron(III) and manganese(IV) by *Shewanella putrefaciens* MR-1. *FEMS Microbiol. Lett.* **114**, 215–222.
Ohlendorf, H. M., Hoffman, D. J., Saiki, M. K., and Aldrich, T. W. (1986). Embryonic mortality and abnormalities of aquatic birds: Apparent impact of selenium from irrigation drain water. *Sci. Total Environ.* **52**, 49–63.
Oremland, R. S., Hollibaugh, J. T., Maest, A. S., Presser, T. S., Miller, L. G., and Culbertson, C. W. (1989). Se(VI) reduction to elemental selenium by anaerobic-bacteria in sediments and culture—Biogeochemical significance of a novel, sulfate-independent respiration. *Appl. Environ. Microbiol.* **55**, 2333–2343.
Palmer, T., Sargent, F., and Berks, B. C. (2005). Export of complex cofactor-containing proteins by the bacterial Tat pathway. *Trends Microbiol.* **13**, 175–180.
Ridley, H., Watts, C. A., Richardson, D. J., and Butler, C. S. (2006). Resolution of distinct membrane-bound enzymes from *Enterobacter cloacae* SLD1a-1 that are responsible for selective reduction of nitrate and Se(VI) oxyanions. *Appl. Environ. Microbiol.* **72**, 5173–5180.
Saffarini, D. A., Blumerman, S. L., and Mansoorabadi, K. J. (2002). Role of menaquinones in Fe(III) reduction by membrane fractions of *Shewanella putrefaciens*. *J. Bacteriol.* **184**, 846–848.
Salmon, K., Hung, S. P., Mekjian, K., Baldi, P., Hatfield, G. W., and Gunsalus, R. P. (2003). Global gene expression profiling in *Escherichia coli* K12—The effects of oxygen availability and FNR. *J. Biol. Chem.* **278**, 29837–29855.
Sharma, V., Hudspeth, M. E., and Meganathan, R. (1996). Menaquinone (vitamin K2) biosynthesis: Localization and characterization of the *menE* gene from *Escherichia coli*. *Gene* **168**, 43–48.
Sharma, V., Meganathan, R., and Hudspeth, M. E. (1993). Menaquinone (vitamin K2) biosynthesis: Cloning, nucleotide sequence, and expression of the *menC* gene from *Escherichia coli*. *J. Bacteriol.* **175**, 4917–4921.
Sharma, V., Suvarna, K., Meganathan, R., and Hudspeth, M. E. (1992). Menaquinone (vitamin K2) biosynthesis: Nucleotide sequence and expression of the *menB* gene from *Escherichia coli*. *J. Bacteriol.* **174**, 5057–5062.
Shineberg, B., and Young, I. G. (1976). Biosynthesis of bacterial menaquinones: The membrane-associated 1,4-dihydroxy-2-naphthoate octaprenyltransferase of *Escherichia coli*. *Biochemistry* **15**, 2754–2758.

Spiro, S. (1994). The FNR family of transcriptional regulators. *Antonie van Leeuwenhoek* **66**, 23–36.
Staskawicz, B., Dahlbeck, D., Keen, N., and Napoli, C. (1986). Molecular characterization of cloned avirulence genes from race 0 and race 1 of *Pseudomonas syringae* pv. glycinea. *J. Bacteriol.* **169**, 5789–5794.
Suvarna, K., Stevenson, D., Meganathan, R., and Hudspeth, M. E. (1998). Menaquinone (vitamin K2) biosynthesis: Localization and characterization of the *menA* gene from *Escherichia coli*. *J. Bacteriol.* **180**, 2782–2787.
Unden, G., Becker, S., Bongaerts, J., Holighaus, G., Schirawski, J., and Six, S. (1995). O_2-sensing and O_2-dependent gene regulation in facultative anaerobic bacteria. *Arch. Microbiol.* **164**, 81–90.
Watts, C. A., Ridley, H., Condie, K. L., Leaver, J. T., Richardson, D. J., and Butter, C. S. (2003). Se(VI) reduction by *Enterobacter cloacae* SLD1a-1 is catalyzed by a molybdenum-dependent membrane-bound enzyme that is distinct from the membrane-bound nitrate reductase. *FEMS Microbiol. Lett.* **228**, 273–279.
Yee, N., Ma, J., Dalia, A., Boonfueng, T., and Kobayashi, D. Y. (2007). Se(VI) Reduction and the precipitation of Se(0) by the facultative bacterium *Enterobacter cloacae* SLD1a-1 are regulated by FNR. *Appl. Environ. Microbiol.* **73**, 1914–1920.

CHAPTER 4

Metagenomics of Dental Biofilms

Peter Mullany,[*,1] **Stephanie Hunter,**[†] and **Elaine Allan**[*]

Contents
    I. Introduction   126
    II. Isolation of Metagenomic DNA   126
    III. Sample Collection and Processing   126
    IV. DNA Extraction   128
    V. Preparation of the Insert DNA for Cloning   128
        A. Partial restriction digestion   129
        B. Mechanical shearing of the DNA   129
    VI. Removal of Human DNA   129
    VII. Metagenomic Library Construction   130
        A. Construction of large insert metagenomic libraries   130
        B. Construction of small insert libraries   132
        C. Phage display   132
    VIII. Limitations of Metagenomics   133
    IX. Conclusions and Future Perspectives   134
    References   134

---

[*] Division of Microbial Diseases, Eastman Dental Institute, University College London, London WC1X 8LD, United Kingdom
[†] Institute of Structural and Molecular Biology, University College London, London, WC1E 6BT, United Kingdom
[1] Corresponding author: Division of Microbial Diseases, Eastman Dental Institute, University College London, London WC1X 8LD, United Kingdom

I. INTRODUCTION

Up to 1000 different bacterial species have been found in the human oral cavity. The interactions between these organisms and the human host are important in the development of, and protection from, some of the commonest human diseases such as dental caries and periodontal disease (Kazor et al., 2003; Kumar et al., 2003; Wilson, 2005). This environment can also act as a reservoir for mobile elements that may encode clinically important antibiotic resistance genes or other genes important in bacterial evolution (Roberts and Mullany, 2006; Ready et al., 2006; Villedieu et al., 2007). A major barrier to fully understanding the roles that oral bacteria have within this complex environment is that only 50% of the oral flora can be cultivated in the laboratory. One way through which the noncultivatable flora can be investigated is to use metagenomics.

This approach, in addition to allowing an investigation of community interactions and antibiotic resistance, also allows any particular ecological niche to be mined for products such as new enzymes, which may be useful for biotechnological applications.

Metagenomics requires that all the bacterial DNA from the oral cavity is isolated. The DNA can then be investigated in several ways (see Table 4.1), most of which are aimed at description and quantitation of the metagenomic DNA. This chapter deals with the techniques available for functional analysis of the oral metagenome. This requires that the DNA is cloned into vectors that allow the construction of large insert libraries, such as BACs and fosmids, or into specialized expression vectors such as phagemids for phage display. This chapter will review the methods available for isolating, cloning, and expressing the oral metagenome.

II. ISOLATION OF METAGENOMIC DNA

The isolation of good quality high molecular weight DNA is a starting point for a functional metagenomic study. Plaque and saliva samples are taken from the oral cavity as described below.

III. SAMPLE COLLECTION AND PROCESSING

Approximately 5 ml of saliva is acquired using 1g sterile paraffin wax (Raymond A Lamb Ltd., Eastbourne, East Sussex, UK) to stimulate saliva and dislodge plaque. Samples are collected into a sterile container by expectoration (care must be taken not to contaminate the samples with bacteria from the lips and chin). An equal volume of Reduced Transport Fluid (RTF) (Syed and Loesche, 1972) is added to each sample in a

TABLE 4.1 A summary of current techniques used to study complex microbial ecosystems

Method	Uses	Limitations
16S rDNA Sequencing/16S rRNA clone libraries	Phylogenetic identification	Laborious, subject to PCR/cloning biases
DGGE/TGGE/TTGE	Monitoring of community shifts; rapid comparative analysis	Subject to PCR biases and co-migration of bands (disorting diversity); semiquantitiative; identification requires clone library
T-RFLP	Monitoring of community shifts; rapid comparative analysis; very sensitive; potential for high throughput	Subject to PCR biases; semiquantitiative; identification requires clone library
SSCP	Monitoring of community shifts; rapid comparative analysis	Subject to PCR biases; semiquantitative; identification requires clone library
RT (real time)-PCR	Quantification of bacteria in an environmental sample. Promising for species present at low concentrations	Requires sequence information
FISH	Detection; enumeration; comparative analysis possible with automation. In combination with flow cytometry can sort uncultivable from cultivables to access to sequence information	Requires sequence information; laborious at species level; high detection levels; relies on cell permeability

(*continued*)

TABLE 4.1 (continued)

Method	Uses	Limitations
Dot-blot hybridization	Detection; estimates relative abundance	Requires sequence information; laborious at species level
Quatitative PCR	Detection; estimates relative abundance	Laborious
Diversity microarrays	Detection; estimates relative abundance	In early stages of development; expensive
Shotgun libraries	Access to sequence information of the whole metegenome	Subject to cloning biases; high detection level; generates sequences encoding unknown functions
Non-16S rRNA profiling	Monitoring of community shifts; rapid comparative analysis	Identification requires additional 16S rRNA-based approaches

Adapted from Zoetendal et al. (2004); additional information from Gafan and Spratt (2005).

microbiological safety cabinet to avoid contamination. The saliva is vortexed for 30 s to resuspend any settled plaque or planktonic cells and the samples are then pooled.

IV. DNA EXTRACTION

Two aliquots of pooled sample are prepared, and the DNA is extracted from one of these aliquots by the Puregene Gram-positive DNA isolation protocol and from the other by the Puregene Gram-negative DNA isolation protocol (Gentra Systems) according to the manufacturer's instructions. The extracted DNA is subsequently pooled.

V. PREPARATION OF THE INSERT DNA FOR CLONING

The method used for the preparation of DNA depends on the type of libraries to be constructed, that is, large or small inserts. In this work, we have used both partial digestion with a restriction enzyme and also controlled shearing of the DNA to prepare inserts for library construction (Diaz-Torres et al., 2006; Seville and Mullany, unpublished).

A. Partial restriction digestion

The DNA is subjected to partial digestion with a suitable restriction enzyme; we have found HindIII to be particularly useful (Seville and Mullany, unpublished).

B. Mechanical shearing of the DNA

We recommend sonicating 0.2 ml of DNA (at a concentration of 250 ng/ml) for 5 s on ice at 80% power by using an ultrasonic homogenizer (IKA-WERKE). The length of time in the sonicator will vary depending on whether large or small insert libraries are being constructed and should be determined empirically. The ends of the DNA can be repaired by treatment with 2 U/mg of mung bean nuclease in a final volume of 100 µl at 37 °C for 1 h to produce blunt ends. The resulting DNA fragments are then separated by agarose gel electrophoresis and fragments of the desired length are cut from the gel and purified by commercially available kits for DNA purification from agarose. To facilitate cloning in T vectors, 3' A overhangs can be created by incubating the DNA fragments at 75 °C for 1 h with 1 U/mg of *Taq* DNA polymerase in a final volume of 100 µl of 1×*Taq* buffer containing 2 mM dATP. Prior to ligation into the vector, the DNA is purified using Qiagen PCR Purification kit.

VI. REMOVAL OF HUMAN DNA

In our experience, plaque and saliva contains a large amount of human DNA (Seville and Mullany, unpublished), which will inevitably be cloned into the metagenomic libraries and complicate the screening procedures. Therefore efforts should be made to remove human DNA. All methods have their limitations, but we have investigated the following. Firstly, osmotic lysis of eukaryotic cells will release the eukaryotic DNA, which may then be removed by the addition of Benzonase^{TM}, (Novagen), a potent nonspecific nuclease that can degrade both DNA and RNA. The nuclease is then inactivated before lysing the bacteria in the sample. We have achieved significant reduction in the amount of human DNA present in oral samples using this method (Hunter and Ward, unpublished) and suggest that it be used for other sample types following optimization of the osmotic lysis buffer (Tris-HCl pH8). The concentration of the buffer (we used 20 mM) used to lyse eukaryotic cells should be optimized for different sample types.

Care should be taken with selective lysis methods as a number of more delicate oral bacterial species (e.g. Mycoplasma spp. and L-forms of streptococci) may be sensitive to such procedures. An alternative method which eliminates this problem is to use differential centrifugation to pellet the heavier eukaryotic cells. We have also found this approach to be useful in eliminating a significant quantity of DNA (Hunter and Ward, unpublished). A limitation of this method, however, is the parallel removal of bacterial cells that are tightly bound to the human cells. The use of detergents, proteases, and sugars could be investigated in order to disrupt bacteria–human cell interactions before differential centrifugation. In our work, we have used combinations of selective lysis and differential centrifugation and shown that DNAs from representatives of most bacterial species from the oral cavity are present within our libraries (unpublished).

VII. METAGENOMIC LIBRARY CONSTRUCTION

The choice of vector and methods used for library construction depend on the overall aim of the experiment. For a comprehensive metagenomic study, we recommend that both large and small insert vectors be used.

A. Construction of large insert metagenomic libraries

Large insert libraries are required to construct an ordered metagenomic library and to isolate large genetic elements such as conjugative transposons, pathogenicity islands, and large metabolic operons. Several different types of vectors have been used for large insert cloning and these are listed in Table 4.2. The preferred method for making large insert metagenomic libraries is to use bacterial artificial chromosomes (BAC) or fosmids. BACs were developed by Shizuya and colleagues in 1992 as a bacterial cloning system for mapping and analysis of complex genomes as part of the effort to create a high-resolution map of the human genome. The BAC is based around the mini-F plasmid pMB0131, which contains genes that regulate its own replication (OriS and repE) and control its copy number (par operon, which is involved in the exclusion of multiple f factors (Easter *et al.*, 1998) thus allowing the stable maintenance of inserted DNA. The addition of a chloramphenicol resistance gene as a selectable marker, a cloning segment composed of cosN sites from bacteriophage λ, loxP sites from P1, and a multiple cloning site flanked by promoters completed the vector (Shizuya, 1992, 2001). The vector was used to clone inserts of more than 300 kb (Shizuya *et al.*, 1992).

Not all genes will express well in *E. coli* and if expression from the large insert library is a requirement of the project, alternative host vector

TABLE 4.2 Metagenomic libraries and the novel characteristics and/or genes found during their screening

Vector	Metagenome/sample	Novel genes/characterizations	Reference
BAC	Marine	Novel bacteriorhodopsin	Beja et al. (2000)
BAC	Soil	Kanamycin resistance encoded by a novel aminoglycoside gene	Riesenfeld (2004)
Small insert	Soil	Eight novel aminoglycoside resistance genes and a novel tetracycline resistance gene	Riesenfeld (2004)
BAC	Soil	Novel indirubis-like antimicrobial compound	MacNeil (2001)
BAC	Soil	Turbomycin A and B synthesis characterization	Gillespie et al. (2002)
Fosmid	Soil	Novel copper-containing nitrate reductase	Treusch et al. (2004)
Cosmid	Soil	Novel Violacein-like antibacterial pigment	August (2000)
Cosmid	Soil	Putative stereoselective amidase, 1. cellulases, α-amylase, 1,4-α-glucan branching enzyme, 2. pectate lyases	Voget et al. (2003)
Cosmid	Soil	Long chain N-Acyltyrosine synthases	Brady et al. (2004)
pBluescriptSK+	Soil	4-hydroxybutyrate utilization	Henne et al. (1999)
Fosmid	Soil	Uncultured Crenarchaeote	Quaiser (2002)
TOPO-XL	Human saliva/plaque	Tetracycline resistance (tet(37))	Diaz-Torres (2003)

systems need to be considered. Vectors that have been designed for use in different host are illustrated in Table 4.2. Other systems to improve expression in *E. coli* have been used such as cloning between oppositely reading promoters to force bidirectional transcription (Lammle *et al.*, 2007) and the random integration of strong transcriptional signals by using transposons that generate transcriptional fusions (Leggewie *et al.*, 2006).

B. Construction of small insert libraries

Plasmids that have been used for the construction of small insert metagenomic libraries are shown in Table 4.2.

C. Phage display

Phage display technologies work by inserting DNA fragments into the genome of filamentous bacteriophage. Insert DNA is fused to a gene encoding a phage coat protein and is expressed as a fusion peptide on the phage surface. It is a powerful technique that provides a direct link between the peptide of interest and the encoding DNA sequence and it allows the screening of very large numbers of unique peptides and proteins (10^9 or more). Using a shotgun approach (pioneered by Jacobsson and Frykberg (1995, 1996)), it is possible to use this technique to screen metagenomic samples for novel adhesion mechanisms and host–pathogen interactions as previously demonstrated for genomic samples (Beckmann *et al.*, 2002; Mullen *et al.*, 2007; Nilsson *et al.*, 2004; Williams *et al.*, 2002).

The technique generally uses a phagemid vector, which is a plasmid encoding a surface protein (either pIII or pVIII) of the filamentous bacteriophage Ff. The phagemid also contains both *E. coli* and phage origins of replication, allowing it to be replicated as a plasmid in *E. coli* and to be packaged as a phage upon coinfection of *E. coli* by helper phage. An antibiotic marker is also encoded by the phagemid. Metagenomic DNA is sonicated to give fragments of approximately 0.5–3 kb and end repaired using T4 DNA polymerase. The fragmented DNA is purified using either phenol–chloroform extractions or the QIAquick PCR Purification kit (Qiagen) and ligated into the multiple cloning site of the phagemid vector, which is adjacent to the gene encoding the phage coat protein.

Several phagemid vectors are available, including pG8H6 and pG8SAET, which allow peptide fusions with the pVIII surface protein (Jacobsson and Frykberg, 1996; Jacobsson *et al.*, 2003) and pG3H6 (Jacobsson *et al.*, 1997), which allows peptide fusions with the pIII surface protein. The phagemid is digested to produce blunt ends (the restriction enzyme will depend on the choice of vector) and dephosphorylated to prevent self-ligation. Ligation is carried out using T4 DNA ligase, although the Ready-to-Go Ligation kit (GE Healthcare, Bucks, UK) has routinely produced the best results (Jacobsson et al., 2003; Mullen et al., 2007; Williams et al., 2002).

The phagemid is introduced into E. coli TG1 by electroporation and the cells are left to recover at 37 °C in 10 ml nutrient broth. The culture is then infected with helper phage, such as R408 (Promega, Southampton, UK), to allow preferential packaging of the phagemid DNA using a multiplicity of infection of around 20 (Mullen et al., 2007). Where a pIII fusion is used, the helper phage "Hyperphage" may be used with greater efficiency (Rondot et al., 2001). The infected cells are then grown overnight in a larger volume of nutrient broth (containing antibiotic) and the phage recovered by polyethylene glycol 8000 precipitation (Williams et al., 2002).

The final stage is panning, allowing for the selection of peptides that bind to a ligand of interest. This could be, for example, a component of the saliva that could be involved in primary colonization of oral surfaces such as albumin, fibronectin or IgA (Scannapieco, 1994). The chosen ligand is immobilized on the surface of an immunotube (MaxiSorp^{TM}, Nunc) overnight at 4 °C and subsequently blocked with 2% BSA at room temperature; controls replace the ligand with BSA. After washing, the phage library is added to the immunotube and incubated at room temperature. Unbound phage are removed by washing with PBS and the bound phage eluted using 0.1 M glycine. Phage are neutralized using Tris-HCl buffer at pH 8.0. Further amplification and panning of the eluted phage improve the specificity of the positive results obtained. The sequences of the identified peptides can be used for further analyses. Several bacterial peptides that bind fibronectin have been identified (Beckmann et al., 2002; Mullen et al., 2007; Williams et al., 2002) and some initial data from our work reveals some oral bacterial peptides that will bind IgA (Hunter, Easton, Schulze-Schweifing and Ward, unpublished).

VIII. LIMITATIONS OF METAGENOMICS

Despite all the advances in this field, the concept of finding a certain gene or pathway in an environmental sample using metagenomics depends on several factors: the abundance of the gene in the sample, the length of the gene or operon, the size of the cloned DNA insert, the presence of expression signals that are functional in the host organism and the correct folding of the translated protein in a heterologous host by appropriate trans acting host factors (chaperones, cofactors, protein-modifying enzymes)—all contribute to the likelihood of successful cloning (Gabor et al., 2004).

The minimal requirements for gene expression are an appropriately spaced promoter for transcription, a ribosome binding site (RBS) and a start codon for initiation of translation, both of which need to be compatible with host transcription and translation factors (Gabor et al., 2004).

In metagenomic cloning, expression may occur by

i. independent gene expression (both the RBS and promoter are provided by the insert)
ii. expression as a transcriptional fusion with only the RBS located on the insert
iii. as a translational fusion (the RBS and promoter are provided by the vector), although this is highly unlikely to occur.

Gabor et al. (2004) have taken these factors into account and designed statistical models to determine the proportion of genes that might be expressed in an E. coli host, taking into account the three modes of gene expression listed earlier. Their results suggest that approximately 40% of enzyme activities from a metagenome can be expected to express in E. coli and would therefore be recovered by random cloning. Therefore, the challenge still remains to access all the genetic information from a metagenome.

Perhaps one concern of the wide use of these methods to define environmental microbial diversity is the deposition of sequences in the database which provide no further information than the degree of relatedness of "clones" to cultivable species, although allowing some quantification of diversity, these methods shed no further light on the role of these novel species in the microbiota. There is also concern that the relative ease of obtaining Mb of sequence data is eclipsing the need to continue culture based studies, especially in light of the fact that novel cultivation strategies are possible and have recently been developed for butyrate-producing and cellobiose-degrading bacteria, allowing access to potentially marketable novel biochemical pathways, and a greater understanding of their role within the environment.

IX. CONCLUSIONS AND FUTURE PERSPECTIVES

Functional metagenomics is a powerful tool for analyzing bacteria that cannot yet be cultivated in the laboratory. The major limitation to the technique is that not all genes will express in the heterologous host. However host vector systems are being developed that are aimed at overcoming this problem.

REFERENCES

August, P. R., Grossman, T. H., Minor, C., Draper, M. P., MacNeil, I. A., Pemberton, J. M., Call, K. M., Holt, D., and Osburne, M. S. (2000). Sequence analysis and functional characterisation of the violacein biosynthetic pathway from *Chromobacterium violaceum*. *J. Mol. Microbiol. Biotechnol.* **2,** 513–519.

Beckmann, C., Waggoner, J. D., Harris, T. O., Tamura, G. S., and Rubens, C. E. (2002). Identification of novel adhesins from Group B streptococci by use of phage display reveals that C5a peptidase mediates fibronectin binding. *Infection and Immunity.* **70,** 2869–2876.

Béjà, O., Suzuki, M. T., Koonin, E. V., Aravind, L., Hadd, A., Nguyen, L. P., Villacorta, R., Amjadi, M., Garrigues, C., Jovanovich, S. B., Feldman, R. A., and DeLong, E. F. (2000). Construction and analysis of bacterial artificial chromosome libraries from a marine microbial assemblage. *Environ. Microbiol.* **2,** 516–529.

Brady, S. F., Chao, C. J., and Clardy, J. (2004). Long chain N-Acetyrosine synthases from environmental DNA. *Appl. Environ. Microbiol.* **70,** 6865–6870.

Diaz-Torres, M. L., McNab, R., Spratt, D. A., Villedieu, A., Hunt, N., Wilson, M., and Mullany, P. (2003). Novel tetracycline resistance determinant from the oral Metagenome. *Antimicrob. Agents Chemother.* **47,** 1430–1432.

Diaz-Torres, M. L., Villedieu, A., Hunt, N., McNab, R., Spratt, D. A., Allan, E., Mullany, P., and Wilson, M. (2006). Determining the antibiotic resistance potential of the indigenous oral microbiota of humans using a metagenomic approach. *FEMS. Microbiol. Lett.* **258,** 257–262.

Easter, C. L., Schwab, H., and Helinski, D. R. (1998). Role of the parCBA operon pf the broad-host-range plasmid RK2 in stable plasmid maintenance. *J. Bacteriol.* **180,** 6023–6030.

Gabor, E. M., Alkema, W. B., and Janssen, D. B. (2004). Quantifying the accessibility of the metagenome by random expression cloning techniques. *Environ. Microbiol.* **6,** 879–886.

Gafan, G. P., and Spratt, D. A. (2005). Denaturing gradient gel electrophoresis gel expansion (DGGEGE)–An attempt to resolve the limitations of co-migration in the DGGE of complex polymicrobial communities. *FEMS Microbiol. Lett.* **253,** 303–307.

Gillespie, D. E., Brady, S. F., Bettermann, A. D., Cianciotto, N. P., Liles, M. R., Rondon, M. R., Clardy, J., Goodman, R. M., and Handelsman, J. (2002). Isolation of antibiotics turbomycin A and B from a metagenomic library of soil microbial DNA. *Appl. Environ. Microbiol.* **68,** 4301–4306.

Henne, A., Daniel, R., Schmitz, R. A., and Gottschalk, G. (1999). Construction of environmental DNA libraries in *Escherichia coli* and screening for the presence of genes conferring utilisation of 4-hydroxybutyrate. *Appl. Environ. Microbiol.* **65,** 3901–3907.

Jacobsson, K., and Frykberg, L. (1995). Cloning of ligand-binding domains of bacterial receptors by phage display. *BioTechniques.* **18,** 878–885.

Jacobsson, K., and Frykberg, L. (1996). Phage Display Shot-Gun Cloning of Ligand-Binding Domains of Prokaryotic Receptors Approaches 100% Correct Clones. *BioTechniques.* **20,** 1070–1081.

Jacobsson, K., Jonsson, H., Lindmark, H., Guss, B., Lindberg, M., and Frykberg, L. (1997). Shot-gun phage display mapping of two streptococcal cell-surface proteins. *Microbiol. Res.* **152,** 121–128.

Jacobsson, K., Rosander, A., Bjerketorp, J., and Frykberg, L. (2003). Shotgun Phage Display – Selection for Bacterial Receptins or other Exported Proteins. *Biological Procedures Online* **5,** 123–135.

Kazor, C. E., Mitchell, P. M., Lee, A. M., Stokes, L. N., Loesche, W. J., Dewhirst, F. E., and Paster, B. J. (2003). Diversity of bacterial populations on the tongue dorsa of patients with halitosis and healthy patients. *J. Clin. Microbiol.* **41,** 558–563.

Kumar, P. S., Griffen, A. L., Barton, J. A., Paster, B. J., Moeschberger, M. L., and Leys, E. J. (2003). New bacterial species associated with chronic periodontitis. *J. Dent. Res.* **82,** 338–344.

Lämmle, K., Zipper, H., Breuer, M., Hauer, B., Buta, C., Brunner, H., and Rupp, S. (2007). Identification of novel enzymes with different hydrolytic activities by metagenomic expression cloning. *J. Biotechnnol.* **127,** 575–592.

Leggewie, C., Henning, H., Schmeisser, C., Streit, W. R., and Jaeger, K. E. (2006). A novel transposon for functional expression of DNA libraries. *J. Biotechnnol.* **123,** 281–287.

MacNeil, I. A., Tiong, C. L., Minor, C., August, P. R., Grossman, T. H., Loiacono, K. A., Lynch, B. A., Phillips, T., Narula, S., Sundaramoorthi, R., Tyler, A., Aldredge, T., *et al.* (2001). Expression and isolation of antimicrobial small molecules from soil DNA libraries. *J. Mol. Microbiol. Biotechnol.* **3,** 301–308.

Mullen, L. M., Nair, S. P., Ward, J. M., Rycroft, A. N., Williams, R. J., and Henderson, B. (2007). Comparative functional genomic analysis of Pasteurellaceae adhesins using phage display. *Veterinary Microbiol.* **122,** 123–134.

Nilsson, M., Bjerketorp, J., Wiebensjö, Å., and Ljungh, Å. (2004). A von Willebrand factor-binding protein from *Staphylococcus lugdunensis*. *FEMS. Microbiol. Lett.* **234,** 155–161.

Quaiser, A., Ochsenreiter, T., Klenk, H. P., Kletzin, A., Treusch, A. H., Meurer, G., Eck, J., Sensen, C. W., and Schleper, C. (2002). First insight into the genome of an uncultivated crenarchaeote from soil. *Environ. Microbiol.* **4,** 603–611.

Ready, D., Pratten, J., Roberts, A. P., Bedi, R., Mullany, P., and Wilson, M. (2006). Potential role of *Veillonella* spp. as a reservoir of transferable tetracycline resistance in the oral cavity. *Antimicrob. Agents Chemother.* **50,** 2866–2868.

Riesenfeld, C. S., Goodman, R. M., and Handelsman, J. (2004). Uncultured soil bacteria are a reservoir of new antibiotic resistance genes. *Environ. Microbiol.* **6,** 981–989.

Roberts, A. P., and Mullany, P. (2006). Genetic basis of horizontal gene transfer among oral bacteria. *Periodontology* 2000. **42,** 36–46.

Rondot, S., Koch, J., Breitling, F., and Dübel, S. (2001). A helper phage to improve single-chain antibody presentation in phage display. *Nature Biotechnol.* **19,** 75–78.

Scannapieco, F. A. (1994). Saliva-bacterium interactions in oral microbial ecology. *Crit. Rev. Oral Bio. Med.* **5,** 203–248.

Shizuya, H., Birren, B., Kim, U. J., Mancino, V., Slepak, T., Tachiiri, Y., and Simon, M. (1992). Cloning and stable maintainance of 300 kilobase pair fragments of human DNA in *Escherichia coli* using an F-factor based vector. *Proc. Natl. Acad. Sci. USA* **89,** 8794–8797.

Shizuya, H., and Kouros-Mehr, H. (2001). The development and applications of the bacterial artificial chromosome cloning system. *Keio. J. Med.* **50,** 26–30.

Syed, S. A., and Loesche, W. J. (1972). Survival of Human dental plaque flora in various transport media. *Appl. Microbiol.* **24,** 638–644.

Treusch, A. H., Kletzin, A., Raddatz, G., Ochsenreiter, T., Quaiser, A., Meurer, G., Schuster, S. C., and Schleper, C. (2004). Characterisation of large insert DNA libraries from soil for environmental genomic studies of Archaea. *Environ. Microbiol.* **6,** 970–980.

Villedieu, A., Roberts, A. P., Allan, E., Hussain, H., McNab, R., Spratt, D. A., Wilson, M., and Mullany, P. (2007). Determination of the genetic support for tet(W) in oral bacteria. *Antimicrob. Agents Chemother.* **51,** 2195–2197.

Voget, S., Leggewie, C., Uesbeck, A., Raasch, C., Jaeger, K. E., and Streit, W. R. (2003). Prospecting for novel biocatalysts in a soil metagenome. *Appl. Environ. Microbiol.* **69,** 6235–6242.

Williams, R. J., Henderson, B., Sharp, L. J., and Nair, S. P. (2002). Identification of a fibronectic-Binding Protein from *Staphylococcus epidermis*. *Infection and Immunity.* **70,** 6805–6810.

Wilson, M. (2005). Microbial Inhabitants of Humans: Their Ecology and Role in Health and Disease. Cambridge University Press, Cambridge, UK.

Zoetendal, E. G., Cheng, B., Koike, S., and Mackie, R. I. (2004). Molecular microbial ecology of the gastrointestinal tract: From phylogeny to function. *Curr. Issues Intest. Microbiol.* **5,** 31–47.

CHAPTER 5

Biosensors for Ligand Detection

Alison K. East,*,1 Tim H. Mauchline,† and Philip S. Poole*

Contents		
	I. Introduction	137
	II. Induction Biosensors	139
	A. General considerations for use of induction biosensors	139
	B. Reporter genes	140
	III. Molecular Biosensors	150
	A. Optical biosensors	150
	B. Electrical biosensors	159
	IV. Conclusions and Future Prospects	160
	Acknowledgments	160
	References	160

I. INTRODUCTION

In recent years, there has been an explosion in our knowledge of genome sequences, microarray data, and proteomics. However, changes in transcription or protein profiles in different environments often leads to the question of what factor(s) induces the change. Tools that are able to accurately determine the amount of a compound, its cellular location and changes in its concentration thereafter are crucial in identifying the environmental inducer. Biosensors are such tools, which couple the ability of a biological sensor to precisely distinguish between inducer

---

* Molecular Microbiology, John Innes Centre, Colney Lane, Norwich NR4 7UH, United Kingdom
† School of Biological Sciences, University of Reading, Berkshire RG6 6AJ, United Kingdom
1 Corresponding author: Molecular Microbiology, John Innes Centre, Colney Lane, Norwich NR4 7UH, United Kingdom

ligands, with a robust abiotic method of detection. The term biosensor can be used to describe devices for monitoring, recording, and transmitting information regarding physiological changes in living organisms. A more specific definition of a biosensor and one that will be used in this review is that biosensors are detection devices that are completely or partly composed of biological material. With this definition in mind, biosensors can be split into four basic types depending on the sensing component: molecular, subcellular, cellular, and tissue-based. The most widely used are cellular and molecular biosensors. Cellular biosensors use reporter genes or proteins inside a cell, while molecular biosensors rely on a purified molecule such as a protein. However, these definitions can become blurred as some molecular biosensors can be expressed inside a cell to generate a cellular biosensor. Another way of considering this is that cellular biosensors can be either induction biosensors or molecular biosensors, and we shall consider both these subclasses in this review.

Induction biosensors rely on a ligand being detected by binding to a protein that induces expression of a reporter gene. The reporter gene codes for a protein that has an easily quantifiable activity, often relying on enzymatic and/or optical detection. Initial developments of reporter proteins required the addition of a substrate but many that do not have since been developed, particularly autofluorescent proteins such as green fluorescent protein (GFP) from the bioluminescent jellyfish *Aequorea victoria*. GFP and its variants require no added substrate and are non-invasive, requiring only excitation at a particular wavelength and measurement of fluorescence emission at another (Tsien, 1998).

An exciting development in the biosensor field has been the advent of molecular biosensors or nanosensors (Dattelbaum *et al.*, 2005; De Lorimier *et al.*, 2002; Deuschle *et al.*, 2006; Fehr *et al.*, 2002, 2003, 2005b; Gu *et al.*, 2006; Lager *et al.*, 2003; Looger *et al.*, 2003; Marvin and Hellinga, 2001a,b; Marvin *et al.*, 1997; Okumoto *et al.*, 2005). These are based on direct detection of ligand-binding by a protein, mediated by the accompanying conformational change. Many molecular biosensors are based on the properties of bacterial solute binding proteins (SBPs), part of the ABC and TRAP transporter complex, whose *in vivo* function is to bind molecules for transport into a bacterial cell and are, in some cases, involved as sensors for the chemotactic response in bacteria (Fehr *et al.*, 2002). There are large numbers of naturally occurring, extremely specific SBPs. Many bacteria have hundreds which are able to bind a range of ligands from sugars (monosaccharides, disaccharides, trisaccharides, polyols), amino acids, peptides, metal ions, and amines, making them ideal for the development of a wide range of biosensors (Mauchline *et al.*, 2006).

II. INDUCTION BIOSENSORS

A. General considerations for use of induction biosensors

First, we will outline the principles of how induction biosensors work. The sensing element of an induction biosensor is composed of a transcriptional unit (gene or operon) that contains a promoter sequence of a specific gene that is known to be responsive to a given inducing condition or ligand. The selectivity and to a certain degree the sensitivity of the system is determined by this element. This sequence is fused to a reporter gene, usually in a transcriptional fusion, containing its own transcriptional start codon and ribosome binding site, thus creating a single transcriptional unit. As such, when the promoter responds to an inducing condition and transcription commences, the reporter gene is also transcribed, and subsequently translated into a protein that can be measured and assayed. Reporter genes vary in their sensitivity and so choice of reporter gene is crucial in determining the sensor's sensitivity and limit of detection.

The choice of cell type that houses the promoter–reporter gene fusion is wide, and care must be taken to ensure that the microbe can establish in the environment that is to be probed and that the host cell has the necessary regulatory machinery to respond to inducing ligands/conditions. Classically, bacterial cells have been used as microbial induction biosensors, and they continue to be the main source. However, recently, there have been advances in the use of other micro organisms such as filamentous fungi and yeasts (see review by Baronian, 2004).

In addition to host cell, the type of genetic element that the fusion is present on in the sensing cell is critical for biosensor use. Generally, reporter fusions are cloned in plasmids that replicate in *Escherichia coli*. However, if they do not replicate in the reporter organism then the vector may integrate by recombination with homologous DNA in the host. The integrated fusion will be as a single copy with reduced sensitivity relative to a multi-copy plasmid. Also, if the region to be recombined into the genome is intragenic (lacking both N- and C-termini of a coded protein) then a mutant will be created. If the reporter vector replicates in the host organism, it will be present in multiple copies, although the actual number will vary depending on the plasmid replicon. A multicopy plasmid may disrupt regulation of the gene or operon being examined. For example, when the number of copies of plasmid is higher than the number of negative regulator molecules, the result will be constitutive expression of the reporter gene (Mauchline *et al.*, 2006).

B. Reporter genes

Understanding the fusing of promoters to reporter genes has greatly enhanced our ability to study gene expression both in the laboratory and in the environment (Karunakaran et al., 2005). There has been significant development of reporter gene technology over the last decade and the different systems available for environmental monitoring are described in this section. In addition, the advantages and disadvantages of each type of reporter gene are outlined (Table 5.1).

1. Chloramphenicol acetyltransferase (CAT)

CAT is an enzyme derived from E. coli, the most active variant of which is the type III enzyme. Its endogenous form is a trimer of identical subunits, forming a deep pocket in which chloramphenicol binds. The trimer is stabilized by hydrogen bonds as well as β-pleated sheets that extend from one subunit to the next (Shaw and Leslie, 1991). CAT detoxifies chloramphenicol by covalently attaching an acetyl group to the antibiotic, rendering it unable to bind bacterial ribosomes and thus preventing disruption of protein synthesis (Shaw, 1983). CAT was discovered in the 1960s following a spread of resistance to antibiotics of which chloramphenicol was the first reported (Shaw, 1983). Its use as a reporter gene couples enzymatic assays with thin layer chromatography (TLC) and hinges around the differential migration of acetylated and unacetylated chloramphenicol. Traditionally, radiolabelled chloramphenicol or acetyl-CoA was used to measure the product of the CAT-mediated acetylation reaction and the detection limit of these assays is reported to be as low as 2 pg, with the assay being linear over three orders of magnitude. The health risk associated with using radioisotopes in these assays has prompted the development of an assay that makes use of fluorescent derivatives of chloramphenicol and that can be detected in a similar way with comparable sensitivity (Lefevre et al., 1995).

2. β-galactosidase

The enzyme β-galactosidase (β-gal) catalyzes the hydrolysis of β-galactosides, such as lactose, into two monomers. The gene for the E. coli version of this enzyme is often used as a reporter gene for transcriptional and translational regulation studies despite there is endogenous β-gal activity in many cell types. The structure of β-gal from E. coli was discovered using X-ray crystallography techniques (Jacobson and Matthews, 1992) and it was revealed to be a tetramer of four identical proteins, each composed of 40% β-sheet, 35% α-helix, 13% β-turn, and 12% random coil that interact with the galactose substrate. In addition, monovalent and divalent cations are shown to mediate catalysis by acting as co-factors (Huber et al., 1994). There are several detection methods for assays using β-gal as the reporter

TABLE 5.1 Advantages and disadvantages of different reporter genes

Gene	Advantages	Disadvantages
cat	Sensitive to 2 pg with both fluorescent and radio-labeled assays	Hazardous radioisotopes often used Destructive sampling required Requires addition of substrate Assay has narrow linear range of only up to three orders of magnitude Laborious assays
lacZ	Sensitive to 2 pg with fluorescent substrates, and 2 fg with luminescence based assays, although only 100 pg with colorimetric substrates (Jain and Magrath, 1991) Can be used in anaerobic environments Protein is very stable	Requires addition of substrate Invasive—requires isolation and permeabilization of target cells Endogenous activity in many cells
uidA	Similar sensitivity to *lacZ*, no endogenous activity	See *lacZ*
lux	No substrate required if entire operon included (*luxCDABE*) Sensitive to pg level (Meighen, 1993) Non-destructive, *in situ* monitoring possible Short half-life of LuxAB for real-time monitoring (Hautefort and Hinton, 2000) Lack of endogenous activity in most cells	Requires oxygen Heat labile (Naylor, 1999) Metabolically expensive (Hautefort and Hinton, 2000) Aldehyde substrate is toxic (Gonzalezflecha and Demple, 1994) Variability in reproducibility in anoxic conditions (Camilli, 1996)

(*continued*)

TABLE 5.1 (continued)

Gene	Advantages	Disadvantages
		Photon yield does not allow single cell monitoring
		Short half life of LuxAB can lead to experimental perturbations (Hautefort and Hinton, 2000)
		Photon yield too weak to monitor single cells (Hautefort and Hinton, 2000)
		Narrower linear range than *luc* (Naylor, 1999)
luc	Nondestructive, *in situ* monitoring possible.	Requires oxygen
	Sensitive down to fg level (Billard and DuBow, 1998)	Requires substrate addition
	Lack of endogenous activity in most cells	
	Broad linear range seven to eight orders of magnitude (Naylor, 1999)	
	Not as metabolically as expensive as *lux* (Hakkila et al., 2002)	
gfp	No substrate addition required	Autofluorescence of sample (Hakkila et al., 2002)
	Nondestructive sampling	Requires oxygen for fluorophore activation
	In situ monitoring	Lack of sensitivity compared to *lux*, *inaZ*, *lacZ* (Kohlmeier et al., 2007; Miller et al., 2001)
	Intracellular studies	Slower response time than *lux* and *luc* (Hakkila et al., 2002)
	Different color variants/dual labeling	Some *gfps* sensitive to mildly acidic conditions (Tsien, 1998)
	Variants available with different stabilities	
	Heat stable varieties	
	High photo stability (Tsien, 1998)	
	Quantifiable	
	No endogenous activity	

dsRed	No substrate addition required	Autofluorescence of sample (Hakkila et al., 2002)
	Nondestructive sampling	Oxygen required for fluorophore activation
	Single cell studies	Lack of sensitivity compared to some other reporters
	Dual-labeling	Slower response time than lux and luc (Hakkila et al., 2002)
	Fluorescent timer variant available (Terskikh et al., 2000)	Slow maturation, though variants such as mRFP1 are 10 times quicker to mature (Campbell et al., 2002)
	Different color variants	Emission of contaminating green light could cause complications for dual labeling (Gross et al., 2000)
	No endogenous activity	Low solubility, though new variants have less green contamination and are more soluble (e.g., dsRed T4) (Barolo et al., 2004)
		Can be prone to photo bleaching (Mirabella et al., 2004)
cobA	Reduced autofluorescence compared to AFPs	Substrate addition required to make assay more reproducible
	Similar intensity to *gfp* (Wildt and Deuschle, 1999)	Endogenous activity
	No substrate addition strictly necessary	
inaZ	No substrate addition required	Destructive sampling
	More sensitive than *lacZ* or *gfp* (Miller et al., 2001)	
	Can be used in anaerobic environments	
	No endogenous activity	

and these vary depending on the substrate used for the enzyme; they include colorimetric (e.g., *O*-nitrophenyl β-D-galactopyranoside or ONPG), histochemical (e.g., 5-bromo-4-chloro-3-indolyl β-D-galactoside or X-gal), fluorometry (e.g., 4-methylumbelliferyl-β-D-galactopyransoside or MUG) and fluorescein-di-beta-D-galactoside or FDG), luminescence (e.g., 1,2-dioxetane substrates), and electrochemical assays (e.g., *p*-aminophenyl-β-D-galactopyransoside or PAPG). The reader is directed to an excellent review of these assays, their pros and cons, and limits of detection (Daunert *et al.*, 2000).

3. β-glucuronidase (GUS)

Glucuronidases are a family of enzymes that cleave glucuronides. The *E. coli* GUS (UidA) has a molecular weight of 68,200 and appears to function as a tetramer of four identical subunits (Jain *et al.*, 1996; Jefferson *et al.*, 1986). Various glucuronide substrates are available for GUS assays, which all contain the sugar D-glucopyranosiduronic acid attached by a glycosidic linkage to a hydroxyl group of a chromophore, which can be detected histochemically or by fluorescence. In histochemical detection, cleavage of the substrate 5-bromo-4-chloro-3-indolyl glucuronide (X-Gluc) results in a blue precipitate. For detection by fluorescence the substrate most often used is MUG. Its hydrolysis results in the formation of the fluorochrome 4-methyl umbelliferone and the sugar glucuronic acid. An alternative substrate to MUG is 4-trifluoromethyl umbelliferyl β-D-glucuronic acid (4-TFMUG). This, unlike MUG, allows continuous monitoring of GUS activity because the substrate becomes fluorescent upon hydrolysis at the assay pH, whereas for MUG, the assay must be terminated with a basic solution for detection (Gallagher, 1992). GUS was initially developed as a marker gene system in *E. coli*, but more recently has been used extensively for the detection of chimeric gene expression in plants.

4. Bioluminescence

Many different organisms, including bacteria, unicellular algae, coelenterates, beetles, fish, and fungi, are able to emit light. The bioluminescent systems that these organisms possess have many times evolved independently and so are not often evolutionarily conserved (Wilson and Hastings, 1998) The ecological significance of bioluminescence varies greatly: it being implicated in quorum sensing in bacteria, attraction of prey in some deep-sea fish and of mates in insects, camouflage as well as repulsion, and as a mechanism to control oxygen concentration in the cell (McElroy and Seliger, 1962; Wilson and Hastings, 1998). The development of luminescence biosensor technology has been based largely around bacterial and firefly systems, although other luciferases have been utilized as reporter genes. For example the luciferase from the marine copepod

Gaussia princes has been cloned and used as a reporter (Wiles *et al.*, 2005) as has that of the sea pansy, *Renilla reniformis* (Srikantha *et al.*, 1996).

The genes of the bacterial luciferase operon (*luxCDABE*) originate from marine bacteria including *Vibrio fischeri* and *Vibrio harveyi* as well as the terrestrial bacterium *Photorhabdus luminescens* (Hakkila *et al.*, 2002). Bacterial luciferases (LuxAB) catalyze the oxidation of a reduced riboflavin phosphate ($FMNH_2$) and a long chain fatty aldehyde, with emission of blue-green light (λ_{max} 490 nm). The *V. harveyi* luciferase is an α-β heterodimer, with individual subunits folded into an 8 barrel motif (Fisher *et al.*, 1995, 1996). The catalytic site of this enzyme is located in a pocket in the α subunit (Fisher *et al.*, 1996). The role of the β-subunit in catalysis is unclear despite studies have shown that it is essential for a high quantum yield reaction (Baldwin *et al.*, 1995). Other genes of the *lux* operon (*luxCDE*) encode enzymes that are responsible for the synthesis of the long-chain substrate which is required for the bioluminescence reaction.

The use of *lux* genes as biosensors couples the *luxAB* genes to a promoter of interest and expression of this fusion in host cells. The *luxAB* genes alone are sufficient to monitor luminescence, however an addition of long chain aldehyde as substrate is required. An alternative strategy is to create fusions containing the entire *lux* operon (*luxCDABE*) to a promoter, and in doing so endogenous substrate will be created for the luciferase to bind to and cause luminescence. The visualization of *lux* bioluminescence from environmental samples is achieved with both *in situ* and *in vitro* assays. In destructive assays, a luminometer is used for photon detection. However, for *in vivo* monitoring assays the sample to be viewed must be covered in total darkness, so that the luminescence of the biosensor can be recorded by use of a charge coupled device (CCD) in a dark field image. This, when stacked on top of a bright field image, allows spatial distribution of the bioluminescence in relation to the sample (Darwent *et al.*, 2003).

Firefly luciferase from *Photinus pyralis* encoded by the *lucFF* genes is a polypeptide monomer of 62 kDa. The crystal structure of firefly luciferase shows two distinct domains, a large N-terminal active site domain, a flexible linker domain and a small C-terminal domain facing the N domain (Conti *et al.*, 1996). Firefly luciferase catalyses the emission of yellow-green light (maximum emission 550–575 nm) from the substrates luciferin (benzothiazoyl-thiazole), Mg-ATP, and oxygen, with the additional production of CO_2, AMP and pyrophosphate (Branchini *et al.*, 1999; Hakkila *et al.*, 2002). The use of *luc* as a biosensor couples the *lucFF* genes to a promoter of interest, and like *luxAB* fusions, an application of exogenous substrate (luciferin for LucFF) is required for enzyme activity in non-luciferin producing cells. However, *luc* fusions are an order of magnitude more sensitive than the equivalent *lux* fusions, and have a broader linear range of quantification (Billard and DuBow, 1998; Meighen, 1993; Naylor, 1999).

5. Autofluorescent proteins (AFP)

AFP is the generic name given to proteins that emit fluorescence upon illumination with light within a specific excitation range without the requirement of additional substrate, cofactor or protein (Larrainzar et al., 2005). The most commonly used AFP is GFP, and the original GFP was discovered in the jellyfish, *A. victoria*, while studying the chemiluminescent protein aequorin (Shimomura et al., 1962). It was found that the emission of blue light from aequorin at 470 nm is absorbed by GFP with subsequent emission of green light with an emission peak near 570 nm. As such, GFP acts as an accessory protein to aequorin, and this relationship between GFP and a primary photoprotein (e.g., aequorin) is repeated in other GFP producing coelenterates, such as representatives of *Obelia*, *Phialidium*, and *Renilla* species (Larrainzar et al., 2005). The ecological reason for this is unclear, although it is possible that GFP functions to enhance the quantum efficiency of emission in these organisms, perhaps attracting a mate or to deter predators.

The identification of the structure of the *Aequorea* GFP revealed that it is a monomeric 23 kDa protein that contains a natural chromophore in an internal hexapeptide, which requires O_2 for cyclization (Chalfie et al., 1994; Inouye and Tsuji, 1994). The three-dimensional structure of GFP has been solved, and it has 11 antiparallel β-strands forming a cylinder (or β-can) that surround an inner α-helix where the chromophore is located (Yang et al., 1996). This structure functions to protect the chromophore and confer the stability of the native GFP protein.

In addition to wild type GFP, many GFP derivatives have been created with altered properties. For example, mutations within the fluorophore region Ser65-Tyr66-Gly67, as well as replacements of Tyr-66 with aromatic amino acids have been created to alter the excitation and emission spectra (Cubitt et al., 1995; Heim et al., 1994). Other variants have been developed with increased stability at temperatures at or above 37 °C such as the UV-excitable variant T-Sapphire (Zapata-Hommer and Griesbeck, 2003), those with altered pH sensitivity such as ecliptic pHluorin (Ashby et al., 2004), as well as brighter variants such as GFP-UV. GFP-UV was produced by shuffle mutagenesis, resulting in a 45-fold higher emission than wild type GFP in *E. coli*, mainly due to increased protein solubility (Crameri et al., 1996). Additionally, three classes of red-shifted mutated proteins, where Ser65 was substituted, were named GFPmut1 (commercialized as EGFP), GFPmut2 and GFPmut3 and it was found that fluorescence intensity increased by up to 100-fold in *E. coli*. This was due to enhanced protein solubility and a shift in the absorption spectrum, with maximum excitation occurring between 480 and 501 nm, compared to a maxima of 395 nm (and a minor peak at 475 nm) in wild type GFP (Cormack et al., 1996). This is much closer to the excitation and emission spectra of fluorescein, making the red-shifted derivatives of GFP suitable for FAC sorters and microscopes set up for fluorescein detection. Furthermore, the addition of a protease-targeting signal to GFPmut has led to the creation of a suite of GFPmut proteins with different stabilities (Andersen *et al.*, 1998). Other examples of GFP mutants are those with blue, cyan, yellowish-green and yellow emission spectra (Shaner *et al.*, 2005; Tsien, 1998). The mutations in GFP-UV and GFPmut3 have also been combined to create a red shifted variant GFP+, which has an 320-fold increase in fluorescence compared to wild type GFP in *E. coli* (Scholz *et al.*, 2000).

DsRed, a 28 kDa fluorescent protein from the sea coral *Discosoma striata*, with an emission maximum of 583 nm, is an alternative AFP to GFP. A disadvantage of wild type DsRed is that it is tetrameric and is slow to mature compared to GFP. However, mutant derivatives DsRedT.3 and DsRedT.4, have recently been isolated, which are tetrameric and mature more rapidly than the wild type protein (Bevis and Glick, 2002). Additionally, a more rapidly maturing monomeric variant of DsRed has been developed, called monomeric red fluorescent protein (mRFP1), that matures 10 times faster than DsRed and has excitation and emission peaks at 584 and 607 nm respectively (Campbell *et al.*, 2002). This is significant as tetrameric proteins can be toxic to bacteria, whereas monomeric proteins are usually less so (Shaner *et al.*, 2004). It is also possible to covert dimeric *Aequorea* AFPs (e.g., GFP) into a monomeric form by introduction of a mutation A206K, generally without deleterious effects (Zacharias *et al.*, 2002). However, in our experience GFP still matures much faster than any of the DsRed and mRFP1 derivatives and pilot experiments should be conducted before any widespread application of these proteins.

Further modifications to DsRed include production of the E5 variant that changes fluorescence from red to green over time so can be used as a "fluorescent timer" (Terskikh *et al.*, 2000) as well as this development of a suite of alternative red DsRed proteins are available such as mCherry, tdTomato and mStrawberry (Shaner *et al.*, 2004), as well as the orange DsRed proteins: mOrange (Shaner *et al.*, 2004), mKO (Karasawa *et al.*, 2004) and the far-red DsRed protein mPlum (Wang *et al.*, 2004). The development of such a wide variety of AFPs has made the monitoring of multiple reporter genes in one system highly achievable and it is now possible to distinguish between up to four different AFPs simultaneously *in situ* (Shaner *et al.*, 2005).

AFPs have been used for various environmental studies, such as studying the interactions between rhizobia and their legume hosts (Bringhurst *et al.*, 2001; Gage, 2002; Gage and Long, 1998; Gage *et al.*, 1996). AFPs can also be used in high-throughput screening assays. In one such example, a fusion library to the complete transportome of *Sinorhizobium meliloti* was made, enabling the solute induction profiles of

47% of the ABC uptake systems and 53% of the TRAP transporters in S. meliloti to be identified (Mauchline et al., 2006). These fusions will be of great use for environmental monitoring and provide the biological community with an invaluable resource for understanding the relationships between members of large paralogous families.

6. Uroporphyrinogen III methyltransferase (UMT)

UMT is an important enzyme in the biosynthetic pathway of vitamin B12 and siroheme, in both eukaryotes and prokaryotes (Fan et al., 2006) and it catalyses the S-adenosyl-L-methionine-dependent di-methylation of urogen III to dihydrosirohydrochlorin (precorrin-2). This can be oxidized to the fluorescent molecule sirohydrochlorin, or can be further methylated to trimethylpyrrocorphin, which like sirohydrochlorin, emits a detectable red-orange fluorescence when illuminated with UV light at 300 nm (Roessner and Scott, 1995; Sattler et al., 1995). However, a more recent study suggests excitation at 498 nm, as at this wavelength there is less background interference from endogenous substances (Feliciano et al., 2006). An advantage of using this system in biosensing, is that the enzymatic substrate is ubiquitous (Roessner and Scott, 1995). However, the requirement of uroporphyrinogen for vitamin B-12 and siroheme biosynthesis, means that this sensing system can be limited by substrate availability. Indeed, in one study the addition of 5-aminolevulinic acid, which is the first committed intermediate in heme biosynthesis, resulted in a more reproducible assay (Feliciano et al., 2006).

UMT has been identified in many organisms and has been found to exist in at least two forms in bacteria. The first being encoded by the gene *cysG* that is required for siroheme and thus cysteine synthesis in E. coli (Macdonald and Cole, 1985; Spencer et al., 1993; Warren et al., 1990) as well as siroheme and Vitamin B-12 in *Salmonella typhimurium* (Goldman and Roth, 1993; Jeter et al., 1984). These two enzymes are closely related showing 95% similarity (Sattler et al., 1995). The other form, encoded by *cobA* is responsible for vitamin B-12 synthesis in *Pseudomonas denitrificans* (Blanche et al., 1989) and has also been isolated from other organisms including *Bacillus megaterium* (Robin et al., 1991), *Methanobacterium ivanovii* (Blanche et al., 1991), *Propionibacterium freudenreichii* (Sattler et al., 1995) and *Selenomonas ruminantium* (Anderson et al., 2001). Both forms of the enzyme are able to perform methylation reactions, so producing percorrin-2, but only *cysG* has NAD^+-precorrin-2 oxidase and ferrochelatase activities (Spencer et al., 1993). CysG and CobA can also be distinguished from each other by their physical properties, CobA is 280 amino acids long, whereas CysG is 458 amino acids in length, and as it is only homologous to CobA at the C-terminus, it is believed that the N-terminus is responsible for the extra enzymatic properties of this enzyme (Spencer et al., 1993). The use of biosensors based around this system is in its infancy though it has been reported that the fluorescent intensity of *cobA* fusions is similar to those of *gfp* and that the signal:noise ratio is lower (Wildt and Deuschle, 1999).

7. Ice nucleation proteins

All the reporter genes discussed earlier are detected either by the production of a chromogenic product or by an autofluorescent protein. However, an entirely different mechanism of detection has been developed based on the ability of some Gram-negative plant pathogenic bacteria of the genera *Pseudomonas* (Arny et al., 1976; Orser et al., 1985), *Erwinia* (Lindow et al., 1978; Orser et al., 1985) and *Xanthomonas* (Kim et al., 1987; Orser et al., 1985) to nucleate the crystallization of ice from super-cooled water. These bacteria are abundant on the foliage of many plants and are responsible for initiating much of the frost damage to crops (Deininger et al., 1988). A series of ice nucleation genes have been identified that are responsible for the observed phenotype, and their protein products are localized in the bacterial outer membrane. These include: *inaZ* in *P. syringae* (Turner et al., 1991), *inaW* in *P. fluorescens* (Corotto et al., 1986; Turner et al., 1991) and *iceE* in *E. herbicola* (Guriansherman et al., 1993; Turner et al., 1991). When *inaZ* is introduced into *E. coli* the bacterium is converted from Ina$^-$ (no ice-nucleating activity) to Ina$^+$ (Corotto et al., 1986).

The best studied ice nucleation protein is InaZ, which possesses a central octa-peptide repeated 132 repeats times (Wolber et al., 1986). This internal reiteration is considered to be crucial for ice nucleation and reflects its nonenzymatic function. The protein also has unique N- and C-termini, and is likely to be subject to post-translational modification (Warren and Wolber, 1987).

Populations of Ina$^+$ cells expressing a particular type of ice nucleation gene have a varied ability to nucleate ice formation at different temperatures. Only a small fraction of cells can nucleate at $-4.4\,°C$ or warmer, whereas almost all can nucleate at $-8\,°C$ (Turner et al., 1991). It is believed that cells most efficient for ice nucleation have a larger amount of ice nucleation protein although this relationship is not linear (Southworth et al., 1988). They also possess larger highly stable aggregates of the protein (Govindarajan and Lindow, 1988) and undergo a particular type of posttranslational modification. Ice nucleation structures are classified as A, B or C, with class A structures allowing the most efficient nucleation followed by B and C. The class A structure contains the ice nucleation protein linked to phosphatidylinositol and mannose, probably as a complex mannan and possibly glucosamine, the class B structure is thought to contain the protein and only mannan and glucosamine moieties, and the Class C structure just the protein with a few mannose residues (Turner et al., 1991). These nonprotein components are characteristic of, and similar to those used by eukaryotes to anchor external proteins to cell membranes, and so it is likely that this is also their function in bacteria (Turner *et al.*, 1991).

The development of *ina* as a reporter gene has focused on *inaZ*. The assay to determine ice nucleation activity is known as the droplet freeze assay (Loper and Lindow, 1994, 1996) and has been used mostly to study specific genes in soil and plant environments. Briefly, this involves destructive sampling to separate the bacterial cells from the host environment. Next the cell suspensions or soil slurries are serially diluted and dropped onto paraffin coated aluminium that is floating on an ethanol bath at a temperature of $-7\,°C$ (Jaeger *et al.*, 1999). The number of ice nuclei formed at various time points is recorded and compared to the number of viable cells to ascertain the concentration of inducer in the sample.

III. MOLECULAR BIOSENSORS

A. Optical biosensors

In the previous section we described several reporter proteins, including AFPs, and how their induction can be used in cell-based assays as a biosensor. It is also possible to use fluorescent proteins as purified molecular biosensors that can be used independently of cells. Such reagent-less molecular biosensors, which rely on detection by optical means, have been described and used under a wide range of different conditions (De Lorimier *et al.*, 2002; Deuschle *et al.*, 2006; Fehr *et al.*, 2002, 2003; Lager *et al.*, 2003; Marvin and Hellinga, 2001a,b; Marvin *et al.*, 1997; Miyawaki *et al.*, 1997). Optical biosensors rely on fluorescent indicator proteins (FLIPs), of which there are two types; those which detect ligand binding by measuring the change in environment of a single fluorophore and those which detect changes in fluorescence resonance energy transfer (FRET) between a pair of fluorophores.

1. Single fluorophores

The extensive and elegant work performed by Hellinga and coworkers (Dattelbaum *et al.*, 2005; De Lorimier *et al.*, 2002; Looger *et al.*, 2003; Marvin and Hellinga, 2001a,b; Marvin *et al.*, 1997) involved the development of a suite of FLIP biosensors used for *in vitro* detection of ligands. On the basis of SBPs and their conformational change on binding a ligand, these biosensors use conversion of a carefully selected amino acid into a cysteine residue, usually by site directed mutagenesis (SDM). The covalent thiol-linkage formed between the Cys of a purified genetically modified SBP and a fluorophore dissolved in acetonitrile, produces a fluorophore-labelled protein (Marvin et al., 1997). X-ray crystallography and protein structural data enable the careful selection of the residue(s) converted to Cys, so that a ligand-dependent change in structure usually results in a detectable shift in the fluorescence of a dye sensitive to its microenvironment. Multiple labelling of proteins is possible by using several rounds of modification with reversible thiol protection mechanisms (Smith et al., 2005). As the dye molecules have different properties (e.g., acrylodan is very sensitive to the polarity of the environment (Marvin et al., 1997)), a range of fluorophores (Table 5.2) are often investigated to obtain the optimum signal on substrate binding (De Lorimier et al., 2002; Marvin et al., 1997). These biosensors are most useful for in vitro detection as the chemically labelled proteins are not easily taken up into living cells because of the large and bulky structure of the dyes (see Table 5.2).

Tsien and coworkers have reported a method of site-specific fluorescent labelling of a recombinant protein in living cells (Adams et al., 2002; Griffin et al., 1998). The peptide sequence Cys-Cys-Xaa-Xaa-Cys-Cys (Xaa is any amino acid except cysteine) reacts with a membrane-permeant biarsenical fluorescein dye derivative, FlAsH, to specifically label the protein. The dye is non-fluorescent until it binds with high affinity and specificity to the tetracysteine motif (Griffin et al., 1998). ReAsH is an analogue of FlAsH, derived from resorufin which fluoresces in the red part of the spectrum (Adams et al., 2002). This method permits the opportunity to single out by fluorescent staining, a slightly modified target protein within a live cell by the addition of non-fluorescent dye outside the cell (Griffin et al., 1998). FlAsH and ReAsH work as induction biosensors, where the proteins report gene induction by a ligand, so again this illustrates the overlap between cellular and molecular biosensors.

a. Development of biosensors A biosensor for maltose was developed using maltose binding protein (MBP) of *E. coli* labelled with environmentally sensitive fluorophores (Marvin et al., 1997). In order for the detection of maltose over a wide concentration range, it was necessary to introduce point mutations into the binding site to reduce the affinity of MBP for its ligand (Marvin et al., 1997). A common theme in development of biosensors is that a native SBP will bind its substrate with too high an affinity to be used under physiological conditions.

b. Modifications to change specificity of a biosensor As well as biosensors based on the binding of the natural substrate for a SBP, Hellinga and coworkers have shown it is possible to completely re-model binding sites to bind other molecules of interest. For example MBP was converted to a biosensor for zinc (Marvin and Hellinga, 2001a), or to bind oxygen (Benson et al., 2002). Looger et al. (2003) reported how five different SBPs were re-designed to bind TNT, L-lactate or serotonin with high

TABLE 5.2 Fluorescent dyes

Dye	Binds to	Use	Reference
Acrylodan	Cys	Environmentally sensitive dye, sensitive to polarity	(De Lorimier et al., 2002; Marvin et al., 1997)
IANBD	Cys	Sensitive to quenching by solvent	(Marvin et al., 1997)
CNBD	Cys	Sensitive to quenching by solvent.	(Marvin et al., 1997)
Fluorescein	Cys		(De Lorimier et al., 2002)
Pyrene	Cys		(De Lorimier et al., 2002)
FlAsH	CysCysXaaXaaCysCys (FlAsHTAG)	Specifically labels protein with FlAsHTAG sequence. Fluoresces green	(Griffin et al., 1998)
ReAsH	CysCysXaaXaaCysCys	Fluoresces red.	(Adams et al., 2002)
AsCy3	CysCysLysAlaGluAlaAlaCysCys (Cy3TAG)	Fluoresces red FRET partner with FlAsH	(Cao et al., 2007)

selectivity and affinity. This feat of protein engineering, made possible by an automated design process requiring days of computing, demonstrates that from different starting points (SBPs binding sugars or amino acids), binding sites can be designed for chemically distinct, very different natural and non-natural (TNT) molecules. Although TNT, serotonin, and L-lactate differ in molecular shape, chirality, functional groups, internal flexibility, charge, and water solubility, protein molecules could specifically bind these new substrates with high affinity that were designed and synthesized (each requiring between 5 and 17 amino acid changes) (Looger et al., 2003). Binding to these modified SBPs was detected by a hybrid (named Trz) of the Trg chemotaxis receptor and the EnvZ signal transduction protein (Looger et al., 2003). When solute (e.g., L-lactate or TNT) bound to their redesigned SBP, the protein interacted with Trz to cause autophosphorylation. The phosphate group was transferred to OmpR, which induced the OmpC promoter linked to a LacZ transcriptional fusion. The detection of serotonin and L-lactate is important in clinical chemistry, as elevated levels are indicative of certain medical conditions (Looger et al., 2003). A simple robust detection of TNT would be advantageous on two fronts; detection of unexploded landmines offers a method of improving safety in war zones. In addition, the explosive TNT is also a potent carcinogen so the opportunity to detect TNT plumes present in the sea from unexploded bombs or contaminated soil with fluorescent bacteria would enable decontamination.

The changes described earlier were carried out on the binding site itself, but it has been demonstrated that it is possible to affect the binding affinity by changing residues outside of the binding cleft. This change in substrate affinity is achieved by manipulating the equilibria between the different protein conformations (Marvin and Hellinga, 2001b).

2. Multiple fluorophores

Fluorescent Resonance Energy Transfer (FRET), in which the change in transfer of fluorescent energy between fluorophores is analyzed, requires an overlap between the emission and excitation spectra of a suitable donor/acceptor pair. The Főrster distance (R_0) defines the distance at which transfer is 50% efficient between a pair of fluorophores, and is dependent upon their spectral overlap, the relative orientation of the chromophore transition dipoles and the quantum yield of the donor in the absence of the acceptor (Fehr et al., 2005a). As R_0 is usually between 20 and 60 Å (in the range of protein dimensions), FRET can be used as a "microscopic ruler" (Fehr et al., 2005a) in approximately 1–10 nm (10–100 Å) range (Deuschle et al., 2005b). FRET is used as a highly sensitive indicator of protein conformational change as it is a non-destructive spectroscopic method of optically monitoring the distance apart and relative orientation of the fluorophores (Deuschle et al., 2005b). FRET biosensors are suitable for both *in vitro* and *in vivo* detection of ligands and have been used for real-time monitoring of metabolites in various cells and cellular compartments (Deuschle *et al.*, 2006; Fehr *et al.*, 2002, 2003, 2005b; Gu *et al.*, 2006; Lager *et al.*, 2003; Okumoto *et al.*, 2005).

The first demonstration of FRET used GFP, together with blue fluorescent protein (BFP), as reporters fused at either end of a protease-sensitive peptide. Cleavage of the peptide resulted in the two reporters being irreversibly separated accompanied by loss of FRET (Mitra *et al.*, 1996). Of the GFP variants, BFP is the least bright and most prone to photobleaching (irreversible destruction of the fluorophore on illumination). As BFP must be excited in the ultraviolet range there is an increase in the background noise due to cell autofluorescence and scattering (Zaccolo, 2004). The GFP variants cyan (CFP) and yellow (YFP) are commonly used for FRET as their excitation and emission spectra show the required overlap. Other advantageous properties are that CFP is brighter and less prone to photobleaching than BFP, and variants of YFP show increased photostability and less sensitivity to H^+ and Cl^- ions (Citrine) and increased brightness and speed of maturation (Venus) (Zaccolo, 2004). Fluorescence in the red-shifted part of the spectrum permits greater tissue penetration and minimises background noise from autofluorescence.

A FRET biosensor, developed to detect the reversible binding of Ca^{2+} to chameleon, used a fusion protein composed of calmodulin and the calmodulin-binding peptide M13, linked at the N- and C-termini to two genetically-encoded GFP variants (based on either BFP/GFP or CFP/YFP pairs) (Miyawaki *et al.*, 1997). The observed Ca^{2+}-dependent change in FRET, due to the protein's conformational change on binding the ion, was used to examine the concentration of Ca^{2+} in the cytosol, nucleus and ER of HeLa cells (Miyawaki *et al.*, 1997). Since this initial report, chameleon and its variants have been used to examine Ca^{2+} levels in animal and yeast cells (Nagai *et al.*, 2001) and in plants (Allen *et al.*, 1999; Miwa *et al.*, 2006).

An approach analyzing the FRET between FlAsH and CFP has been demonstrated in determining the interaction between proteins (Hoffmann *et al.*, 2005). As the FlAsH dye molecule is far smaller than any GFP variant (GFP full size protein is 238 amino acids) it is less likely to disrupt the folding of the protein and perturb its interaction with other molecules.

a. Development of SBP-based FRET biosensors Frommer *et al.* have developed a series of FRET biosensors able to detect sugars (Deuschle *et al.*, 2006; Fehr *et al.*, 2002, 2003; Lager *et al.*, 2003), amino acids (Okumoto *et al.*, 2005) and ions (Gu *et al.*, 2006). In this set of nanosensors, the genes encoding two fluorescent proteins are linked at either end to that of an SBP, forming a fusion protein. Although the primary sequences of SBPs with different specificities show little homology, they share high tertiary structure similarity (Fukami-Kobayashi *et al.*, 1999). In the ellipsoidal SBP, the ligand-binding cleft is between two globular domains (Fehr *et al.*, 2003) and on binding a ligand, the substrate-induced Venus flytrap-like hinge twist results in movement of these relative to one another. This conformational change often causes an observable change in FRET when these domains have attached fluorescent moieties. In addition to the genetically encoded FRET biosensors described in detail below, Smith *et al.* (2005) describe how two SBPs (binding maltose and glucose) have been chemically modified and labelled with multiple fluorophores, to form FRET biosensors.

Fehr *et al.* (2002) described a maltose biosensor (FLIPmal) based on *E. coli* MBP. MBP, without its N-terminal signal peptide, was fused between two genes encoding variants of GFP. Enhanced cyan fluorescent protein (ECFP), the donor chromophore, was attached to the N-terminus and enhanced yellow fluorescent protein (EYFP) as an acceptor chromophore at the C-terminus of MBP. Although the biosensor using the whole protein was found to have no observable change in FRET upon ligand binding, a mutant lacking the first five amino acids showed maltose concentration-dependent FRET activity (Fehr *et al.*, 2002). On binding maltose the fluorescent energy was more effectively transferred between these two fluorophores as they are moved closer together in the bound form. FLIPmal is described as a nanosensor with a type II structure (Fukami-Kobayashi *et al.*, 1999) in which the C- and N-termini (and hence the attached fluorophores) are on distal ends of the two lobes relative to the hinge region (Fehr *et al.*, 2002) (Fig. 5.1). In addition to the change in distance between the chromophores, the movement between the "open" and "closed" forms of the SBP includes twisting at the hinge which is thought to affect the relative orientation of the transition dipoles, contributing to the increase in FRET (Fehr *et al.*, 2002). Other examples of nanosensors described as having a type II structure are detectors for glutamate, FLIPE (Okumoto *et al.*, 2005) and for phosphate, FLIPPi (Gu *et al.*, 2006). Rather confusingly and unlike MBP, in these two SBPs the N- and C-termini are both located within the same lobe (Gu *et al.*, 2006; Okumoto *et al.*, 2005).

In contrast, in a type I structure (Fukami-Kobayashi *et al.*, 1999) with the C- and N-termini located on different lobes and proximal relative to the hinge region, on binding substrate the FRET signal decreases as the fluorophores are moved further apart (Fig. 5.1) (Fehr *et al.*, 2003). Examples of nanosensors of type I include those based on the glucose/galactose binding protein, FLIPglu (Fehr *et al.*, 2003) and FLIPglu△13 (Deuschle *et al.*, 2006), on the ribose binding protein, FLIPrib (Lager *et al.*, 2003) and the sucrose nanosensor, FLIPsuc (Lager *et al.*, 2006). Nanosensors based on both types I and II have been used to great effect as outlined below and

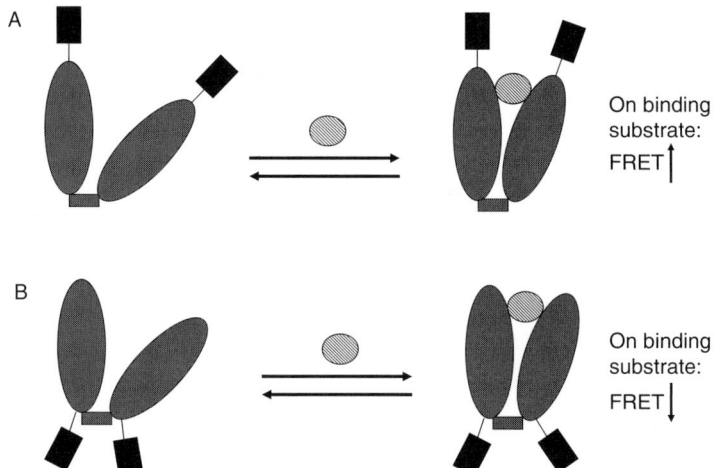

FIGURE 5.1 (A) FLIPmal, a type II nanosensor, with the fluorophores located on different lobes, distal to the hinge region; (B) FLIPglu, type I, with fluorophores on different lobes but proximal to the hinge region. SBP protein is shown in grey, fluorophores in black and substrate is striped.

summarized in Table 5.3. In each case, a series of biosensors has been generated (Table 5.3).

b. Modifications to improve biosensors In designing nanosensors for *in vivo* applications based on ligand-SBP interaction, one consideration is that the affinity of the SBP for substrate is too high for use under physiological conditions. Fehr *et al.* (2002) described how mutants, generated by SDM with lower affinity for maltose, were useful sensors of physiological concentrations of maltose. The various mutants in the FLIPmal series are able to measure maltose over a range of 0.26–2.039 µM (Fehr *et al.*, 2002). Other nanosensors were developed in a similar way using SDM and protein engineering techniques, for example, the FLIPglu series can detect glucose in the range 0.019–5.30 µM (Fehr *et al.*, 2003) and the FLIPrib series detects 0.028 µM–105 mM ribose (Lager *et al.*, 2003) [for a summary see Deuschle *et al.* (2005a)] (see Table 5.3).

In addition to changing the binding affinity, the specificity can be altered to give a more specific, and therefore more useful, nanosensor. This was demonstrated for the glucose nanosensor; FLIPglu-170n was found to bind other molecules (xylose, fructose, lactose and sorbitol) with an affinity which would interfere with glucose detection. However, another member of the FLIPglu series, FLIPglu-600 µ, has increased selectivity for glucose (as well as decreased glucose binding affinity) and was successfully used to monitor glucose, *in vivo*, in both yeast and

TABLE 5.3 SBP-based FRET biosensors using autofluorescent proteins

Nanosensor series	Observed FRET	Concentration range covered by series	Attached fluorophores	SBP type	Position of GFPs	Demonstrated use *in vivo*	Reference
FLIPmal	Maltose-dependent increase	0.26–2.039 μM	eCFP/eYFP	Type II	Different lobes, distal ends relative to the hinge region.	Cytosol of single yeast cells *In vitro*: beer	(Fehr et al., 2002)
FLIPglu	Glucose-dependent decrease	0.019–5,301μM	eCFP/eYFP	Type I	Different lobes, proximal ends relative to the hinge region.	Cytosol of mammalian COS-7 cells	(Fehr et al., 2003)
FLIPrib	Ribose-dependent decrease	0.028 μM–105 mM	eCFP/eYFP	Type I	Different lobes, proximal ends relative to the hinge region.	Cytosol of mammalian COS-7 cells	(Lager et al., 2003)
FLIPE	Glutamate-dependent decrease	10 μM–10 mM	eCFP/Venus	Type II	Same lobe	Cytosol and cell surface of rat hippocampal neurons and PC12 cells *In vitro*: soy sauce, calf serum	(Okumoto et al., 2005)

(*continued*)

TABLE 5.3 (continued)

Nanosensor series	Observed FRET	Concentration range covered by series	Attached fluorophores	SBP type	Position of GFPs	Demonstrated use *in vivo*	Reference
FLIPsuc	Sucrose-dependent decrease[a]	low nm to high mM	eCFP/eYFP	Type I			(Lager et al., 2006)
FLIPPi	P$_i$-dependent decrease	25 nM–170 mM	eCFP/Venus	Type II	Same lobe	Cytosol of mammalian COS-7 and CHO cells	(Gu et al., 2006)
FLIPglu△13	Glucose-dependent decrease	low nm– high mM	eCFP/eYFP or Ares/Aphrodite	Type I	Different lobes, proximal ends relative to the hinge region	*Arabidopsis* root and leaf cells	(Deuschle et al., 2006)

[a] Possibly unexpected, given the high homology with MBP (Lager et al., 2006).

mammalian cells (Fehr *et al.*, 2003). Rational design of the binding site was carried out to a much greater extent in the development of FLIPsuc, making the nanosensor specific for sucrose by reducing sufficiently the binding of other sugar molecules (Lager *et al.*, 2006). In addition to the techniques used to change the affinity for a ligand, careful selection of the initial SBP is important in developing a successful biosensor, as demonstrated for FLIPsuc and FLIPPi series (Gu *et al.*, 2006; Lager *et al.*, 2006). A wealth of naturally occurring bacterial ABC transporter systems are found in organisms growing in an extremely wide range of conditions, and able to bind and transport a plethora of different molecules in a highly specific way. Characterization of such systems is a natural springboard for development of FRET nanosensors with a wide range of specificities, useful in numerous situations.

The successful use of the nanosensors has demonstrated that they are able to produce a good signal:noise ration in animal, plant and yeast cells (Lager *et al.*, 2006). Improvements to this ratio have mainly been achieved by changing the length of the linkers connecting the parts of the fusion protein and thus reorienting the fluorophore dipole (Looger *et al.*, 2005). For example, in FLIPsuc series an improved FRET signal was obtained when the linkers between GFP moieties and the SBP were shortened (Lager *et al.*, 2006) and the FLIPgluΔ13 series improved sensitivity over that of FLIPglu, was achieved by linker truncation (Deuschle *et al.*, 2006). In FLIPPi, as well as reducing the linkers between the parts of the fusion proteins, the C-terminus of Venus and N-terminus of eCFP were reduced in length in order to improve the FRET signal (Gu *et al.*, 2006).

Other ways to improve a nanosensor is by changing the GFP variant used, for example, EYFP variant Venus has reduced pH and chloride sensitivity (Gu *et al.*, 2006) and is therefore more robust under certain conditions. It is also possible to introduce the GFP moiety at different positions in the SBP using rational design, rather than using the N- and C-termini, shown in FLIPglu and FLIPE (Deuschle *et al.*, 2005b).

B. Electrical biosensors

A method to detect binding of a ligand using an electrical biosensor was first described by Benson *et al.* (2001). This method exploits the ligand-induced hinge-bending motion of SBPs, linking the binding to allosterically controlled interactions between a redox-active Ruthenium-labelled protein and an electrode surface. MBP from *E. coli* was tethered via a COOH-terminus His-tag to a gold electrode and a Ru(II) reporter group was introduced site-specifically onto a mutant Cys (introduced by SDM) in a position facing the surface of the electrode. Upon binding maltose, the change in structure of MBP caused the Ru(II) redox group to move away from the electrode, resulting in the current through the electrode decreasing in a maltose concentration-dependent fashion (Benson et al., 2001). In addition, these authors demonstrated that with electrodes using mutant MBPs with a lower affinity for maltose, the observed maltose affinity varied with the solution binding constants of the mutant proteins (Benson et al., 2001). Using this SDM, Benson et al. (2001) showed it was possible to completely remodel MBP to bind Zn(II) and produce an electrical biosensor which responded to Zn(II) rather that to maltose.

To illustrate the robust practical nature of the electrical biosensors, electrodes that responded to glucose and glutamine, based on glucose- and glutamine binding proteins respectively, were shown to be able to measure the concentration of their ligand in complex mixtures of serum (with large protein components) and beer (containing a mixture of competing small molecules) (Benson et al., 2001).

IV. CONCLUSIONS AND FUTURE PROSPECTS

Biosensors have become an essential tool in biological research because they are able to report the presence of ligands at the nanoscale, hence the frequent reference to them as nanosensors. Their role in some systems for detection of cell signalling, such as the use of chameleon for detection of calcium, has been spectacular. In turn, they have a myriad of applications in applied science and technology, such as detection of medically important compounds and environmental pollutants. The route from demonstration of ligand binding to a practical biosensor is not always a simple one, requiring changes in ligand specificity and binding affinity. However, a rational computational design and SDM have resulted in huge progress in this field.

ACKNOWLEDGMENTS

Work in the Poole Laboratory was supported by the BBSRC.

REFERENCES

Adams, S. R., Campbell, R. E., Gross, L. A., Martin, B. R., Walkup, G. K., Yao, Y., Llopis, J., and Tsien, R. Y. (2002). New biarsenical ligands and tetracysteine motifs for protein labeling *in vitro* and *in vivo*: Synthesis and biological applications. *J. Am. Chem. Soc.* **124,** 6063–6076.

Allen, G. J., Kwak, J. M., Chu, S. P., Llopis, J., Tsien, R. Y., Harper, J. F., and Schroeder, J. I. (1999). Cameleon calcium indicator reports cytoplasmic calcium dynamics in *Arabidopsis* guard cells. *Plant. J.* **19,** 735–747.

Andersen, J. B., Sternberg, C., Poulsen, L. K., Bjorn, S. P., Givskov, M., and Molin, S. (1998). New unstable variants of green fluorescent protein for studies of transient gene expression in bacteria. *Appl. Environ. Microbiol.* **64,** 2240–2246.

Anderson, P. J., Entsch, B., and McKay, D. B. (2001). A gene, *cobA* plus *hemD*, from *Selenomonas ruminantium* encodes a bifunctional enzyme involved in the synthesis of vitamin B-12. *Gene* **281,** 63–70.

Arny, D. C., Lindow, S. E., and Upper, C. D. (1976). Frost sensitivity of *Zea mays* increased by application of *Pseudomonas syringae*. *Nature* **262,** 282–284.

Ashby, M. C., Ibaraki, K., and Henley, J. M. (2004). It's green outside: Tracking cell surface proteins with pH-sensitive GFP. *Trends Neurosci.* **27,** 257–261.

Baldwin, T. O., Christopher, J. A., Raushel, F. M., Sinclair, J. F., Ziegler, M. M., Fisher, A. J., and Rayment, I. (1995). Structure of bacterial luciferase. *Curr. Opin. Struct. Biol.* **5,** 798–809.

Barolo, S., Castro, B., and Posakony, J. W. (2004). New *Drosophila* transgenic reporters: Insulated P-element vectors expressing fast-maturing RFP. *Biotechniques* **36,** 436–442.

Baronian, K. H. R. (2004). The use of yeast and moulds as sensing elements in biosensors. *Biosens. Bioelectron.* **19,** 953–962.

Benson, D. E., Conrad, D. W., de Lorimer, R. M., Trammell, S. A., and Hellinga, H. W. (2001). Design of bioelectronic interfaces by exploiting hinge-bending motions in proteins. *Science* **293,** 1641–1644.

Benson, D. E., Haddy, A. E., and Hellinga, H. W. (2002). Converting a maltose receptor into a nascent binuclear copper oxygenase by computational design. *Biochemistry* **41,** 3262–3269.

Bevis, B. J., and Glick, B. S. (2002). Rapidly maturing variants of the *Discosoma* red fluorescent protein (DsRed). *Nat. Biotechnol.* **20,** 83–87.

Billard, P., and DuBow, M. S. (1998). Bioluminescence-based assays for detection and characterization of bacteria and chemicals in clinical laboratories. *Clin. Biochem.* **31,** 1–14.

Blanche, F., Debussche, L., Thibaut, D., Crouzet, J., and Cameron, B. (1989). Purification and characterization of S-adenosyl-L-methionine-uroporphyrinogen III methyltransferase from *Pseudomonas denitrificans*. *J. Bacteriol.* **171,** 4222–4231.

Blanche, F., Robin, C., Couder, M., Faucher, D., Cauchois, L., Cameron, B., and Crouzet, J. (1991). Purification, characterization, and molecular cloning of S-adenosyl-L-methionine: Uroporphyrinogen III Methyltransferase from *Methanobacterium ivanovii*. *J. Bacteriol.* **173,** 4637–4645.

Branchini, B. R., Magyar, R. A., Murtiashaw, M. H., Anderson, S. M., Helgerson, L. C., and Zimmer, M. (1999). Site-directed mutagenesis of firefly luciferase active site amino acids: A proposed model for bioluminescence color. *Biochemistry.* **38,** 13223–13230.

Bringhurst, R. M., Cardon, Z. G., and Gage, D. J. (2001). Galactosides in the rhizosphere: Utilization by *Sinorhizobium meliloti* and development of a biosensor. *Proc. Natl. Acad. Sci. USA.* **98,** 4540–4545.

Camilli, A. (1996). Noninvasive techniques for studying pathogenic bacteria in the whole animal. *Trends Microbiol.* **4,** 295–296.

Campbell, R. E., Tour, O., Palmer, A. E., Steinbach, P. A., Baird, G. S., Zacharias, D. A., and Tsien, R. Y. (2002). A monomeric red fluorescent protein. *Proc. Natl. Acad. Sci. USA* **99,** 7877–7882.

Cao, H., Xiong, Y., Wang, T., Chen, B., Squier, T. C., and Mayer, M. U. (2007). A red Cy3-based biarsenical fluorescent probe targeted to a complementary binding peptide. *J. Am. Chem. Soc.* **129,** 8672–8673.

Chalfie, M., Tu, Y., Euskirchen, G., Ward, W. W., and Prasher, D. C. (1994). Green fluorescent protein as a marker for gene expression. *Science* **263,** 802–805.

Conti, E., Franks, N. P., and Brick, P. (1996). Crystal structure of firefly luciferase throws light on a superfamily of adenylate-forming enzymes. *Structure* **4,** 287–298.

Cormack, B. P., Valdivia, R. H., and Falkow, S. (1996). FACS-optimized mutants of the green fluorescent protein (GFP). *Gene* **173,** 33–38.

Corotto, L. V., Wolber, P. K., and Warren, G. J. (1986). Ice nucleation activity of *Pseudomonas fluorescens*: Mutagenesis, complementation analysis and identification of a gene product. *EMBO J.* **5,** 231–236.

Crameri, A., Whitehorn, E. A., Tate, E., and Stemmer, W. P. C. (1996). Improved green fluorescent protein by molecular evolution using DNA shuffling. *Nat. Biotechnol.* **14,** 315–319.

Cubitt, A. B., Heim, R., Adams, S. R., Boyd, A. E., Gross, L. A., and Tsien, R. Y. (1995). Understanding, improving and using green fluorescent proteins. *Trends Biochem. Sci.* **20,** 448–455.

Darwent, M. J., Paterson, E., McDonald, A. J. S., and Tomos, A. D. (2003). Biosensor reporting of root exudation from *Hordeum vulgare* in relation to shoot nitrate concentration. *J. Exp. Bot.* **54,** 325–334.

Dattelbaum, J. D., Looger, L. L., Benson, D. E., Sali, K. M., Thompson, R. B., and Hellinga, H. W. (2005). Analysis of allosteric signal transduction mechanisms in an engineered fluorescent maltose biosensor. *Protein. Sci.* **14,** 284–291.

Daunert, S., Barrett, G., Feliciano, J. S., Shetty, R. S., Shrestha, S., and Smith-Spencer, W. (2000). Genetically engineered whole-cell sensing systems: coupling biological recognition with reporter genes. *Chem. Rev.* **100,** 2705–2738.

De Lorimier, R. M., Smith, J. J., Dwyer, M. A., Looger, L. L., Sali, K. M., Paavola, C. D., Rizk, S. S., Sadigov, S., Conrad, D. W., Loew, L., and Hellinga, H. W. (2002). Construction of a fluorescent biosensor family. *Protein. Sci.* **11,** 2655–2675.

Deininger, C. A., Mueller, G. M., and Wolber, P. K. (1988). Immunological characterization of ice nucleation proteins from *Pseudomonas syringae*, *Pseudomonas fluorescens*, and *Erwinia herbicola*. *J. Bacteriol.* **170,** 669–675.

Deuschle, K., Chaudhuri, B., Okumoto, S., Lager, I., Lalonde, S., and Frommer, W. B. (2006). Rapid metabolism of glucose detected with FRET glucose nanosensors in epidermal cells and intact roots of Arabidopsis RNA-silencing mutants. *Plant. Cell.* **18,** 2314–2325.

Deuschle, K., Fehr, M., Hilpert, M., Lager, I., Lalonde, S., Looger, L. L., Okumoto, S., Persson, J., Schmidt, A., and Frommer, W. B. (2005a). Genetically encoded sensors for metabolites. *Cytometry* **64,** 3–9.

Deuschle, K., Okumoto, S., Fehr, M., Looger, L. L., Kozhukh, L., and Frommer, W. B. (2005b). Construction and optimization of a family of genetically encoded metabolite sensors by semirational protein engineering. *Protein. Sci.* **14,** 2304–2314.

Fan, J., Wang, D. Q., Liang, Z., Guo, M., Teng, M. K., and Niu, L. W. (2006). Maize uroporphyrinogen III methyltransferase: Overexpression of the functional gene fragments in *Escherichia coli* and one-step purification. *Protein Expr. Purif.* **46,** 40–46.

Fehr, M., Frommer, W. B., and Lalonde, S. (2002). Visualization of maltose uptake in living yeast cells by fluorescent nanosensors. *Proc. Natl. Acad. Sci. USA.* **99,** 9846–9851.

Fehr, M., Lalonde, S., Lager, I., Wolff, M. W., and Frommer, W. B. (2003). *In vivo* imaging of the dynamics of glucose uptake in the cytosol of COS-7 cells by fluorescent nanosensors. *J. Biol. Chem.* **278,** 19127–19133.

Fehr, M., Okumoto, S., Deuschle, K., Lager, I., Looger, L. L., Persson, J., Kozhukh, L., Lalonde, S., and Frommer, W. B. (2005a). Development and use of fluorescent nanosensors for metabolite imaging in living cells. *Biochem. Soc. Trans.* **33,** 287–290.

Fehr, M., Takanaga, H., Ehrhardt, D. W., and Frommer, W. B. (2005b). Evidence for high-capacity bidirectional glucose transport across the endoplasmic reticulum membrane by genetically encoded fluorescence resonance energy transfer nanosensors. *Mol. Cell. Biol.* **25,** 11102–11112.

Feliciano, J., Liu, Y., and Daunert, S. (2006). Novel reporter gene in a fluorescent-based whole cell sensing system. *Biotech. Bioeng.* **93,** 989–997.

Fisher, A. J., Raushel, F. M., Baldwin, T. O., and Rayment, I. (1995). Three-dimensional structure of bacterial luciferase from *Vibrio harveyi* at 2.4 Å resolution. *Biochemistry* **34,** 6581–6586.

Fisher, A. J., Thompson, T. B., Thoden, J. B., Baldwin, T. O., and Rayment, I. (1996). The 1.5-Å resolution crystal structure of bacterial luciferase in low salt conditions. *J. Biol. Chem.* **271**, 21956–21968.

Fukami-Kobayashi, K., Tateno, Y., and Nishikawa, K. (1999). Domain dislocation: A change of core structure in periplasmic binding proteins in their evolutionary history. *J. Mol. Biol.* **286**, 279–290.

Gage, D. J. (2002). Analysis of infection thread development using Gfp- and DsRed-expressing *Sinorhizobium meliloti*. *J. Bacteriol.* **184**, 7042–7046.

Gage, D. J., and Long, S. R. (1998). Alpha-galactoside uptake in *Rhizobium meliloti*: Isolation and characterization of *agpA*, a gene encoding a periplasmic binding protein required for melibiose and raffinose utilization. *J. Bacteriol.* **180**, 5739–5748.

Gage, D. J., Bobo, T., and Long, S. R. (1996). Use of green fluorescent protein to visualize the early events of symbiosis between *Rhizobium meliloti* and alfalfa (*Medicago sativa*). *J. Bacteriol.* **178**, 7159–7166.

Gallagher, S. R. (1992). "GUS Protocols. Using the GUS Gene as a Reporter of Gene Expression." Academic Press, San Diego, CA.

Goldman, B. S., and Roth, J. R. (1993). Genetic structure and regulation of the *cysG* gene in *Salmonella typhimurium*. *J. Bacteriol.* **175**, 1457–1466.

Gonzalezflecha, B., and Demple, B. (1994). Intracellular generation of superoxide as a by-product of *Vibrio harveyi* luciferase expressed in *Escherichia coli*. *J. Bacteriol.* **176**, 2293–2299.

Govindarajan, A. G., and Lindow, S. E. (1988). Size of bacterial ice nucleation sites measured *in situ* by radiation inactivation analysis. *Proc. Natl. Acad. Sci. USA* **85**, 1334–1338.

Griffin, B. A., Adams, S. R., and Tsien, R. Y. (1998). Specific covalent labeling of recombinant protein molecules inside live cells. *Science* **281**, 269–272.

Gross, L. A., Baird, G. S., Hoffman, R. C., Baldridge, K. K., and Tsien, R. Y. (2000). The structure of the chromophore within DsRed, a red fluorescent protein from coral. *Proc. Natl. Acad. Sci. USA* **97**, 11990–11995.

Gu, H., Lalonde, S., Okumoto, S., Looger, L. L., Scharff-Poulsen, A. M., Grossman, A. R., Kossmann, J., Jakobsen, I., and Frommer, W. B. (2006). A novel analytical method for *in vivo* phosphate tracking. *FEBS Lett.* **580**, 5885–5893.

Guriansherman, D., Lindow, S. E., and Panopoulos, N. J. (1993). Isolation and characterization of hydroxylamine-induced mutations in the *Erwinia Herbicola* ice nucleation gene that selectively reduce warm temperature ice nucleation activity. *Mol. Microbiol.* **9**, 383–391.

Hakkila, K., Maksimow, M., Karp, M., and Virta, M. (2002). Reporter genes *lucFF*, *luxCDABE*, *gfp*, and *dsred* have different characteristics in whole-cell bacterial sensors. *Anal. Biochem.* **301**, 235–242.

Hautefort, I., and Hinton, J. C. D. (2000). Measurement of bacterial gene expression *in vivo*. *Philos. Trans. R. Soc. Lond. B Biol. Sci.* **355**, 601–611.

Heim, R., Prasher, D. C., and Tsien, R. Y. (1994). Wavelength mutations and post-translational autoxidation of green fluorescent protein. *Proc. Natl. Acad. Sci. USA* **91**, 12501–12504.

Hoffmann, C., Gaietta, G., Bünemann, M., Adams, S. R., Oberdorff-Maass, S., Behr, B., Vilardaga, J. P., Tsien, R. Y., Ellisman, M. H., and Lohse, M. J. (2005). A FlAsH-based FRET approach to determine G protein-coupled receptor activation in living cells. *Nat. Methods* **2**, 171–176.

Huber, R. E., Gupta, M. N., and Khare, S. K. (1994). The active site and mechanism of the beta-galactosidase from *Escherichia coli*. *Int. J. Biochem.* **26**, 309–318.

Inouye, S., and Tsuji, F. I. (1994). *Aequorea* green fluorescent protein. Expression of the gene and fluorescence characteristics of the recombinant protein. *FEBS Lett.* **341**, 277–280.

Jacobson, R. H., and Matthews, B. W. (1992). Crystallization of beta-galactosidase from *Escherichia coli*. *J. Mol. Biol.* **223**, 1177–1182.

Jaeger, C. H., Lindow, S. E., Miller, S., Clark, E., and Firestone, M. K. (1999). Mapping of sugar and amino acid availability in soil around roots with bacterial sensors of sucrose and tryptophan. *Appl. Environ. Microbiol.* **65**, 2685–2690.

Jain, S., Drendel, W. B., Chen, Z. W., Mathews, F. S., Sly, W. S., and Grubb, J. H. (1996). Structure of human beta-glucuronidase reveals candidate lysosomal targeting and active-site motifs. *Nat. Struct. Biol.* **3**, 375–381.

Jain, V. K., and Magrath, I. T. (1991). A chemiluminescent assay for quantitation of beta-galactosidase in the femtogram range: Application to quantitation of beta-galactosidase in *lacZ*-transfected Cells. *Anal. Biochem.* **199**, 119–124.

Jefferson, R. A., Burgess, S. M., and Hirsh, D. (1986). Beta-glucuronidase from *Escherichia coli* as a gene-fusion marker. *Proc. Natl. Acad. Sci. USA* **83**, 8447–8451.

Jeter, R. M., Olivera, B. M., and Roth, J. R. (1984). *Salmonella typhimurium* synthesizes cobalamin (Vitamin-B12) *de novo* under anaerobic growth conditions. *J. Bacteriol.* **159**, 206–213.

Karasawa, S., Araki, T., Nagai, T., Mizuno, H., and Miyawaki, A. (2004). Cyan-emitting and orange-emitting fluorescent proteins as a donor/acceptor pair for fluorescence resonance energy transfer. *Biochem. J.* **381**, 307–312.

Karunakaran, R., Mauchline, T. H., Hosie, A. H. F., and Poole, P. S. (2005). A family of promoter probe vectors incorporating autofluorescent and chromogenic reporter proteins for studying gene expression in Gram-negative bacteria. *Microbiol.* **151**, 3249–3256.

Kim, H. K., Orser, C., Lindow, S. E., and Sands, D. C. (1987). *Xanthomonas campestris* pv. translucens strains active in ice nucleation. *Plant Dis.* **71**, 994–997.

Kohlmeier, S., Mancuso, M., Tecon, R., Harms, H., van der Meer, J. R., and Wells, M. (2007). Bioreporters: *gfp* versus *lux* revisited and single-cell response. *Biosens. Bioelectron.* **22**, 1578–1585.

Lager, I., Fehr, M., Frommer, W. B., and Lalonde, S. (2003). Development of a fluorescent nanosensor for ribose. *FEBS Lett.* **553**, 85–89.

Lager, I., Looger, L. L., Hilpert, M., Lalonde, S., and Frommer, W. B. (2006). Conversion of a putative Agrobacterium sugar-binding protein into a FRET sensor with high selectivity for sucrose. *J. Biol. Chem.* **281**, 30875–30883.

Larrainzar, E., O'Gara, F., and Morrissey, J. P. (2005). Applications of autofluorescent proteins for *in situ* studies in microbial ecology. *Ann. Rev. Microbiol.* **59**, 257–277.

Lefevre, C. K., Singer, V. L., Kang, H. C., and Haugland, R. P. (1995). Quantitative non-radioactive CAT assays using fluorescent BODIPY(R) 1-deoxychloramphenicol substrates. *Biotechniques* **19**, 488–493.

Lindow, S. E., Arny, D. C., and Upper, C. D. (1978). *Erwinia herbicola*—Bacterial ice nucleus active in increasing frost injury to corn. *Phytopathology* **68**, 523–527.

Looger, L. L., Dwyer, M. A., Smith, J. J., and Hellinga, H. W. (2003). Computational design of receptor and sensor proteins with novel functions. *Nature* **423**, 185–190.

Looger, L. L., Lalonde, S., and Frommer, W. B. (2005). Genetically encoded FRET sensors for visualizing metabolites with subcellular resolution in living cells. *Plant Physiol.* **138**, 555–557.

Loper, J. E., and Lindow, S. E. (1994). A biological sensor for iron available to bacteria in their habitats on plant surfaces. *Appl. Environ. Microbiol.* **60**, 1934–1941.

Loper, J. E., and Lindow, S. E. (1996). "Reporter Gene Systems Useful in Evaluating *in situ* Gene Expression by Soil- and Plant-associated Bacteria." American Society for Microbiology, Washington, DC.

Macdonald, H., and Cole, J. (1985). Molecular cloning and functional analysis of the *cysG* and *nirB* genes of *Escherichia coli* K12, 2 closely linked genes required for NADH-dependent nitrite reductase activity. *Mol. Gen. Genet.* **200**, 328–334.

Marvin, J. S., Corcoran, E. E., Hattangadi, N. A., Zhang, J. V., Gere, S. A., and Hellinga, H. W. (1997). The rational design of allosteric interactions in a monomeric protein and its applications to the construction of biosensors. *Proc. Natl. Acad. Sci. USA* **94**, 4366–4371.

Marvin, J. S., and Hellinga, H. W. (2001a). Conversion of a maltose receptor into a zinc biosensor by computational design. *Proc. Natl. Acad. Sci. USA* **98,** 4955–4960.

Marvin, J. S., and Hellinga, H. W. (2001b). Manipulation of ligand binding affinity by exploitation of conformational coupling. *Nat. Struct. Mol. Biol.* **8,** 795–798.

Mauchline, T. H., Fowler, J. E., East, A. K., Sartor, A. L., Zaheer, R., Hosie, A. H. F., Poole, P. S., and Finan, T. M. (2006). Mapping the *Sinorhizobium meliloti* 1021 solute-binding protein-dependent transportome. *Proc. Natl. Acad. Sci. USA* **103,** 17933–17938.

McElroy, W. D., and Seliger, H. H. (1962). "Origin and Evolution of Bioluminescence." Academic, New York.

Meighen, E. A. (1993). Bacterial bioluminescence: Organization, regulation, and application of the *lux* genes. *FASEB J.* **7,** 1016–1022.

Miller, W. G., Brandl, M. T., Quinones, B., and Lindow, S. E. (2001). Biological sensor for sucrose availability: Relative sensitivities of various reporter genes. *Appl. Environ. Microbiol.* **67,** 1308–1317.

Mirabella, R., Franken, C., van der Krogt, G. N. M., Bisseling, T., and Geurts, R. (2004). Use of the fluorescent timer DsRED-E5 as reporter to monitor dynamics of gene activity in plants. *Plant Physiol.* **135,** 1879–1887.

Mitra, R. D., Silva, C. M., and Youvan, D. C. (1996). Fluorescence resonance energy transfer between blue-emitting and red-shifted excitation derivatives of the green fluorescent protein. *Gene* **173,** 13–17.

Miwa, H., Sun, J., Oldroyd, G. E. D., and Downie, J. A. (2006). Analysis of calcium spiking using a cameleon calcium sensor reveals that nodulation gene expression is regulated by calcium spike number and the developmental status of the cell. *Plant J.* **48,** 883–894.

Miyawaki, A., Llopis, J., Heim, R., McCaffery, J. M., Adams, J. A., Ikura, M., and Tsien, R. Y. (1997). Fluorescent indicators for Ca^{2+} based on green fluorescent proteins and calmodulin. *Nature* **388,** 882–887.

Nagai, T., Sawano, A., Park, E. S., and Miyawaki, A. (2001). Circularly permuted green fluorescent proteins engineered to sense Ca^{2+}. *Proc. Natl. Acad. Sci. USA* **98,** 3197–3202.

Naylor, L. H. (1999). Reporter gene technology: The future looks bright. *Biochem. Pharmacol.* **58,** 749–757.

Okumoto, S., Looger, L. L., Micheva, K. D., Reimer, R. J., Smith, S. J., and Frommer, W. B. (2005). Detection of glutamate release from neurons by genetically encoded surface-displayed FRET nanosensors. *Proc. Natl. Acad. Sci. USA* **102,** 8740–8745.

Orser, C., Staskawicz, B. J., Panopoulos, N. J., Dahlbeck, D., and Lindow, S. E. (1985). Cloning and expression of bacterial ice nucleation genes in *Escherichia coli*. *J. Bacteriol.* **164,** 359–366.

Robin, C., Blanche, F., Cauchois, L., Cameron, B., Couder, M., and Crouzet, J. (1991). Primary structure, expression in *Escherichia coli*, and properties of S-adenosyl-L-methionine: Uroporphyrinogen-III methyltransferase from *Bacillus megaterium*. *J. Bacteriol.* **173,** 4893–4896.

Roessner, C. A., and Scott, A. I. (1995). Fluorescence-based method for selection of recombinant plasmids. *Biotechniques* **19,** 760–764.

Sattler, I., Roessner, C. A., Stolowich, N. J., Hardin, S. H., Harrishaller, L. W., Yokubaitis, N. T., Murooka, Y., Hashimoto, Y., and Scott, I. (1995). Cloning, sequencing, and expression of the uroporphyrinogen III methyltransferase *cobA* gene of *Propionibacterium freudenreichii* (*shermanii*). *J. Bacteriol.* **177,** 1564–1569.

Scholz, Q., Thiel, A., Hillen, W., and Niederweis, M. (2000). Quantitative analysis of gene expression with an improved green fluorescent protein. *Eur. J. Biochem.* **267,** 1565–1570.

Shaner, N. C., Campbell, R. E., Steinbach, P. A., Giepmans, B. N. G., Palmer, A. E., and Tsien, R. Y. (2004). Improved monomeric red, orange and yellow fluorescent proteins derived from *Discosoma* sp. red fluorescent protein. *Nat. Biotechnol.* **22,** 1567–1572.

Shaner, N. C., Steinbach, P. A., and Tsien, R. Y. (2005). A guide to choosing fluorescent proteins. *Nat. Methods* **2,** 905–909.

Shaw, W. V. (1983). Chloramphenicol acetyltransferase: Enzymology and molecular biology. *CRC Crit. Rev. Biochem.* **14**, 1–46.

Shaw, W. V., and Leslie, A. G. W. (1991). Chloramphenicol acetyltransferase. *Annu. Rev. Biophys. Biophys. Chem.* **20**, 363–386.

Shimomura, O., Johnson, F. H., and Saiga, Y. (1962). Extraction, purification and properties of aequorin, a bioluminescent protein from the luminous hydromedusan, Aequorea. *J. Cell Comp. Physiol.* **59**, 223–239.

Smith, J. J., Conrad, D. W., Cuneo, M. J., and Hellinga, H. W. (2005). Orthogonal site-specific protein modification by engineering reversible thiol protection mechanisms. *Protein Sci.* **14**, 64–73.

Southworth, M. W., Wolber, P. K., and Warren, G. J. (1988). Nonlinear relationship between concentration and activity of a bacterial ice nucleation protein. *J. Biol. Chem.* **263**, 15211–15216.

Spencer, J. B., Stolowich, N. J., Roessner, C. A., and Scott, A. I. (1993). The *Escherichia coli cysG* gene encodes the multifunctional protein, siroheme synthase. *FEBS Lett.* **335**, 57–60.

Srikantha, T., Klapach, A., Lorenz, W. W., Tsai, L. K., Laughlin, L. A., Gorman, J. A., and Soll, D. R. (1996). The sea pansy *Renilla reniformis* luciferase serves as a sensitive bioluminescent reporter for differential gene expression in *Candida albicans*. *J. Bacteriol.* **178**, 121–129.

Terskikh, A., Fradkov, A., Ermakova, G., Zaraisky, A., Tan, P., Kajava, A. V., Zhao, X., Lukyanov, S., Matz, M., Kim, S., Weissman, I., and Siebert, P. (2000). Fluorescent timer: Protein that changes color with time. *Science* **290**, 1585–1588.

Tsien, R. Y. (1998). The green fluorescent protein. *Ann. Rev. Biochem.* **67**, 509–544.

Turner, M. A., Arellano, F., and Kozloff, L. M. (1991). Components of ice nucleation structures of bacteria. *J. Bacteriol.* **173**, 6515–6527.

Wang, L., Jackson, W. C., Steinbach, P. A., and Tsien, R. Y. (2004). Evolution of new nonantibody proteins via iterative somatic hypermutation. *Proc. Natl. Acad. Sci. USA* **101**, 16745–16749.

Warren, G. J., and Wolber, P. K. (1987). Heterogeneous ice nucleation by bacteria. *Cryo-Letters* **8**, 204–217.

Warren, M. J., Roessner, C. A., Santander, P. J., and Scott, A. I. (1990). The *Escherichia coli cysG* gene encodes S-adenosylmethionine-dependent Uroporphyrinogen III methylase. *Biochem. J.* **265**, 725–729.

Wildt, S., and Deuschle, U. (1999). cobA, a red fluorescent transcriptional reporter for *Escherichia coli*, yeast, and mammalian cells. *Nat. Biotechnol.* **17**, 1175–1178.

Wiles, S., Ferguson, K., Stefanidou, M., Young, D. B., and Robertson, B. D. (2005). Alternative luciferase for monitoring bacterial cells under adverse conditions. *Appl. Environ. Microbiol.* **71**, 3427–3432.

Wilson, T., and Hastings, J. W. (1998). Bioluminescence. *Annu. Rev. Cell Dev. Biol.* **14**, 197–230.

Wolber, P. K., Deininger, C. A., Southworth, M. W., Vandekerckhove, J., Vanmontagu, M., and Warren, G. J. (1986). Identification and purification of a bacterial ice-nucleation protein. *Proc. Natl. Acad. Sci. USA* **83**, 7256–7260.

Yang, F., Moss, L. G., and Phillips, G. N. (1996). The molecular structure of green fluorescent protein. *Nat. Biotechnol.* **14**, 1246–1251.

Zaccolo, M. (2004). Use of chimeric fluorescent proteins and fluorescence resonance energy transfer to monitor cellular responses. *Circ. Res.* **94**, 866–873.

Zacharias, D. A., Violin, J. D., Newton, A. C., and Tsien, R. Y. (2002). Partitioning of lipid-modified monomeric GFPs into membrane microdomains of live cells. *Science* **296**, 913–916.

Zapata-Hommer, O., and Griesbeck, O. (2003). Efficiently folding and circularly permuted variants of the Sapphire mutant of GFP. *BMC Biotechnol.* **3**.

CHAPTER 6

Islands Shaping Thought in Microbial Ecology

Christopher J. van der Gast[*,1]

Contents		
	I. Introduction	167
	II. The Importance of Islands	169
	A. Biogeographic islands for studies of bacterial diversity	169
	B. Island biogeography	170
	III. Species–Area Relationships	170
	A. Microbial biogeography	171
	B. Island size and bacterial diversity	172
	IV. Beta Diversity	174
	A. Smaller islands are less stable than larger ones	175
	B. Species–time relationships	175
	C. Distance–decay relationships	177
	V. Opposing Perspectives on Community Assembly	177
	VI. Conclusions	179
	Acknowledgments	180
	References	180

I. INTRODUCTION

In the classic science-fiction novel *"The War of the Worlds,"* recently reworked into a blockbuster movie, H.G. Wells depicted a story of how Earth was invaded by sinister Martians intent on the destruction of mankind (Wells, 1898). In the end, the twist to the tale was that the

---

[*] NERC Centre for Ecology and Hydrology, Oxford, OX1 3SR, United Kingdom
[1] Corresponding author: NERC Centre for Ecology and Hydrology, Oxford, OX1 3SR, United Kingdom

Martian invasion was not thwarted by man with all his technological know-how but instead the Martians were stopped in their tracks by the humble bacteria that are everywhere on this planet. Although a work of fantasy, it does or should provoke thought into the relative importance of bacteria to both mankind and the Earth itself. Bacteria comprise one of the three domains of life the [other two are Archaea and Eukaryota (which includes all of the animals and plants)] and are ubiquitous in the environment, living in every conceivable habitat on Earth. Although invisible to the naked eye, bacteria and their communities are central to health, agriculture, sustainable cities, and play key roles in many of the Earth's geochemical cycles. There are 10^{30} microbial individuals on Earth when compared with 10^{21} stars in the universe, making the microbial world immense (Curtis and Sloan, 2005). Furthermore, with approximately 1×10^9 bacterial individuals in 1 g of soil, you would only have to sample 7 g to surpass the current global population of approximately 6.6×10^9 humans. In terms of bacterial diversity, it is estimated that there are 70 species per milliliter in sewage works, 160 per milliliter in oceans, and up to 38,000 per gram of soil (Curtis et al., 2002).

In general ecology, the communities of animals and plants, and the patterns therein, are relatively easy to observe (though the underlying mechanisms or causes for those patterns remain obscure). However, bacterial communities are hard to observe and it is only recently that the technologies in bacterial ecology have advanced enough that the long established patterns in general ecology, such as the species–area relationship (SAR), can be investigated (Bell et al., 2005a; Horner-Devine et al., 2004; van der Gast et al., 2005, 2006). Despite sophisticated biotechnological exploitation of bacteria, it is fair to comment that, on many levels, bacterial ecology has yet to deliver its promise. Bacterial ecology has long been in a state of accumulating situation-bound statements of limited predictive ability, rendering little insight to either researchers or practitioners (Prosser et al., 2007). The potential of exploiting theories, models and principles from general ecology in microbiology could well provide invaluable insights into how bacterial communities organize and change in time and space. In time, this increased knowledge in bacterial community ecology will help us better understand and predict changes in the natural environment, allow improved manipulation of industrial processes, and give improved protection of human health. The application of theory is severely lacking in bacterial ecology where, paradoxically, it is required most (Prosser et al., 2007). Established ecological theory may provide the answer but, itself, must be tested using microbial systems to demonstrate its generality. At present, the process of how bacterial communities assemble and develop is poorly understood. This coming together of hundreds or thousands of different bacterial taxa composed of millions of individuals in a given space or time is known as community assembly (Curtis et al., 2003).

Here I will emphasize the importance of using island systems to test and apply ecological principles so as to better understand bacterial community assembly, and bacterial spatial and temporal scaling.

II. THE IMPORTANCE OF ISLANDS

In microbial ecology, it makes logical sense to test ecological theory initially in island systems. Since Darwin's visit to the Galápagos Islands, biologists have been using islands as microcosms to address key questions in ecology and evolution (Krebs, 1994). Diamond and May (1976) stated that islands have proven well suited to provoking or testing theoretical ideas in general ecology as they contain tractable communities with definite boundaries, coming in different sizes and remoteness. Islands have played a key role in the development of ecological theories, from the initial proposition of evolution and biogeography with Darwin's finches on the Galápagos Islands (Darwin, 1859), through Mayr's demonstration of the role of geographic isolation in speciation (Mayr, 1963), to the theory of island biogeography pioneered by MacArthur and Wilson (1967). In addition, studying groups of islands enables the investigator to view a simpler microcosm of the vastly more complex terrestrial or oceanic systems (MacArthur and Wilson, 1967).

A. Biogeographic islands for studies of bacterial diversity

Rosenzweig (1995) defines a biogeographic island as "a self-contained region whose species originate entirely by immigration from outside the region," and on the basis of this definition a diverse range of islands have been used to study ecology and evolution, from classical oceanic islands to oases, trees, lakes, and mountain tops (Rosenzweig, 1995). In microbial ecology, island systems must be at spatial scales relevant to the lifestyles of bacteria, so that instead of oceanic islands, we must think smaller, such that tree-holes, wastewater treatment bioreactors or even the lungs of cystic fibrosis patients can be used as biogeographic islands. This is because sampling efforts still fall short of describing the extent of microbial diversity in complex contiguous environments such as soil or sediments. Although bacterial species–area relationships have been recently described in studies of contiguous soil and sediment habitats of increasing area (Horner-Devine et al., 2004; Noguez et al., 2005), as in most other studies of such complex environments, most descriptions of bacterial communities have been at spatial scales of human convenience and are probably not relevant for the biology of such organisms. Without detracting from those studies but instead clarifying the point, soil bacteria and other soil-borne microbes are not continuously distributed throughout soil but are organized into microcolonies associated with soil crumbs and particles (Nunan *et al.*, 2002, 2003). In fact, at the 159th meeting of the Society for General Microbiology (2006), Ian Young described in his lecture on spatial modelling of microbial communities in soil, how a few grams of soil can have a surface area equivalent to five tennis courts. Therefore, it is fair to comment that previous studies have overlooked the relevant and important microspatial scale in complex terrestrial systems. This also suggests that most soil bacteria and other microbes exist in microisland communities and that the basic unit of study that should be addressed first in terrestrial habitats for microbes should be in fact these microislands before one could reasonably extend work to macrospatial scales (e.g., from centimeters to kilometers).

B. Island biogeography

From the various theories and principles developed in general ecology, it is the equilibrium theory of island biogeography that has generated the most interest in microbial ecology as a framework for better understanding bacterial community assembly (Curtis *et al.*, 2003; Graham and Curtis, 2003; Graham and Smith, 2004; van der Gast *et al.*, 2006). Pioneered by MacArthur and Wilson (1967), the underpinning features of the theory are that (a) the number of species found on an island is determined by the size of an island, the distance or relative isolation of an island from an immigration (mainland) source and thus is a balance between immigration and extinction; (b) the immigration/extinction balance is dynamic, where species are continually going extinct and being replaced through immigration or births by individuals of new species or species already present on the island; (c) immigration and extinction rates vary with island size and isolation, for example, smaller islands will have less species than larger islands. Before attempting to apply the island biogeography equilibrium model to bacterial island systems it would be prudent to test underlying issues that contribute to the theory such as do larger islands house more bacterial taxa than smaller ones (species–area relationships)? The species–area relationship is an underlying issue of island biogeography and as MacArthur and Wilson stated, "theories, like islands, are often reached by stepping stones. The species–area curves are such stepping stones" (MacArthur and Wilson, 1967).

III. SPECIES–AREA RELATIONSHIPS

In 1859, H. C. Watson proposed a basic yet fundamental ecological model, now termed the species–area relationship (SAR), and the relationship between species richness and island or area size is one of the few generalizations in ecology (Rosenzweig, 1995). The SAR has been extensively used for estimating population distributions, numbers and species diversity for animals and plants (MacArthur and Wilson, 1967; Preston, 1960, 1962a,b; Rosenzweig, 1995). Several mechanisms explain how the number of species in a particular area results from the balance between the immigration and colonization of new species and the extinction of extant species (Bell et al., 2005a). The size of the area influences the rate of colonization and extinction and so has an effect on biodiversity. Alternatively, if species are adapted to a particular habitat, then larger areas or islands will likely contain more habitats (increased habitat heterogeneity) and therefore contain more species. Finally, a species–area relationship will appear with increased sampling effort, because the number of species discovered increases with sampling effort (Rosenzweig, 1995). The relationship between diversity and island or area size is well described by the power law, $S = cA^z$, where S is the number of species, A is the area sampled, c is an empirically derived taxon- and location-specific constant, and z is the slope of the log–log line (Arrhenius, 1921). The value for z has been shown to be consistent across animal and plant species but differs between islands ($z \sim 0.2$–0.35) and nonisolated sample areas of contiguous habitat ($z \sim 0.1$–0.17) (Rosenzweig, 1995).

A. Microbial biogeography

Although, the SAR has been used in ecology to describe spatial diversity patterns in contiguous pieces of biota and on islands for many phyla within the plants and animals, microorganisms lay claim to 2 of the 3 domains of life (Archaea and Bacteria) and their diversity dominates the Eukaryota with the protests to protests. Until recently their small size and abundance had precluded the detailed study of their distributions and ecological relationships. However, it may prove that microbial diversity is fundamentally different from those of general ecology as has been suggested, based upon studies of microbial eukaryotic species (Finlay, 2002; Finlay et al., 2004; Finlay and Clarke, 1999). Finlay and colleagues examined the global abundance of free living ciliate species. Their data strongly implied that although there was high local diversity, globally, all species of ciliates were ubiquitous ($z = 0.043$). They proposed that this was due to the small body size of the ciliates and that their ubiquity was due to global dispersal by wind and water currents. Finlay and colleagues have reasoned that as bacteria have much smaller body sizes than ciliates and are many orders of magnitude more abundant, they are therefore even more likely to be globally dispersed and so less likely to be biogeographically restricted unlike most larger organisms (Finlay, 2002; Finlay et al., 2004).

This view would support the old microbiological tenet of "everything is everywhere, but, the environment selects" (Baas Becking, 1934). Indeed the issue of whether microorganisms are globally dispersed or biogeographically restricted has been the subject of high profile debate in the last few years (Bell et al., 2005b; Fenchel and Finlay, 2005; Hughes Martiny et al., 2006; Whitfield, 2005). There is now an increasing body of evidence to support the proposition that bacteria do have a recognizable biogeography as demonstrated in recent studies of population genetics and community ecology studies. Evidence of bacterial endemicity has been demonstrated in studies of fluorescent *Pseudomonas* strains in soil (Cho and Tiedje, 2000), purple nonsulfur bacteria in freshwater marsh sediments (Oda et al., 2003), and thermophilic cyanobacteria in hot springs (Papke et al., 2003). From a community ecology perspective, two studies of contiguous habitats have shown that bacterial diversity is area dependent. Horner-Devine and colleagues (2004) found a bacterial taxa–area relationship ($z \sim 0.04$) in a contiguous sampling study of a New England (USA) salt marsh sediment. Likewise, Noguez et al. (2005) observed that richness in bacterial taxa scaled with increasing area ($z \sim 0.42$–0.47) in a study of tropical deciduous forest soil in western Mexico. To date, as there have been only two studies of bacterial species–area relationships in contiguous habitats, it is difficult to draw any conclusions about the scaling exponents that they observed. The disparity between the z values may be (in part), as mentioned previously, that both studies overlooked the microspatial (local) scale of diversity ($<g$) which may have well skewed their findings.

B. Island size and bacterial diversity

The slope of the SAR is expected to be steeper on discrete islands ($z \sim 0.2$–0.35), than on single contiguous habitats ($z \sim 0.1$–0.17). In a series of three studies, we predicted that the slope of the SAR for insular bacterial communities would be similar to that found for communities of larger organisms. In the first study we sampled bacterial communities colonizing metal-cutting fluids from machines of increasing sump tank size, taking these to be analogous to islands of variable size (van der Gast et al., 2005). Estimating the area available for colonization was not possible because of the irregular shapes of the machines. Therefore, we used volume to indicate the size of island for the sump tanks system. The species–area, or in this case the species–volume relationship slopes were typical of z values described for island archipelagos of approximately 0.25 (Fig. 6.1A). In the second study, we tested the island biogeography species–area relationship for bacterial communities colonizing water-filled tree-holes, of increasing size, in the buttress root systems of mature European beech trees (*Fagus sylvatica*) (Bell et al., 2005a). This was chosen

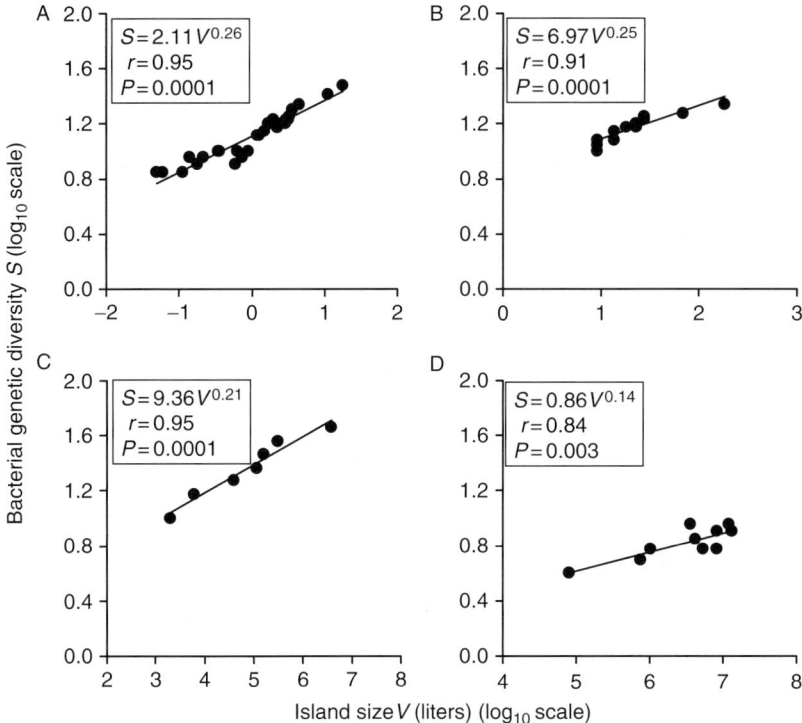

FIGURE 6.1 The relationship between island size and bacterial taxa for (A) water-filled tree-holes, (B) metal-cutting sump tanks, (C) WWTP bioreactors, and (D) high-mountain lakes. Also given is the power law ($S = cV^z$), the correlation coefficient (r), and significance (P). Note in (D) the statistical outlier, Lake Caldera, has been removed. When the outlier is included in the analyses then; $z = 0.08$, $r = 0.60$, and $P = 0.56$.

as it was a natural system that facilitated determination of both surface area and volume. We found a linear relationship between tree-hole surface area (A) and taxa richness ($S = 3.30A^{0.28}$, $r = 0.62$, $P < 0.001$). However, regression analyses revealed that volume (V) was the main predictor for bacterial taxa in the tree-hole systems studies ($S = 2.11V^{0.26}$, $r = 0.95$, $P < 0.0001$) (Fig. 6.1B). In a third study, we sampled several wastewater treatment plant (WWTP) bioreactors of increasing size located across the UK (van der Gast et al., 2006). In this study, we observed that the slope ($z \sim 0.21$) of the relationship between island size (WWTP bioreactor volume) and taxa richness was similar to that found for larger organisms in insular island systems (Fig. 6.1C). Reche et al. (2005) assessed whether the size of ecosystem determined aquatic bacterial richness in 11 high-mountain lakes from Sierra Nevada (Spain). Using lake area as the predictor they observed a slope (z) of the SAR was 0.104 ± 0.051 ($r = 0.562$, $P = 0.072$) (Reche et al., 2005).

They also added data from other published data sets (Lindström and Leskinen, 2002; Zwart et al., 2002) where $z = 0.152 \pm 0.042$ ($n = 21$ lakes, $r = 0.643, P = 0.002$). There were some controversies as Lindström et al. (2007) argued that the results derived from the added published data sets were not comparable because of differences in molecular techniques employed across the three separate studies and therefore determination of operational taxonomic units used to infer bacterial richness would possibly bias the results (Lindström et al., 2007; Reche et al., 2007). It is possible to compare the Sierra Nevada (Spain) lake data with our own by estimating the volume of those lakes (from the surface area and depth data provided) (Fig. 6.1D).

It can be observed from Fig. 6.1 that larger islands will support more bacterial taxa than smaller ones as predicted by MacArthur and Wilson (1967) for larger organisms such as animals and plants. The slope values were approximately within the range, described by MacArthur and Wilson (1967), as being typical for larger organisms. It is notable that this taxa–volume relationship holds across substantial changes in scale, although the scaling exponent (z) begins to decline when larger island types are compared. Also, volume appears to be a better measure for island size (Fig. 6.1A and D) and therefore a better predictor of spatial scaling of bacterial taxa in aquatic islands systems. It is not surprising that volume is better predictor over area as these are aquatic systems where the bacteria inhabit three-dimensional space. Also, the routes for immigration are not the same as that for animals or plants colonizing an oceanic island. If we consider WWTP bioreactors, apart from initial sludge seed, the main source for bacterial immigrants getting into a reactor would be the feed water, and immigration from the surrounding air would, in contrast, be negligible, thus making the area redundant as a measure of island size for such systems.

IV. BETA DIVERSITY

So far I have concentrated on bacterial taxa diversity or richness, which is a measure of alpha diversity. Formally, alpha diversity is the diversity or richness of a defined assemblage or habitat (Magurran, 2004). However, when comparing two or more spatially (or temporally) separated assemblages or habitats we move from alpha to beta diversity. Beta diversity is in essence a measure of the extent to which two or more spatial units differs (Magurran, 2004). Green et al. (2004) stated that ecologists studying animals and plants have long recognized that beta diversity is central to understanding the forces responsible for the magnitude and variability of diversity. Microbial beta diversity patterns are largely unknown (Green et al., 2004), which could offer valuable insights into the relative influence of dispersal limitation and environmental change in the structure of bacterial communities.

A. Smaller islands are less stable than larger ones

Theoretical predictions for island biogeography state that smaller islands have lower variable (less stable) diversity, whereas larger islands have greater and more stable diversity over time and this has chiefly been explained because of larger islands having more refuge opportunities than smaller ones (Curtis *et al.*, 2003). Therefore, smaller islands have more variable, less stable bacterial populations. This was observed from two studies (in 2001 and 2002) of three metal-cutting fluid sump tanks (small, medium, and large islands) sampled every week for 8 weeks (van der Gast *et al.*, 2005). The cluster analyses revealed, in the 2001 study, that the bacterial assemblages in the large sump tank (island) were 98% ± 2% similar ($n = 8$ in all cases). Similarity for the bacterial communities in the medium and small islands was 80% ± 9% and 58% ± 29%, respectively. In the 2002 study, similarity for bacterial community was large island, 90% ± 8%; medium, 83% ± 6%; and small 63% ± 10%. In brief, the cluster analyses revealed that large island samples were more similar temporally. In contrast, the smaller islands showed less similarity across time, with the medium size islands falling in between the observed properties of large and small islands (van der Gast *et al.*, 2005).

B. Species–time relationships

The manner in which species richness changes with time has received even less attention than that of SARs (Rosenzweig, 1995). Originally proposed by Preston (1960), the species–time relationship (STR) describes how the observed species richness of a community in a fixed area increases with the length of time over which the community is monitored (Preston, 1960; White, 2004; White *et al.*, 2006). The species–area power law can be modified to describe the relationship between species richness and time, T (Preston, 1960). For clarity, the scaling exponent z is changed to w, so that the STR power law becomes $S = cT^w$ (Adler and Lauenroth, 2003). Preston (1960) surmised that the STR should mimic the SAR, following a straight line in log–log space.

Temporal turnover has been defined as 'the number of species eliminated and replaced per unit time' and is a core concept to the theory of island biogeography (MacArthur and Wilson, 1967). If temporal turnover can be modelled by the STR then the scaling exponent (w) of the STR can be assumed to reflect turnover and is thus implicit in community assembly and colonization (Magurran, 2004). Reanalyzing the earlier-mentioned temporal metal-cutting fluid sump tank data, it was possible to plot the STR for each of the large, medium, and small sump tank islands (Fig. 6.2). Both large islands (Fig. 6.2A and D) have low temporal scaling exponents ($w < 0.1$), which suggest that bacterial species turnover is also low on these islands, which is in agreement with the earlier-mentioned cluster analyses. Conversely, the small islands (Fig. 6.2C and F) had much steeper STR relationships ($w \geq 0.23$) indicating, as observed previously, that turnover is much greater and that smaller islands have less stable bacterial populations.

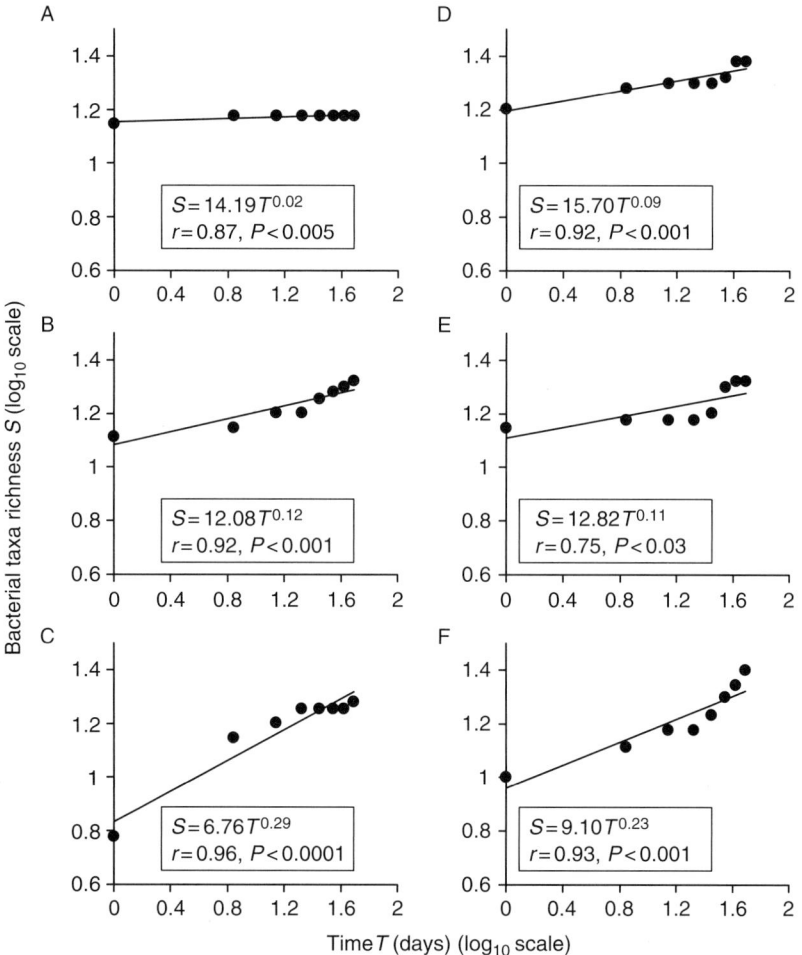

FIGURE 6.2 Bacterial species–time relationships for large (A, D), medium (B, E) and small (C, F) metal-cutting fluid sump tank islands from the 2001 (1st column) and 2002 (2nd column) studies. Also given are the species–time power law ($S = cT^w$), the correlation coefficient (r), and significance (P).

C. Distance–decay relationships

Another prediction of island biogeography is that islands close together in an archipelago will have more similar community composition than those that are farther away. A way to visualize this is to use distance–decay relationships where community composition decays with increasing geographical distance (Green et al., 2004). Both Horner-Devine et al. (2004) and Green et al. (2004) used distance–decay relationships to approximate for SARs in contiguous terrestrial habitats for bacteria and fungi, respectively. From the island studies already detailed we have been able to directly observe the bacterial SARs, therefore there is no need to perform a similar approximation. But, testing the distance–decay relationship for the water-filled tree-holes and metal-cutting fluid sump tank studies revealed that island community composition does decay with distance (Fig. 6.3). Interestingly, we did not find a significant inverse relationship between geographical distance and bacterial community similarity from the WWTP bioreactor study ($r = 0.34$, $P = 0.20$). There are a few possible explanations for this. First, the source community for each bioreactor would originate from the town or populated area local to the WWTP. This would mean that the bioreactors were not part of an archipelago, unlike the tree-holess or sump tanks, and therefore would not share a common single source region for colonizing. Secondly, the geographic isolation of islands arising from the large distances between WWTP bioreactors sampled in the study (greatest distance between bioreactors was 1100 km).

V. OPPOSING PERSPECTIVES ON COMMUNITY ASSEMBLY

At present, the process of how bacterial communities assemble and develop is poorly understood (Curtis et al., 2003). Emerging from the fields of community ecology and biogeography, there are two differing perspectives on the organization of ecological communities, which are the niche-assembly perspective or the dispersal-assembly perspective, respectively (Hubbell, 1997). Deterministic community assembly or the niche-assembly perspective is marked by a coming together in space or time of a set of species with distinct functions and ecological niches. The determinism that underpins the niche-assembly perspective is two fold; first, an avoidance of serious competition (i.e., coexistence), and second, improved or optimal ecosystem function due to selection from the regional community of the most competitive species for the available ecological niches in the local community. A deterministically assembled community, typically, is conserved in composition and structure and only changes when niches within the habitat are altered or the entire habitat is

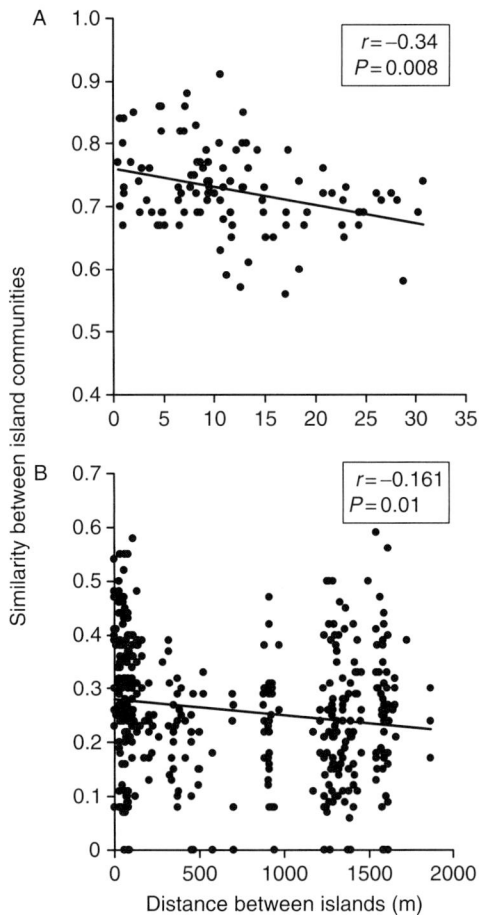

FIGURE 6.3 The distance–decay relationships for (A) metal-cutting fluid sump tanks (visit 1) and (B) water-filled tree-holes derived using the Mantel test. Also given are the Mantel statistic (*r*) and significance (*P*) following 9999 random iterations.

perturbed and changed. Conversely, stochastic community assembly or the dispersal community perspective asserts that communities are formed on a 'first come first served basis,' where species are largely thrown together by chance, history, and random dispersal (Hubbell, 2001). In addition, a randomly assembled community is typically, highly dynamic with constantly changing composition. Historically, bacterial communities have been thought to be shaped by deterministic factors, manifested in the 'everything is everywhere, but, the environment selects' tenet (Baas Becking, 1934). Although, neutral dispersal community models, inspired by Hubbell's unified neutral theory of biodiversity and biogeography, are now

TABLE 6.1 Raup and Crick probability-based index of pair-wise similarity (S_{RC}) frequency percentages between bacterial community samples within five Islands datasets

Study	% S_{RC} values		
	<0.05	>0.05 and < 0.95	>0.95
Sump tanks (visit 1)	1	61	38
Sump tanks (visit 2)	4	63	33
Sump tanks (visit 3)	0	70	30
Tree-holes	6	88	6
WWTP bioreactors	9	83	8

being developed for microbial communities (Sloan et al., 2006; Woodcock et al., 2006, 2007). It is important to note, as Hubbell also highlighted, that although dispersal community theories do not have to be neutral, most are, including the theory of island biogeography (Hubbell, 2001).

Using the Raup and Crick probability-based similarity index (S_{RC}) we assessed whether colonization was driven by stochastic (dispersal) or niche-considerations for the tree-holes, metal fluid-cutting sump tank, and WWTP bioreactor studies (Raup and Crick, 1979). This was achieved by recording the proportions of island communities taken pair-wise, whose compositional similarities were not more or less significant than could occur by chance (Table 6.1) (Rowan et al., 2003; van der Gast et al., 2005). Therefore, S_{RC} values of > 0.95 and < 0.5 signify differences, which are not random assortments of the same species. For the five islands studies (Table 6.1), the proportion of such pair-wise similarities ranged between 61 and 88%, indicating a strong role for dispersal in colonization and community structure.

VI. CONCLUSIONS

Patterns in bacterial diversity will give us a better understanding on how bacteria are distributed spatially and temporally. The studies of biogeography and spatial scaling outlined in this article are but a start. In time, this will in turn provide important information about the underlying mechanisms that regulate microbial diversity and the complexity of the ecosystems they inhabit. Utilizing, testing and adapting theories and principles developed in general ecology can help us to achieve just that. We need to move away from being in a position of collecting situation bound statements of limited predictive ability and therefore a 'horizontal gain of information' (de Lorenzo, 2002; Prosser et al., 2007). By coupling a theoretical approach with the ever improving molecular methodologies available in time should put us onto a firm predictive footing to better understand bacterial diversity and the services they provide.

ACKNOWLEDGMENTS

The impetus for this article came from topical discussions following my talk on '*island size and bacterial diversity*' at the 159th Society of General Microbiology meeting (York, September 2006). I am grateful to Duane Ager (CEH Oxford), Tom Bell (University of Oxford), and Andy Lilley (King's College London) without which this work would not have been possible. I also thank the Natural Environment Research Council (NERC) and the Engineering and Physical Sciences Research Council (EPSRC) for funding.

REFERENCES

Adler, P. B., and Lauenroth, W. K. (2003). The power of time: Spatiotemporal scaling of species diversity. *Ecol. Lett.* **6,** 749–756.
Arrhenius, O. (1921). Species and area. *J. Ecol.* **9,** 95–99.
Baas Becking, L. G. M. (1934). Geobiologie of Inleiding Tot de Milieukunde. W. P. Van Stockum & Zoon, The Hague, the Netherlands (in Dutch).
Bell, T., Ager, D., Song, J. I., Newman, J. A., Thompson, I. P., Lilley, A. K., and van der Gast, C. J. (2005a). Larger islands house diverse bacterial taxa. *Science* **308,** 1884.
Bell, T., Newman, J. A., Thompson, I. P., Lilley, A. K., and van der Gast, C. J. (2005b). Bacteria and island biogeography. *Science* **309,** 1998–1999.
Cho, J. C., and Tiedje, J. M. (2000). Biogeography and degree of endemicity of fluorescent *Pseudomonas* strains in soil. *Appl. Environ. Microbiol.* **66,** 5448–5456.
Curtis, T. P., Head, I. M., and Graham, D. W. (2003). Theoretical ecology for engineering biology. *Environ. Sci. Technol.* **37,** 64–70.
Curtis, T. P., and Sloan, W. T. (2005). Exploring microbial diversity – A vast below. *Science* **309,** 1331–1333.
Curtis, T. P., Sloan, W. T., and Scannell, J. W. (2002). Estimating prokaryotic diversity and its limits. *Proc. Natl. Acad. Sci. USA* **99,** 10494–10499.
Darwin, C. (1859). On the Origin of Species by Means of Natural Selection, or the Preservation of Favoured Races in the Struggle for Life. John Murray, London.
de Lorenzo, V. (2002). Towards the end of experimental (micro)biology? *Environ. Microbiol.* **4,** 6–9.
Diamond, J. M., and May, R. M. (1976). Island biogeography and the design of natural reserves. In "Theoretical Ecology: Principles and Applications." (R. M. May, ed.), pp. 163–186. Blackwell Scientific, Oxford, UK.
Fenchel, T., and Finlay, B. J. (2005). Bacteria and island biogeography. *Science* **309,** 1997.
Finlay, B. J. (2002). Global dispersal of free-living microbial eukaryote species. *Science* **296,** 1061–1063.
Finlay, B. J., and Clarke, K. J. (1999). Ubiquitous dispersal of microbial species. *Nature* **400,** 828.
Finlay, B. J., Esteban, G. F., and Fenchel, T. (2004). Protist diversity is different? *Protist.* **155,** 15–22.
Graham, D. W., and Curtis, T. P. (2003). Ecological theory and bioremediation. In "Bioremediation: A critical review." (I. M. Head, I. Singleton and M. G. Milner, eds.), pp. 61–92. Horizon Scientific Press, Wymondham, Norfolk.

Graham, D. W., and Smith, V. H. (2004). Designed ecosystem services: Application of ecological principles in wastewater treatment engineering. *Front. Ecol. Environ.* **2,** 199–206.

Green, J. L., Holmes, A. J., Westoby, M., Oliver, I., Briscoe, D., Dangerfield, M., Gillings, M., and Beattie, A. J. (2004). Spatial scaling of microbial eukaryote diversity. *Nature* **432,** 747–750.

Horner-Devine, M. C., Lage, M., Hughes, J. B., and Bohannan, B. J. M. (2004). A taxa-area relationship for bacteria. *Nature* **432,** 750–753.

Hubbell, S. P. (1997). A unified theory of biogeography and relative species abundance and its application to tropical rainforest and coral reefs. *Coral. Reefs.* **16,** S9–S21.

Hubbell, S. P. (2001). "The Unified Neutral Theory of Biodiversity and Biogeography." Princeton University Press, Princeton, NJ.

Hughes Martiny, J. B., Bohannan, B. J. M., Brown, J. H., Colwell, R. K., Fuhrman, J. A., Green, J. L., Homer-Devine, M. C., Kane, M., Krumins, J. A., Kuske, C. R., Morin, P. J., Naeem, S., *et al.* (2006). Microbial biogeography: Putting microorganisms on the map. *Nature Rev. Microbiol.* **4,** 102–112.

Krebs, C. J. (1994). "Ecology: The Experimental Analysis of Distribution and Abundance." HarperCollins, New York.

Lindström, E. S., Eiler, A., Langenheder, S., Bertilsson, S., Drakare, S., Ragnarsson, H., and Tranvik, L. J. (2007). Does ecosystem size determine aquatic bacterial richness? Comment. *Ecology* **88,** 252–253.

Lindström, E. S., and Leskinen, E. (2002). Do neighbouring lakes share common taxa of bacterioplankton? Comparison of 16S rDNA fingerprints from three geographic regions. *Microbial. Ecol.* **44,** 1–9.

MacArthur, R. H., and Wilson, E. O. (1967). "The Theory of Island Biogeography." Princeton University Press, Princeton, NJ.

Magurran, A. E. (2004). "Measuring Biological Diversity." Blackwell Publishing, Oxford, UK.

Mayr, E. (1963). "Animal Species and Evolution." Harvard University Press, Cambridge, MA.

Noguez, A. M., Arita, H. T., Escalante, A. E., Forney, L. J., Garcia-Oliva, F., and Souza, V. (2005). Microbial macroecology: Highly structured prokaryotic soil assemblages in a tropical deciduous forest. *Global. Ecol. Biogeogr.* **14,** 241–248.

Nunan, N., Wu, K., Young, I. M., Crawford, J. W., and Ritz, K. (2002). In situ spatial patterns of soil bacterial populations, mapped at multiple scales, in an arable soil. *Microbial. Ecol.* **44,** 296–305.

Nunan, N., Wu, K. J., Young, I. M., Crawford, J. W., and Ritz, K. (2003). Spatial distribution of bacterial communities and their relationships with the micro-architecture of soil. *FEMS Microbiol. Ecol.* **44,** 203–215.

Oda, Y., Star, B., Huisman, L. A., Gottschal, J. C., and Forney, L. J. (2003). Biogeography of the purple nonsulfur bacterium *Rhodopseudomonas palustris. Appl. Environ. Microbiol.* **69,** 5186–5191.

Papke, R. T., Ramsing, N. B., Bateson, M. M., and Ward, D. M. (2003). Geographical isolation in hot spring cyanobacteria. *Environ. Microbiol.* **5,** 650–659.

Preston, F. W. (1960). Time and space and the variation of species. *Ecology* **41,** 611–627.

Preston, F. W. (1962a). The canonical distribution of commoness and rarity: Part I. *Ecology* **43,** 185–215.

Preston, F. W. (1962b). The canonical distribution of commoness and rarity: Part II. *Ecology* **43,** 410–432.

Prosser, J. I., Bohannan, B. J. M., Curtis, T. P., Ellis, R. J., Firestone, M. K., Freckleton, R. P., Green, J. L., Green, L. E., Killham, K., Lennon, J. J., Osborn, A. M., Solan, M., *et al.* (2007). The role of ecological theory in microbial ecology. *Nature Rev. Microbiol.* **5,** 384–392.

Raup, D., and Crick, R. E. (1979). Measurement of faunal similarity in paleontology. *J. Paleontol.* **53,** 1213–1227.

Reche, I., Pulido-Villena, E., Morales-Baquero, R., and Casamayor, E. O. (2005). Does ecosystem size determine aquatic bacterial richness? *Ecology* **86**, 1715–1722.

Reche, I., Pulido-Villena, E., Morales-Baquero, R., and Casamayor, E. O. (2007). Does ecosystem size determine aquatic bacteria richness? Reply. *Ecology* **88**, 253–255.

Rosenzweig, M. L. (1995). "Species Diversity in Space and Time." Cambridge University Press, Cambridge, UK.

Rowan, A. K., Snape, J. R., Fearnside, D., Barer, M. R., Curtis, T. P., and Head, I. M. (2003). Composition and diversity of ammonia-oxidising bacterial communities in wastewater treatment reactors of different design treating identical wastewater. *FEMS Microbiol. Ecol.* **43**, 195–206.

Sloan, W. T., Lunn, M., Woodcock, S., Head, I. M., Nee, S., and Curtis, T. P. (2006). Quantifying the roles of immigration and chance in shaping prokaryote community structure. *Environ. Microbiol.* **8**, 732–740.

van der Gast, C. J., Lilley, A. K., Ager, D., and Thompson, I. P. (2005). Island size and bacterial diversity in an archipelago of engineering machines. *Environ. Microbiol.* **7**, 1220–1226.

van der Gast, C. J., Jefferson, B., Reid, E., Robinson, T., Bailey, M. J., Judd, S. J., and Thompson, I. P. (2006). Bacterial diversity is determined by volume in membrane bioreactors. *Environ. Microbiol.* **8**, 1048–1055.

Wells, H. G. (1898). "The war of the worlds." William Heinemann, London.

White, E. P. (2004). Two-phase species-time relationships in North American land birds. *Ecol. Lett.* **7**, 329–336.

White, E. P., Adler, P. B., Lauenroth, W. K., Gill, R. A., Greenberg, D., Kaufman, D. M., Rassweiler, A., Rusak, J. A., Smith, M. D., Steinbeck, J. R., Waide, R. B., and Yao, J. (2006). A comparison of the species-time relationship across ecosystems and taxonomic groups. *Oikos* **112**, 185–195.

Whitfield, J. (2005). Biogeography: Is everything everywhere?. *Science* **310**, 960–961.

Woodcock, S., Curtis, T. P., Head, I. M., Lunn, M., and Sloan, W. T. (2006). Taxa-area relationships for microbes: The unsampled and unseen. *Ecol. Lett.* **9**, 805–812.

Woodcock, S., van der Gast, C. J., Bell, T., Lunn, M., Curtis, T. P., Head, I. M., and Sloan, W. T. (2007). Neutral assembly of bacterial communities. *FEMS Microbiol. Ecol.* **62**, 171–180.

Zwart, G. G., Crump, B. C., Kamst-van Agterveld, M. P., Hargen, F., and Han, S. K. (2002). Typical freshwater bacteria: Analysis of 16S rRNA from plankton of lakes and rivers. *Aquat. Microbial. Ecol.* **28**, 141–155.

CHAPTER 7

Human Pathogens and the Phyllosphere

John M. Whipps,[1] Paul Hand, David A. C. Pink, and Gary D. Bending

Contents		
	I. Introduction	183
	II. Food Poisoning Outbreaks Associated with Consumption of Fresh and Minimally Processed Fresh Vegetables, Salads, and Fruit	185
	III. Sources of Human Pathogens on Plants	186
	IV. Ecology of Human Pathogens in Relation to Phyllosphere Contamination	189
	A. Ecology and survival in sewage, manure, soil, and water	189
	B. Ecology and survival in association with plants	190
	C. Ecology and survival during processing	203
	V. Conclusions and Future	208
	Acknowledgments	209
	References	209

I. INTRODUCTION

Historically, human pathogens have not been considered as part of the microbial population of the aerial parts of plants termed the phyllosphere. However, with increased consumption of fresh and minimally processed fruit and vegetables as part of a growing awareness of the need to

Warwick HRI, University of Warwick, Wellesbourne, Warwick, CV35 9EF, United Kingdom
[1] Corresponding author: Warwick HRI, University of Warwick, Wellesbourne, Warwick, CV35 9EF, United Kingdom

improve human health, particularly in Europe and North America, the occurrence and importance of human pathogens on fresh produce has to be considered. This is of even more significance because a clear relationship has been found between increased food-related disease outbreaks and greater consumption of fresh produce (Beuchat, 2002; Sivapalasingam *et al.*, 2004). This reflects changing production practices, including increased use of prepackaged and minimally processed fresh foods, awareness of the potential for fresh produce-related food poisoning by physicians, improved detection and monitoring of human pathogens, increasing numbers of immunocompromised patients who are more susceptible to diseases than the rest of the population, and the emergence of new pathogens and strains of pathogens, particularly multidrug resistant isolates (Brandl, 2006a,b; Sagoo *et al.*, 2003; Suslow, 2002; Tauxe, 1997). Because of these concerns, the biology and ecology of human pathogens associated with many aspects of production and consumption of fresh fruit and vegetables have been reviewed in considerable detail over the last few years (Aruscavage *et al.*, 2006; Beuchat, 1996a,b, 1998, 2002; Beuchat and Ryu, 1997; Brandl, 2006a,b; Burnett and Beuchat, 2000; de Roever, 1998; Farber and Peterkin, 1991; Lin *et al.*, 1996; Long *et al.*, 2002; Monaghan, 2006; Nguyen-the and Carlin, 1994; Sivapalasingam *et al.*, 2004; Steele and Odumeru, 2004; Suslow, 2002; Tauxe *et al.*, 1997; Warriner *et al.*, 2003b). Consequently, this overview provides a general background to the whole area of human pathogens in the phyllosphere of fresh produce. This provides the correct context to highlight new or most appropriate approaches for control of human pathogens in the phyllosphere of fresh and minimally processed food. Special emphasis is placed on salad crops, particularly lettuce, where there have been notable problems with outbreaks of food poisoning in the past (Ackers *et al.*, 1998; Hilborn *et al.*, 1999; Islam *et al.*, 2004a,b; Monaghan, 2006; Nuorti *et al.*, 2004) and where there are new concerns such as the occurrence and dissemination of antibiotic-resistant bacteria with potential for horizontal transfer of antibiotic resistance genes to human pathogens (Boehme *et al.*, 2004; Rodríguez *et al.*, 2006). Nevertheless, in spite of these outbreaks of fresh produce-related food poisoning, fresh and minimally processed fruit and vegetables (including those from organic crops) have usually been found to be of satisfactory microbiological quality in developed countries (Johannessen *et al.*, 2002; Johnston *et al.*, 2005; Little *et al.*, 1999; Loncarevic *et al.*, 2005; McMahon and Wilson, 2001; Mukherjee *et al.*, 2004; Sagoo *et al.*, 2001, 2003) although there may still be concerns with human pathogen contamination of vegetables in some countries (Froder *et al.*, 2007; Ibenyassine *et al.*, 2007; Machado *et al.*, 2007). Moreover, occurrence of foodborne diseases linked with the consumption of fresh produce is low (<6%) when compared to total recorded food poisoning outbreaks (Long *et al.*, 2002).

II. FOOD POISONING OUTBREAKS ASSOCIATED WITH CONSUMPTION OF FRESH AND MINIMALLY PROCESSED FRESH VEGETABLES, SALADS, AND FRUIT

Several microorganisms have been associated with outbreaks of food poisoning related to consumption of fresh or minimally processed produce, including bacteria (*Aeromonas hydrophila*, *Bacillus cereus*, *Campylobacter* spp., *Clostridium* spp., *Escherichia coli*, *Listeria monocytogenes*, *Salmonella* spp., *Shigella* spp., *Vibrio cholerae*, and *Yersinia enterocolytica*), viruses (hepatitis A virus and norovirus (formerly norwalk/norwalk-like virus)) and protozoa (*Cyclospora cayetanensis*) (Aruscavage *et al.*, 2006; de Roever, 1998; McMahon and Wilson, 2001; Nuorti *et al.*, 2004; Rangel *et al.*, 2005; Sivapalasingam *et al*, 2004; Steele and Odumeru, 2004; Taormina *et al.*, 1999). In the USA, lettuce, melon, seed sprouts and fruit juice were the most important individual sources associated with foodborne illness between 1973–1997 (Sivapalasingam *et al.*, 2004). Consumption of other fresh produce implicated in food poisoning outbreaks include tomato (Cummings *et al.*, 2001), fresh coriander (cilantro) (Campbell *et al.*, 2001), parsley (Naimi *et al.*, 2003), spinach (CDC, 2004, 2006), spring onions (Tauxe *et al.*, 1997), carrot (CDC, 1994), and cabbage (Schlech *et al.*, 1983). In the United Kingdom, some significant outbreaks of food poisoning associated with consumption of fresh or minimally processed produce were linked to consumption of imported lettuce contaminated with *Shigella sonnei* (Kapperud *et al.*, 1995) or *Salmonella* Newport PT33 (Sagoo *et al.*, 2003), lettuce prepared in fast food restaurants contaminated with *Salmonella enterica* serotype Typhimurinum PT104 (Horby *et al.*, 2003), and beansprouts colonized by *Salmonella* Saint-Paul (O'Mahony *et al.*, 1990). In a review of food poisoning outbreaks on several single fresh product items (not mixed salads) in the USA from 1990–2004, *E. coli* strains were the most common agent in most (48%) of cases associated with leafy vegetables, followed by *Salmonella enterica* (30%), and then *Shigella* and *Campylobacter* spp. (11% each) (Brandl, 2006a). Interestingly, fruit and seed sprout related infections were predominantly due to *S. enterica* (70% and 60%, respectively), indicating that some specificity may exist between plant structure and these pathogens. These results confirm the generally held view that the majority of food poisoning outbreaks associated with the consumption of fresh and minimally processed vegetables are due to *E. coli* and *Salmonella* spp. (Sivapalasingam *et al.*, 2004). The emergence of *E. coli* O157:H7, an enterotoxigenic serotype that can cause haemolytic uremic syndrome and lead to death in children, the elderly, and the immunocompromised (Rangel *et al.*, 2005), is of particular importance as this was the predominant serotype of *E. coli* found in leafy vegetables in the USA for over 14 years (Brandl, 2006a).

Note that several bacterial species, such as *Pseudomonas aeruginosa*, *Burkholderia cepacia* and *Erwinia* spp., that commonly occur in the phyllosphere and rhizosphere of crop plants can also cause opportunistic infections in humans who are debilitated, immunocompromised, or suffering from cystic fibrosis (Berg *et al.*, 2005; Cao *et al.*, 2001; Parke and Gurian-Sherman, 2001; Starr and Chatterjee, 1972), but these infections are unlikely to have arisen from food consumption. *Ps. aeruginosa* is known to cause gastrointestinal infections (Ohara and Itoh, 2003) but this would depend upon a critical level of pathogen to be present and generally relate to the presence of other mitigating factors such as antibiotic treatment. Thus overall, such bacterial species are unlikely to have an impact on the normal healthy human population.

Surveys of fresh and minimally processed fruit and vegetables regularly record the presence of several human pathogens although their presence is sporadic and varies between crops, location of production, and season (Beuchat, 1996a,b, 1998; Johnston *et al.*, 2005; Lin *et al.*, 1996; Nguyen-the and Carlin, 1994; Odumeru *et al.*, 1997; Sagoo *et al.*, 2003). In ecological terms, it is perhaps not unexpected that *Clostridium* spp., *B. cereus*, and *Listeria monocytogenes* are found as they occur naturally in soil in the absence of feces (Beuchat and Ryu, 1997; de Roever, 1998; Fenlon, 1985). However, *E. coli* and *Salmonella* spp. are enteric organisms of fecal origin and are thought to be limited to soils that have been contaminated with feces as they are considered transient in nature (de Roever, 1998). Nevertheless, it is somewhat surprising given the regular recovery of these organisms from fresh and minimally processed foods that more outbreaks of food poisoning do not occur. In part, this may reflect the number of cells of the pathogen that have to be ingested to cause disease (infectious dose), which may be quite large for some pathogens, although infections can be caused by as few as 800 *Campylobacter* cells, 100 cells of *E. coli* O157:H7 and *Shigella* spp., and fewer than 100 particles of norovirus virus (Black *et al.*, 1988; Jaquette *et al.*, 1996; Jay *et al.*, 2005; Wachsmuth and Morris, 1989). Sources of human pathogens in fresh and minimally processed produce and procedures to minimize this contamination and subsequent proliferation are considered in the following sections.

III. SOURCES OF HUMAN PATHOGENS ON PLANTS

Some human pathogens, such as *Listeria* spp., are natural soil microorganisms (Beuchat and Ryu, 1997) and can be expected to spread onto plants directly from the soil or by water splash. Recently, *Clostridia* populations were also found in soil and as endophytes in the perennial grass *Miscanthus sinensis*, suggesting a natural ecosystem distribution for this genus too (Miyamoto *et al.*, 2004). However, enteric pathogens such as *E. coli* and *Salmonella* spp. are usually associated with fecal contamination and numerous studies have investigated the potential sources of such contamination both preharvest in the field and postharvest in the food chain. During the preharvest phase, the potential exists for persistent pathogen populations to establish on harvested crops and then the risk to be amplified postharvest either by further direct contamination or by proliferation of existing pathogen populations during processing and handling procedures.

In the field, one potential source of human pathogen contamination related to agronomic practice involves application of inadequately composted or raw animal manures or sewage (Beuchat and Ryu, 1997; de Roever, 1998; Jones and Martin, 2003; Santamaria and Toranzos, 2003). Application of such organic matter is widely practised worldwide and provides an important source of nutrients as well as a disposal method, particularly for organic farmers (Ingham *et al.*, 2004; Mukherjee *et al.*, 2004) even though manure may contain a variety of human pathogenic bacteria (de Freitas *et al.*, 2003; Jones, 1980; Losinger *et al.*, 1995; Pell, 1997; Wang *et al.*, 1996; Wells *et al.*, 2001; Zschock *et al.*, 2000). Composting manure has been shown to reduce populations of *E. coli* and *Salmonella* but effectiveness is dependent on composting time, temperature, and thoroughness of the composting procedure (Fukushima *et al.*, 1999; Jones and Martin, 2003; Lung *et al.*, 2001; Turner, 2002) and so risk of soil contamination with human pathogens with composted material still remains in some circumstances. In the United Kingdom, specific standards for composting including those containing animal manures are in place to ensure compost safety and such good agricultural practices should address this risk (BSI, 2005). Contamination in the field may also occur because of water run-off from nearby animal pastures and exposure to contaminated feces of wild animals (Ackers *et al.*, 1998; Hilborn *et al.*, 1999; Rice *et al.*, 1995). Pathogens may also be spread by insects (Iwasa *et al.*, 1999; Janisiewicz *et al.*, 1999; Sela *et al.*, 2005) and can be associated with dust in the air (Beuchat and Ryu, 1997). The increasing interest in applications of compost teas (extracts produced by soaking composted material in aerated or nonaerated water and collecting the material for use) as alternatives for agrochemicals also provide a risk of carrying human pathogens if their regrowth occurred during the soaking procedure (Duffy *et al.*, 2004; Ingram and Millner, 2007). Addition of nutrients to the water appears to enhance the risk of human pathogen growth.

Another potential source of contamination in the field relates to irrigation water (Steele and Odumeru, 2004; Steele *et al.*, 2005; Tyrrel *et al.*, 2006). In general, groundwater from aquifers is of good microbiological quality, surface water from ponds, lakes, and rivers is of good microbiological quality unless contaminated with surface run-off (Wachtel *et al.*, 2002a), and human wastewater (sewage) is of very poor quality, but is still used in countries where water availability is limited (Hamilton *et al.*, 2006; Ibenyassine *et al.*, 2006, 2007; Melloul *et al.*, 2001; Steele and Odumeru, 2004). The risk associated with using water from a range of sources and microbiological qualities for irrigation of salads and vegetables has been assessed and the need for improved guidelines recognized (Hamilton *et al.*, 2006; Tyrrel *et al.*, 2006). Certainly in lettuce plants, spraying with water contaminated with *E. coli* resulted in recovery of the pathogen on foliage 30 days later (Solomon *et al.*, 2003), and both surface irrigation and spray irrigation with suspensions of. *E. coli* O157:H7 lead to recovery of the pathogen from lettuce tissue albeit at a lower level than that from the irrigation treatment (Solomon *et al.*, 2002a). The lettuce remained contaminated with *E. coli* O157:H7 even after washing, indicating that irrigation of food crops with water of unknown microbiological quality should be avoided (Solomon *et al.*, 2002a). This would also include water used for preparing pesticide solutions as *Salmonella*, *E. coli* O157:H7, *Listeria monocytogenes*, and *Shigella* have all been shown to survive, and in some cases, grow in a fungicide solution (chlorothalonil as Bravo) and when these suspensions were sprayed onto leaves and fruits of tomato, *Salmonella* and *E. coli* were still found to survive (Guan *et al.*, 2001, 2005). The final source of human pathogen contamination preharvest is human handling where inappropriate hygiene and sanitation occur (Beuchat, 1996a; Bidawid *et al.*, 2000).

The period encompassing harvest, transport, processing, and distribution through to sale has several sources of potential contamination in common with the preharvest period such as human handling, animals, insects, and dust (Beuchat, 1996a). However, additional sources of contamination include harvesting equipment; transport containers (from field to packing shed); wash and rinse water; sorting, packing, cutting, and other processing equipment; ice; transport vehicles; improper storage (temperature and physical conditions allowing proliferation); crosscontamination (other foods in storage, preparation, and display areas); improper display temperature (allowing proliferation); improper handling after wholesale or retail purchase (hygiene and sanitation) (Beuchat, 1996a; Beuchat and Ryu, 1997; Holah *et al*, 2004).

In the United Kingdom and Europe, legislation has taken off the responsibility for fresh produce safety from government and moved that responsibility into the supply chain, and retailers are driving food procedures through use of their own or other recognized standards (CFA, 2002; Monaghan, 2006). For example, Marks and Spencer have instigated careful controls on the minimum number of months between soil exposure to potential human pathogen contaminants as a result of application of sewage sludge, animal manures, and composts, which differ between crops, and are supported by strict microbiological testing (Monaghan, 2006). The National Organic Program in the United States has recommended a minimum of 120 days between manure application and harvest (Anonymous, 2000) although the time required is still a matter of debate (Franz *et al.*, 2005; Ingham *et al.*, 2004, 2005). There are also Voluntary guidelines in the United States to minimize microbial food safety hazards (Anonymous, 1998). Interestingly, the approach of predicting potential problems with consuming crops irrigated with reclaimed water and microbial risks with consuming food crops in general are being addressed using a range of different quantitative microbial risk assessment models (Hamilton *et al.*, 2006; Koseki and Isobe, 2005a,b). Utilizing this approach togther with the principal of hazard analysis critical control point (HACCP), which recognizes risks and identifies appropriate control points that should be managed (de Roever, 1998; Monaghan, 2006), should enhance fresh produce safety.

IV. ECOLOGY OF HUMAN PATHOGENS IN RELATION TO PHYLLOSPHERE CONTAMINATION

In the earlier sections, identification of potential sources of human pathogen contamination in the food chain have been discussed but the ability of these microorganisms to occupy these environments and their adaptations to thrive in these environments, particularly in plants, were not considered in detail. Hence, the basic ecology of human pathogens from "farm to fork" is examined in the following sections.

A. Ecology and survival in sewage, manure, soil, and water

Properly treated sewage and effluent should be pathogen free once leaving the treatment works but viruses including hepatitis A virus, norovirus virus, and enterovirus, and bacteria such as *Listeria* spp. and *Salmonella* spp, have been found in sewage sludge and effluent (Bagdasaryan, 1964; Jones and Martin, 2003). In the United Kingdom, under the "safe sludge matrix" agreement, land growing food crops can only receive sewage sludge that has been treated to conventional or higher standards, and no raw sewage or untreated sewage sludge may be applied (ADAS, 2001). Nevertheless, the application of raw sewage or sewage sludge to soil may still go on in other parts of the world on pasture or on soils for growing crops, with concomitant risks of human pathogen contamination of soil.

The most common sources of human pathogen contamination of soil arise from animal wastes in the form of slurry or manure. Depending on the source and time of year, feces may naturally contain between 10^2 and 10^5 cfu/g *Escherichia coli* and between 10^2 and 10^7 cfu/g *Salmonella* spp. (Himathongkham *et al.*, 1999), slurry between 10 and 10^4 cfu/g *E. coli* and

Yersinia spp. (Kearney *et al.*, 1993), and manure between 10^2 and 10^7 cfu/g *Salmonella* spp. (Pell, 1997). The survival capability of human pathogens in fecally derived material have been determined (mostly in artificially inoculated feces, slurries, and manures) in numerous studies, Once again, depending on numerous factors such as pathogen, inoculum level, source of fecally derived material, temperature, pH, and aeration, a huge range of survival patterns have been observed (Table 7.1). Consequently, it is difficult to make direct comparisons but generally, human pathogens decrease in populations over storage time and this is enhanced with aeration and higher temperatures (Kearney *et al.*, 1993; Kudva *et al.*, 1998; Table 7.1). Similarly, large number of studies have been conducted on the survival of human pathogens in soil following application of slurries, manures, and composts either naturally or artificially infested with human pathogens (Table 7.2). As with survival in slurries and composts, survival in soil decreases over time with observations for *E. coli* survival ranging from as short as 10 days up to 13 years in ryegrass plots (Sjogren, 1995; Taylor and Burrows, 1971). Much of this variation reflects crop type, general climatic conditions, inoculum form and level, soil type, and associated soil physicochemical characteristics such as pH, temperature, exchange capacity, water holding capacity, organic matter content, and nutrient levels, along with biological factors including microbial competition and predation (Campo *et al.*, 2007; Franz *et al.*, 2005; Ibekwe *et al.*, 2004; Jiang *et al.*, 2002; Nicholson *et al.*, 2005; Unc and Goss, 2004). The presence of plants *per se* can also enhance survival (Gagliardi and Karns, 2002; Ibekwe *et al.*, 2004).

Water may also act as a source of human pathogens and a number of survival times for *E. coli* 0157 have been recorded. For example, 21 days in pond water at 13 °C (Porter *et al.*, 1997), 22 days in river water at 18 °C (Maule, 1999), 40 days in contaminated drinking water at 20 °C (Geldrich *et al.*, 1992), and 91 days in lake water at 8 °C but only 14–21 days at 13 °C (Wang and Doyle, 1998). However, survival may be longer as detection by traditional methods may underestimate *E. coli* 0157:H7 because of low sensitivity or the possibility that the bacteria may enter a viable but nonculturable state (Maule, 2000; Wang and Doyle, 1998). Importantly, modern molecular techniques are increasing sensitivity and speed of detection of human pathogens in the environment (Ibekwe *et al.*, 2004; Ibenyassine *et al.*, 2006; Keeling *et al.*, 2007; Moganedi *et al.*, 2007; Perry *et al.*, 2007).

B. Ecology and survival in association with plants

There have been a very large number of studies on the behavior of human pathogens on plants over the last 15 years, extending the area from simple observation of presence or absence of pathogens to more detailed

TABLE 7.1 Examples of survival of human pathogens in feces, slurries and manure

Pathogen	Material	Observation	Reference
Escherichia coli	Feces	Survival 70 days; 5 °C	Wang et al., 1996
		Survival 56 days; 22 °C	
		Survival 49 days; 37 °C	
		Survival > 54 days; 18 °C; decreasing from 10^7 to 10^6 cfu/g	Maule, 1997
		Survival 21 months; (only 47–120 d if aerobically mixed)	Kudva et al., 1998
		No survival on dry surface of feces	
		Survival > 99 days; field; decreasing from 10^8 to 10^3 cfu/g	Bolton et al., 1999
		Survival 28 days; 10 °C; (no decrease)	
	Slurry	Survival 11 weeks	Rankin and Taylor, 1969
		Survival > 9 days; 18 °C; decreasing from 10^8 to 10^2 cfu/g	Maule, 1997
		Survival 5 days; 23 °C and 37 °C; (decreasing from 10^7 to 10 cfu/g)	Kudva et al., 1998
		Survival 28 days; 4 °C; (decreasing from 10^7 to 10^5 cfu/g)	
		Decrease faster at 37 °C and slowest at 4 °C	Himathongkham et al., 1999
		Survival 3 months	Nicholson et al., 2005
	Manure	Decrease faster at 37 °C and slowest at 4 °C	Himathongkham et al., 1999
		Survival 84–133 days	Franz et al., 2005

(continued)

TABLE 7.1 (continued)

Pathogen	Material	Observation	Reference
Salmonella	Slurry	Survival 11 weeks	Rankin and Taylor, 1969
		Survival 19–33 weeks	Findlay, 1972
		Survival 27–286 days	Jones, 1980
		Survival 286 days; (90% reduction in 1 month)	Strauch and Ballarini, 1994
		Decrease faster at 37 °C and slowest at 4 °C	Himathongkham et al., 1999
		Survival 3 months	Nicholson et al., 2005
	Manure	Survival 183–204 days	Forshell and Ekesbo, 1993
		Survival > 133 days	Franz et al., 2005
Campylobacter	Manure	Survival 3 months	Nicholson et al., 2005
Listeria	Manure	Survival 6 months	Nicholson et al., 2005

TABLE 7.2 Examples of survival of human pathogens in soil

Pathogen	Inoculum material	Observation	Reference
Escherichia coli	Cells	Survival 130 days	Maule, 1999
		Survival 25–41 days; three soil types	Gagliardi and Karns, 2002
	Feces	Survival 57–99 days	Bolton et al., 1999
	Slurry	Survival 10 days	Taylor and Burrows, 1971
		Survival at least 7 days	Johannessen et al., 2004
	Manure	Survival 42–56 days; 5 °C	Jiang et al., 2002
		Survival 34–152 days; 15 °C	
		Survival 103–193 days; 21 °C	
		Survival 9 weeks (decreasing from 10^5 to 10^2 cfu/g)	Natvig et al., 2002
		Survival 132–168 days; (decreasing from 10^4 to 10^3 cfu/g in 90 days)	Ingham et al., 2004
		Survival at least 7 days	Johannessen et al., 2004
		Survival 2–56 days	Franz et al., 2005
		Survival > 17 weeks; (decreasing from 10^4 to 10^2 cfu/g in 7 weeks)	Ingham et al., 2005
		Survival 1 month	Nicholson et al., 2005
		Survival 69–92 days; decrease faster at 4 °C than ambient temperature	Mukherjee et al., 2006
	Compost	Survival 154–217 days	Islam et al., 2004a
		Survival 154–196 days	Islam et al., 2005
		Survival at least 7 days	Johannessen et al., 2004

(continued)

TABLE 7.2 (continued)

Pathogen	Inoculum material	Observation	Reference
	Water	Survival 45 days	Ibekwe et al., 2004
		Survival > 5 months	Islam et al., 2004a
		Survival 154–196 days	Islam et al., 2005
	Bacterial culture	Survival 13 years	Sjogren, 1995
Salmonella	Slurry	Survival 21 days	Baloda et al., 2001
	Manure	Survival 9 weeks (decreasing from 10^5 to 10^2 cfu/g)	Natvig et al., 2002
		Survival 56 days	Franz et al., 2005
		Survival 1 month	Nicholson et al., 2005
		Survival 161–213 days	Islam et al., 2004b
	Water	Survival 161–213 days	Islam et al., 2004b
Campylobacter	Manure	Survival 1 month	Nicholson et al., 2005
Listeria	Manure	Survival > 1 month	Nicholson et al., 2005

population dynamic studies (Table 7.3). Most studies have focussed on the behavior of *E. coli* (largely *E. coli* O157:H7) and *Salmonella* sprayed or applied directly onto foliage of plants in some way, or applied on seeds, roots, or into soil. Frequently, these studies have been carried out under glasshouse or controlled environment conditions, often at unrealistically high inoculum levels, although some studies have utilized field sites naturally contaminated with human pathogens as experimental systems. Perhaps understandably, lettuce has been the model plant for most experimentation but there have been studies of other salad and herb crops grown for leaves such as coriander, parsley, and rocket, those grown for fruits including tomato and pepper, those grown for roots such as carrot, onion and radish, as well as sprouting seeds. With so many different experimental systems it is difficult to make general statements concerning population dynamics and survival of human pathogens on crop plants, and some results may reflect artificially high levels of inoculum used (Warriner *et al.*, 2003b). Nevertheless, when applied directly to foliage by whatever means, both *E. coli* and *Salmonella* can survive there for extensive periods of time, with maximum values found on parsley in the field of 177 and 231 days, respectively (Islam *et al.*, 2004a,b). In a single study, six different human pathogens including bacteria and viruses were all found to survive for 14 days on the phyllosphere of cantaloupe, lettuce, and pepper under controlled environment conditions (Stine *et al.*, 2005), demonstrating that human pathogens can survive in this ecological niche on important crop plants for commercially relevant periods. The presence of human pathogens in the form of aggregates with other bacteria or in protected or environmentally favorable niches such as depressions in leaf veins, similar to common epiphytic bacteria, may explain part of this survival capability (Brandl and Mandrell, 2002). Nevertheless, survival may be much shorter in some environments, particularly under conditions of high insolation and ultraviolet radiation (Vaz da Costa-Vargas *et al.*, 1991).

Significantly, when these human pathogens are applied to soil or seed, they are also regularly found in aerial parts of plants as well as on roots and seed, showing that spread from the soil to the foliage is possible. Detailed microscopical and analytical studies by several workers have shown that the spread from roots to shoots can occur through plant-based surface or internal means of transmission (Cooley *et al.*, 2006; Franz *et al.*, 2007; Guo *et al.*, 2002; Ibekwe *et al.*, 2004; Jablasone *et al.*, 2005; Kutter *et al.*, 2006; Solomon *et al.*, 2002a,b; Wachtel *et al.*, 2002a; Warriner *et al.*, 2003a, 2005). For example, following application to lettuce seeds in an *in vitro* system, clusters of cells of *E. coli* O157:H7 were found to colonize the detached seed coat, and the whole length of the root, particularly the junction of the root and shoot (Cooley *et al.*, 2006). Similarly, following application to *Arabidopsis thaliana* seed in an aseptic system, cells of *E. coli*

TABLE 7.3 Examples of studies of population dynamics of human pathogens on plants

Inoculum form	Pathogen	System	Plant	Observation	Reference
Inoculation of foliage	*Escherichia coli*	Spray irrigation with effluent from water treatment plant; field	Lettuce	Survival on leaves at least 11 days after irrigation stopped	Vaz da Costa-Vargas et al., 1991
		Cells sprayed onto foliage; field	Ryegrass	Survival on leaves 41 days	Sjogren, 1995
	E. coli O157:H7	Cells sprayed onto foliage; glasshouse	Lettuce	Survival on leaves 20 days	Solomon et al., 2002a
		Cells sprayed onto foliage; CE	Lettuce	Detected on leaves for 30 days	Solomon et al., 2003
		Cells sprayed onto foliage; field	Lettuce	Detected on leaves for 77 days	Islam et al., 2004a
			Parsley	Detected on leaves for 177 days	
		Cells sprayed onto foliage; field	Tomato	Detected on leaves for at least 26 days (decreasing from 10^6 to 10 cfu/g)	Guan et al., 2005
	Salmonella	Spray irrigation with effluent from water treatment plant; field	Lettuce	Survival on leaves only 5 days after irrigation stopped	Vaz da Costa-Vargas et al., 1991
		Cells injected into stem or brushed onto flowers; glasshouse	Tomato	Pathogen spread to ripe fruit	Guo et al., 2001
		Shoot immersion; CE	Coriander	Pathogen increased in population on leaves over 6 days in humid conditions (10^4 to 10^6 cfu/g)	Brandl and Mandrell, 2002

Inoculation of seed	E. coli	Cells sprayed onto plants; field	Lettuce	Detected on leaves for 63 days	Islam et al., 2004b
			Parsley	Detected on leaves for 231 days	
		Cells sprayed on to plants; field	Tomato	Detected on leaves for at least 56 days	Guan et al., 2005
		Cells on seed; agar; CE	Mungbean	Growth occurred over 4 days	Warriner et al., 2003c
		Cells on seed; soil; glasshouse	Spinach	Detected on roots and leaves for 42 days	Warriner et al., 2003a
	E. coli O157:H7	Cells on seed; hydroponics; CE	Radish	Detected endophytically in hypocotyls after 7 days	Itoh et al., 1998
		Cells on seed; soil; CE	*Arabidopsis thaliana*	Detected for 30 days on plant and recovered on new seed	Cooley et al., 2003
		Cells on seed; soil; CE	Lettuce	Detected at least 12 days in leaf (decreasing from 10^6 to 10^3 cfu/g) Detected for at least 3 days in root (decreasing from 10^8 to 10^7 cfu/g)	Cooley et al., 2006
		Cells on seed; agar; CE	Cress, lettuce, radish, spinach	Detected at 10^5–10^7 cfu/g in all plants and persisted until maturity (49 days). Became transiently internalized.	Jablasone et al., 2005
	E. coli P36	Cells on seed; soil; CE	Celery	Detected for only 30 days	Warriner et al., 2005
			Coriander	Detected on leaves for at least 42 days	

(continued)

TABLE 7.3 (continued)

Inoculum form	Pathogen	System	Plant	Observation	Reference
	Salmonella		Lettuce	Detected on leaves for at least 42 days	Warriner et al., 2003c
			Spinach	Detected on leaves for at least 42 days	
			Watercress	Detected on leaves for only 10 days	
		Cells on seed; agar; CE	Mungbean	Growth occurred over 4 days	Cooley et al., 2003
		Cells on seed; soil; CE	Arabidopsis thaliana	Detected for 30 days on plant and recovered on new seed	
		Cells on seed; agar; CE	Cress, lettuce, radish, spinach	Detected at $10^5 - 10^7$ cfu/g in all plants and persisted until maturity (49 days). Became transiently internalized in lettuce and radish.	Jablasone et al., 2005
		Cells on seed in soil	Cowpea	Detected in plants for 45 days and in hay subsequently	Singh et al., 2007
	Listeria monocytogenes	Cells on seed; agar; CE	Cress, lettuce, radish, spinach	Detected at $10^5 - 10^7$ cfu/g in all plants and persisted until maturity (49 days). Did not become internalized.	Jablasone et al., 2005
Inoculation of roots or soil	E. coli	Sewage spill on to field soil	Cabbage	Roots contaminated but no pathogen recovered on foliage	Wachtel et al., 2002b

Organism	Treatment	Plant	Observation	Reference
	Cells in hydroponic solution; glasshouse	Spinach	Detected in roots and leaves after 16 days	Warriner et al., 2003a
	Slurry, manure and compost added to soil; field	Lettuce	Sporadic, low levels of recovery on leaves	Johannessen et al., 2004
	Manure added to soil; field	Carrot, lettuce, radish	Pathogen recovered sporadically	Ingham et al., 2004, 2005
	Cells onto seedlings in sand; gnotobiotic; CE	Barley	Colonized roots but not endophytically	Kutter et al., 2006
E. coli O157:H7	Cells into soil or hydroponic solution; CE	Lettuce	All plant parts contaminated with pathogen	Wachtel et al., 2002a
	Cells into soil microcosm ± manure	Alfalfa	Detected 47–96 days on roots	Gagliardi and Karns, 2002
		Ryegrass	Detected for 92 days on roots	
	Cells sprayed onto soil; glasshouse	Lettuce	Transferred to foliage	Solomon et al., 2002a
	Cells irrigated on to soil containing manure; CE	Lettuce	Pathogen spread from roots to leaves in 9 days	Solomon et al., 2002b
	Cells into compost; field	Lettuce	Detected for 77 days	Islam et al., 2004a
		Parsley	Detected for 177 days	
	Cells into soil microcosms; CE	Spinach	Pathogen colonized roots but did not transfer to shoots; mechanical or nematode damage did not influence colonization	Hora et al., 2005

(continued)

TABLE 7.3 (continued)

Inoculum form	Pathogen	System	Plant	Observation	Reference
		Cells into soil; CE	Lettuce	Detected in roots and phyllosphere for 45 days	Ibekwe et al., 2004
		Cells irrigated on to soil or incorporated with compost; field	Onion	Detected for 74 days on bulbs	Islam et al., 2005
			Carrot	Detected for 168 days on roots	
		Cells into manure in soil; glasshouse	Lettuce	No pathogen detection in roots or leaves after 3 or 7 weeks	Johannessen et al., 2005
		Cells into manure in soil; glasshouse	Lettuce	No detection after 21 days on leaves	Franz et al., 2005
		Cells into soil or hydroponic solution; CE	Lettuce	No detection in leaf from hydroponic system; in soil spread to leaves with internalization.	Franz et al., 2007
	Salmonella	Wastewater irrigated onto soil; field	Lettuce, parsley, pimento, tomato	All contaminated	Melloul et al., 2001
		Cells in hydroponic solution; CE	Tomato	Pathogen entered roots and spread to stems, cotyledons and leaves in 9 days	Guo et al., 2002
		Pathogen and manure added to soil; field	Radish, rocket	Sporadically recovered on both plants but depended on time of year and interval between application and planting	Natvig et al., 2002
		Cells onto seedling roots in agar; CE	Alfalfa	Endophytic colonization of roots and hypocotyls	Dong et al., 2003
			Lettuce		Franz et al., 2005

	Pathogen and manure added to soil; glasshouse		No detection after 21 days on leaves	
	Cells onto seedlings in sand; gnotobiotic; CE	Barley	Colonized roots and spread to leaves with internalization	Kutter et al., 2006
	Cells into soil or hydroponic solution; CE	Lettuce	Spread to leaves with internalization.	Franz et al., 2007
	Cells into soil	Cowpea	Detected in plants for 45 days and in hay subsequently	Singh et al., 2007
Listeria spp.	Cells onto seedlings in sand; gnotobiotic; CE	Barley	Colonized root hair zone but not other parts of the roots; did not spread to leaves or become internalized	Kutter et al., 2006

CE: controlled environment.

O157:H7 and *Salmonella* were found within the root tissue, particularly the root tip and the branch points of the lateral roots. The pathogens then moved to the crown and meristematic region, occurring at specific sites such as the base of petioles, on floral buds and eventually open flowers (Cooley *et al.*, 2003). On the leaves, colonization was concentrated in shallow depressions or near the veins, reflecting behavior of many other bacteria on the phyllosphere (Lindow and Brandl, 2003). Use of motility minus mutants indicated that flagellae were required for movement on the plant (Cooley *et al.*, 2003). Experiments in soil showed that *E. coli* O157:H7 could colonize subsurface locations in the root, hypocotyl and leaf of lettuce (Solomon *et al.*, 2002b) and that in *A. thaliana* both *E. coli* O157:H7 and *Salmonella* could be recovered from seeds and chaff after the original seed-inoculated plant had died (60 days post germination) (Cooley *et al.*, 2003), demonstrating the potential for vertical transmission of these pathogens in plants, and emphasizing the need to obtain pathogen-free seed for crop cultivation (Warriner *et al.*, 2003b). Certainly, the bacterial plant pathogen *Xanthomonas campestris* pv. *vitians* can invade and move within the vascular system of lettuce and result in seed infested with the pathogen (Barak *et al.*, 2002), illustrating that this transmission process can occur in natural conditions. Nevertheless, in some crop plants spread of human pathogens from roots to leaves may not occur (Hora *et al.*, 2005; Jablasone *et al.*, 2005; Kutter *et al.*, 2006; Wachtel *et al.*, 2002b; Warriner *et al.*, 2003a). For example, under identical gnotobiotic conditions, *Listeria* spp. survived on plant surfaces but failed to become endophytic and spread to shoots whereas *Salmonella enterica* did spread to the phyllosphere (Jablasone *et al.*, 2005; Kutter *et al.*, 2006). Interestingly, in soil microcosms, *E. coli* O157:H7 failed to colonize leaves of spinach following root colonization irrespective of mechanical or nematode damage (Hora *et al.*, 2005). This may reflect plant species and conditions used and emphasizes that behavior of human pathogens on plants cannot be easily generalized.

Survival on foliage from soil or seed inoculum may be shorter than when applied directly to the foliage (Franz *et al.*, 2005; Johannessen *et al.*, 2005) and reflect the inoculum dose used in experimentation (Solomon *et al.*, 2002b) but may still provide an avenue for plants contaminated with human pathogens to be harvested. Transfer of human pathogens has also been recorded from tomato flowers or stems to the fruit (Guo *et al.*, 2001). There is also evidence that contamination of fresh fruit and vegetables with *Salmonella* is positively associated with bacterial soft rots (Wells and Butterfield, 1997), indicating an amelioration of conditions enhancing the ability of human pathogens to survive or grow. Nevertheless, co-inoculation of intact spinach leaves with *E. coli* O157:H7 with the plant pathogen *Pseudomonas syringae* resulted in localized necrosis, but the colonization by *E. coli* was not affected (Hora *et al.*, 2005). Interestingly, some studies have demonstrated that depending on temperature and humidity, specific microbial populations on the spermosphere, rhizosphere, and phyllosphere can either enhance or decrease survival and growth of human pathogens on the plant (Brandl and Mandrell, 2002; Cooley et al., 2003, 2006; Gagliardi and Karns, 2002; Jablasone et al., 2005; Johannessen et al., 2005; O'Brien and Lindow, 1989), suggesting that applications of competing bacteria or procedures that encourage growth of competing bacteria on the plant may be used to reduce the incidence of contamination of crops by human pathogens.

One interesting aspect of the ecology of human pathogens such as E. coli, Salmonella enterica, and species of Shigella and Yersinia is that as members of the Enterobacteriaceae, they are closely related to the soft-rotting plant pathogen Erwinia carotovora subsp. atroseptica (Pectobacterium atrosepticum). Comparative genomics of the Enterobacteriaceae along with 14 plant-associated bacteria has shown that E. carotovora has acquired many coding sequences with greater sequence identity to plant associated bacteria than have animal associated enterobacterial species, probably by horizontal gene transfer of large regions or "islands" of DNA (Toth et al., 2006). They include genes encoding plant cell wall degrading enzymes, phytotoxins, capsular polysaccharides, and components of the Type III secretion system, several of which are known as potentially important for microbial colonization of the phyllosphere (Lindow and Brandl, 2003). Significantly, E. carotovora also has genes important in other bacteria for colonization of roots and competition for nutrients from root exudates in the absence of disease. These include those coding for production of all six types of bacterial secretion systems, methyl accepting chemotaxis proteins, ABC transporters and regulators, siderophores, lipopolysaccharides, quorum sensing systems, nitrogen fixation, and adhesion. As many of these genes were probably obtained through horizontal gene transfer, the potential exists for this process to occur in enterobacterial human pathogens, which could allow enhanced colonization of plants by human pathogens in future. The potential for such enhanced environmental capability through horizontal gene transfer of key genes associated with phyllosphere colonization deserves further study.

C. Ecology and survival during processing

Field-harvested plants ready for processing are naturally colonized by microorganisms and because of the importance of the raw consumption of lettuce numerous studies have examined the microbial populations present on freshly harvested lettuce leaves (Delaquis et al., 1999; Ercolani, 1976; Garg et al., 1990; King et al., 1991; Kondo et al., 2006; Li et al., 2001b; Liao and Fett, 2001; Magnuson et al., 1990; Nascimento et al., 2003; Nicholl et al., 2004; Pieczarka and Lorbeer, 1975). Although the microbial populations on lettuce leaves may vary in size depending on time of year, rainfall, irrigation, and drying (Fonseca, 2006; Kondo et al., 2006; Pieczarka and Lorbeer, 1975), some general qualitative and quantitative observations on the microbial population can be made. Both Total aerobic mesophilic bacteria ranging from 10^4 to 10^7 cfu/g and psychrophilic bacteria ranging from 10^4 to 10^6 cfu/g are generally dominated by *Pseudomonas* spp., with frequent occurrence of *Enterobacter*, *Erwinia*, and *Serratia* spp. Pectolytic bacteria, believed to be important in spoilage of lettuce, can form 10% of the total aerobic bacteria represented largely by *Pseudomonas* spp. along with *Bacillus, Erwinia, Flavobacterium*, and *Xanthomonas* spp. Coliform bacteria are found at levels between 10^3 and 10^5 cfu/g, and lactic acid bacteria, typified by *Leuconostoc*, with occasional occurrence of *Lactobacillus* and *Streptococcus*, between 10 and 10^5 cfu/g. Yeasts are found between 10^3 and 10^5 cfu/g, with most common occurrence of *Candida, Cryptococcus, Pichia, Torulospora*, and *Trichophyton* and filamentous fungi (moulds) found less frequently at $< 10^3$ cfu/g, dominated by *Aspergillus* and *Penicillium* spp. As only 0.1–3% of total bacterial populations can be grown under laboratory conditions (Amann et al., 1995), culture independent molecular-based analyses have also been used recently to characterize the phyllosphere microbial population of lettuce from the field (Handschur et al., 2005; Ibekwe and Grieve, 2004; Rudi et al., 2002). These studies have confirmed the presence of members of *Pseudomonas* along with *Acidobacterium, Agrobacterium, Bacillus, Enterobacter*, and lactic acid bacteria but additional groups such as members of *Oxalobacter* and *Microbacteriaceae* have been detected for which selective media are not commonly used or are not available. In general, populations of mesophilic bacteria are generally 1–2 log cfu/g higher on the outer leaves compared with those on the inside of the lettuce (Maxcy, 1978).

For processing, lettuce are trimmed and cored, often in the field, and then transported to a processing facility where heads are cut, sliced or shredded, and then washed, and treated in some way to reduce populations of potential spoilage microorganisms and any human pathogens. The lettuce is then dried of surface moisture and packaged, which, for iceberg lettuce, often involves use of a modified or controlled atmosphere using low oxygen and/or high carbon dioxide, and stored at low temperatures to reduce physiological browning of the tissue that occurs in response to the cutting process, and to reduce potential microbial development (Bolin and Huxsoll, 1991; Heimdal et al., 1995; López-Gálvez et al., 1996). However, attempts to reduce browning by dipping fresh cut lettuce in water for brief periods at 45 °C or 55 °C, which inhibits phenyl ammonia lyase (Loaiza-Valarde et al., 1997), may actually result in more rapid growth of some human pathogens subsequently (Li et al., 2001a, 2002). Thus, the production of minimally processed lettuce and salads is a balance to maintain good appearance, taste, and other sensory properties along with good microbiological quality (Delaquis *et al.*, 2004).

Water washes can reduce the natural microbial population to a small extent (<1 log cfu/g) (Delaquis *et al.*, 2004; Lang *et al.*, 2004; Nascimento *et al.*, 2003) and standard decontamination procedures with solutions containing approximately 20–200 µg/ml free chlorine for various lengths of time can reduce the population by a further 1–3 log cfu/g but cannot eliminate either the natural microbial population or human pathogens completely (Beuchat and Brackett, 1990; Delaquis *et al.*, 1999; Lang *et al.*, 2004; Li *et al.*, 2001b; Seo and Frank, 1999). As there is no linear relationship between free chlorine levels and decontamination achieved, there is little advantage to use of high free-chlorine levels as this affects organoleptic properties of the produce, potentially rendering it unmarketable. Indeed there is evidence that spray treatment with deionized water was equally effective in killing or removing *E. coli* O157:H7 on lettuce as solutions containing 200 µg/ml free chlorine (Beuchat, 1999). This reflects the presence of both natural microbial populations and human pathogens in protected niches where, for example, hydrophobicity of the leaf surface may prevent access by the decontamination solution, as well as those cells present within the stomata, associated with trichomes, in cracks in the cuticle, within the cut surfaces or as aggregates or biofilms, and as endophytes (Carmichael *et al.*, 1999; Hassan and Frank, 2003; Scolari and Vescovo, 2004; Seo and Frank, 1999; Takeuchi and Frank, 2000, 2001; Takeuchi *et al.*, 2000). In addition, there are clear differences between several human pathogens and members of the natural microbial population in their ability to attach to intact surfaces and cut edges of lettuce, and their degree of subsequent epiphytic and endophytic development, and this is another factor that has to be considered when developing successful decontamination treatments (Hassan and Frank, 2004; Seo and Frank, 1999; Takeuchi *et al.*, 2000). For *E. coli* O157:H7, attachment to lettuce during the processing stage does not seem to require active bacterial processes or specific cell surface moieties but reflects simple physical entrapment, which is temperature dependent (Solomon and Matthews, 2006; Takeuchi and Frank, 2000), but this may not be the same for pathogens and other bacteria on different food produce (Brandl, 2006b; Garrood *et al.*, 2004). For example, hepatitis A virus can survive for longer on lettuce after inoculation and storage than on carrot and fennel (Croci *et al.*, 2002). Recently, mutants of *S. enterica* with reduced adhesion to alfalfa sprouts were identified including one mutant that was blocked in curli (fimbriae-like structures) formation, needed for binding to and infection of animal cells (Barak *et al.*, 2005). This may indicate that these curli have a role in attachment to both plant and animal cells (Brandl, 2006b). Certainly, understanding the interrelationships between binding, survival, and susceptibility to sanitizing procedures will aid in developing the safety of minimally processed fruit and vegetables (Takeuchi and Frank, 2000).

Those microorganisms present after the decontamination treatment are then free to grow, potentially stimulated by increased nutrient availability from the cut surfaces. Depending on the decontamination process used, and the length and temperature of storage, aerobic mesophilic bacteria can regularly increase on minimally processed lettuce from 10^3 to 10^6 cfu/g or greater although other groups appear to change to a lesser extent (Abdul-Raouf *et al.*, 1993; King *et al.*, 1991; Li *et al.*, 2001b, 2002; Nicholl *et al.*, 2004). Nevertheless, even though the pre-existing microbial population provides the inoculum for this growth, specific microorganisms or groups of microorganisms such as psychrophiles preferentially grow under these conditions and their composition can reflect differences in origin of the crop as well as the storage conditions (Handschur *et al.*, 2005; Magnuson *et al.*, 1990; Rudi *et al.*, 2002). Importantly, if present, human pathogens can also survive and potentially proliferate during storage providing a hazard to consumption of minimally processed vegetables (Beuchat, 1999). Not surprisingly, therefore, great efforts are being made to improve decontamination treatments, including searching for new chemicals, exploring chemical concentrations and exposure times, along with different forms of radiation, sanitising gases, and empirically manipulating the environmental conditions used during the decontamination process without compromising appearance or other sensory properties of lettuce (Beuchat, 1998; Beuchat *et al.*, 2004; Bialka and Demirci, 2007; Delaquis *et al*, 1999, 2004; Hassenberg *et al.*, 2007; Huang *et al.*, 2006; Kim *et al.*, 2006; Kondo *et al.*, 2006; Koseki *et al.*, 2004; Li *et al.*, 2001b; Nascimento *et al.*, 2003; Oh *et al.*, 2005; Rodgers *et al.*, 2004; Singh *et al.*, 2002; Weissinger *et al.*, 2000; Yuk *et al.*, 2006). This is still a major challenge under intense study.

A novel area relating to ecology of human pathogens during processing concerns their control through use of natural microorganisms. Microorganisms, in general, grow and compete for attachment sites and nutrients in the phyllosphere both before harvest and during production of minimally processed lettuce, and these phenomena could be utilized to reduce the levels of human pathogens. For example, in *in vitro* studies, *Enterobacter asburiae* decreased the survival of *E. coli* O157:H7 by 20–30 times on lettuce foliage after 10 days when grown from co-inoculated lettuce seed and reflected early inhibition of *E. coli* O157:H7 in the rhizosphere (Cooley *et al.*, 2006). Thus the possibility exists for management of crop growth to encourage such competing bacteria like *E. asburiae* to reduce incidence of human pathogens on the crop (Cooley *et al.*, 2006) or alternatively, utilizing the bacteria directly as a biocontrol agent applied to seed (Jablasone *et al.*, 2005) or to the foliage in the field. However, some care must be taken with these approaches as in the same study, another bacterium, *Wasteria paucula*, applied in the same way enhanced the survival of *E. coli* O157:H7 by sixfold on lettuce foliage, emphasizing the need for careful ecological studies before any such approach is utilized.

Similarly, selection and use of antagonistic bacteria, particularly lactic acid bacteria, as biocontrol agents against human pathogens in minimally processed lettuce and other vegetables has started to be explored (Kostrzynska and Bachand, 2006). Application of such bacteria or other microorganisms as "biopreservatives" to fresh cut salads to extend shelf life and safety could represent a considerable opportunity if the toxicological and environmental impact along with appropriate economic benefit could be achieved (Liao and Fett, 2001). For example, 37 bacterial isolates with *in vitro* inhibitory activities against *E. coli* O157:H7, *Listeria monocytogenes, Salmonella*, or *Staphylococcus aureus* were obtained from lettuce shreds from minimally processed vegetables (Schuenzel and Harrison, 2002) and a single *Pseudomonas fluorescens* isolate obtained from Romaine lettuce showed *in vitro* inhibition of *E. coli, L. monocytogenes*, and *Erwinia carotovora* (Liao and Fett, 2001). Some studies have also extended this work to inhibition on plant material. For instance, *Lactobacillus casei* added to mechanically damaged escarole lettuce inhibited *E. coli* O157:H7, *Listeria monocytogenes, Aeromonas hydrophila*, and *Staphylococcus aureus* (Scolari and Vescovo, 2004) but a *Lactobacillus delbruckeii* subsp *lactis* strain capable of producing hydrogen peroxide, when added to chopped vegetables including lettuce had no effect on *E. coli* O157:H7 and *Listeria monocytogenes*. In this case, the cut vegetables contained enough catalase activity to destroy the hydrogen peroxide and prevent antagonistic action against the pathogens (Harp and Gilliland, 2003). An alfalfa-derived microbial flora inoculated onto alfalfa seeds reduced growth of introduced *Salmonella* more effectively and for longer than a single isolate of *Pseudomonas fluorescens*, indicating not only that different modes of action were occurring but that multistrain inocula could be a useful approach to control human pathogens on alfalfa sprouts (Matos and Garland, 2005). Similarly, the native microflora from fresh peeled baby carrots has also been reported to inhibit human pathogens in storage (Liao, 2007). Such bacterial–bacterial interactions generally involve production of inhibitory compounds or competition for nutrients, space, or colonization sites (Whipps, 2001) but recently, the concept of utilizing bacteria that interfere with quorum sensing systems in other bacteria for biocontrol of both human and plant pathogens has been mooted (Rasmussen and Givskov, 2006) and could be an avenue worth examining in the fresh produce situation. These studies again reinforce the need to understand the ecology of the microbial interactions with human pathogens in the phyllosphere and during processing.

In addition to the use of bacteria for biocontrol of human pathogens on fresh produce, opportunities exist for the exploitation of bacteriophages for the same purpose (Greer, 2005; Hudson *et al.*, 2005). Potential for control of *E. coli* O157:H7, *Listeria monocytogenes*, and *Salmonella* with bacteriophages has been demonstrated in culture and in some instances on fresh produce (Carey-Smith *et al.*, 2006; Kudva *et al.*, 1999; Leverentz *et al.*, 2001, 2003). Nevertheless, their use is limited by the apparent need for a threshold concentration of host before replication can proceed, poor lysis at suboptimal temperatures, narrow host range, phage resistant mutants, and the potential for the transduction of genetic information from one bacterial strain to another (Greer, 2005; Hudson *et al.*, 2005). Ecological concerns about bacteriophage use in nature as well as during processing in addition to general consumer acceptance of the use of bacteriophages may also limit their widespread use.

Another way of controlling human pathogen populations on the phyllosphere of crop plants that seems to have been overlooked is to breed plants with lower ability to support human pathogen survival and growth. This may reflect the inherent ability of different human pathogens to colonize plant tissues or the ability to influence populations of competing microorganisms on plant tissues, both of which have been discussed earlier. In theory, genetically modified plants could be used to control microbial populations on the phyllosphere but this is very unlikely to be adopted commercially in Europe. Therefore, the potential to use plant breeding to reduce incidence of human pathogens also deserves study. In addition, other than a brief observation based on an axenic study of alfalfa seedling colonization by *Salmonella* (Dong *et al.*, 2003), there appear to be no studies of any cultivar effects of any plant on human pathogen development in association with the phyllosphere of rhizosphere and this also needs to be investigated further.

V. CONCLUSIONS AND FUTURE

It is clear that human pathogens, particularly *E. coli* and *S. enterica*, can survive in the phyllosphere of plants in the environment. Depending on the environmental conditions, plant species and inoculum level, these pathogens can spread both epiphytically and endophytically and pose a risk associated with the consumption of fresh fruit and vegetables. Many processes are already in place to minimize the risk of contamination but several avenues for research may allow some novel approaches for control of these pathogens on fresh produce. For example, there is a need to examine the role of genetic variation in plants for colonization of the phyllosphere (and rhizosphere) by human pathogens *per se* and to extend these studies to identify those plant characteristics that could be realistically manipulated by plant breeding to decrease the pathogen colonization. Studies of the effects and properties of plant cultivars in relation to human pathogen colonization could be valuable rapid approach to gain information in this area. In addition, the agricultural practices that may affect plant growth such that human pathogens cannot grow or are out-competed by the natural microbial population deserves further study. Understanding the ecology and factors controlling the attachment of both natural microbial populations and human pathogens to intact and cut plant surfaces and the sensitivity of these microorganisms to sanitizing procedures during processing still warrants further research even though this is currently a major focus of study. This could lead to a reduction in microorganisms on fresh produce and lead to a decrease in the risk of foodborne illness. Another direct intervention strategy worthy of further consideration involves the development of biological control procedures against human pathogens in the crop as well as during processing. Finally, with the increasing availability of genomic information concerning many enteric pathogens, the potential for the development of enhanced environmental capability of human pathogens thorough horizontal gene transfer of key genes associated with phyllosphere colonization merits further examination.

ACKNOWLEDGMENTS

We thank the Department for Environment Food and Rural Affairs (Project FO0305) for financial support.

REFERENCES

Abdul-Raouf, U. M., Beuchat, L. R., and Ammar, M. S. (1993). Survival and growth of *Escherichia coli* O157:H7 on salad vegetables. *Appl. Environ. Microbiol.* **59**, 1999–2006.

Ackers, M. L., Mahon, B. E., Leahy, E., Goode, B., Damrow, T., Hayes, P. S., Bibb, W. F., Rice, D. H., Barrett, T. J., Hutwagner, L., Griffin, P. M., and Slutsker, L. (1998). An outbreak of *Escherichia coli* O157:H7 infections associated with leaf lettuce consumption. *J. Infect. Dis.* **177**, 1588–1593.

ADAS (2001). "The Safe Sludge Matrix: Guidelines for the Application of Sewage Sludge to Agricultural Land,". 3rd ed. ADAS, Wolverhampton, UK.

Amann, R. I., Ludwig, W., and Schleifer, K. H. (1995). Phylogenetic identification and *in situ* detection of individual microbial cells without cultivation. *Microbiol. Rev.* **59**, 143–169.

Anonymous (1998). Guide to Minimize Microbial Food Safety Hazards for Fresh Fruits and Vegetables. http://vm.cfsan.fda.gov/~dms/prodguide.html.

Anonymous (2000). National Organic Program 7 CFR part 205:203, federal register **65**, pp. 805–80684. USDA, Washington, DC.

Aruscavage, D., Lee, K., Miller, S., and LeJeune, J. T. (2006). Interactions affecting the proliferation and control of human pathogens on edible plants. *J. Food Sci.* **71**, R89–R99.

Bagdasaryan, G. A. (1964). Survival of viruses of the enterovirus group (poliomyelitis, ECHO, Coxsackie) in soil and on vegetables. *J. Hygiene Epidemiol. Microbiol. Immunol.* **8,** 497–505.

Baloda, S. B., Christensen, L., and Trajcevska, S. (2001). Persistence of a *Salmonella enterica* serovar typhimurium DT12 clone in a piggery and in agricultural soil amended with *Salmonella*-contaminated slurry. *Appl. Environ. Microbiol.* **67,** 2859–2862.

Barak, J. D., Gorski, L., Naraghi-Arani, P., and Charkowski, A. O. (2005). *Salmonella enterica* virulence genes are required for bacterial attachment to plant tissue. *Appl. Environ. Microbiol.* **71,** 5685–5691.

Barak, J. D., Koike, S. T., and Gilbertson, R. L. (2002). Movement of *Xanthomonas campestris* pv. *vitians* in the stems of lettuce and seed contamination. *Plant Pathol.* **51,** 506–512.

Berg, G., Eberl, L., and Hartmann, A. (2005). The rhizosphere as a reservoir for opportunistic human pathogenic bacteria. *Environ. Microbiol.* **7,** 1673–1685.

Beuchat, L. R. (1996a). Pathogenic microorganisms associated with fresh produce. *J. Food Prot.* **59,** 204–216.

Beuchat, L. R. (1996b). *Listeria monocytogenes*: Incidence on vegetables. *Food Control* **7,** 223–228.

Beuchat, L. R. (1998). Surface decontamination of fruits and vegetables eaten raw: A review Food Safety Unit, World Health Organisation, WHO/FSF/98.2. http://www.who.int/foodsafety/publications/fs_management/en/surface_decon.pdf.

Beuchat, L. R. (1999). Survival of enterohemorrhagic *Escherichia coli* O157:H7 in bovine feces applied to lettuce and the effectiveness of chlorinated water as a disinfectant. *J. Food Prot.* **62,** 845–849.

Beuchat, L. R. (2002). Ecological factors influencing survival and growth of human pathogens on raw fruits and vegetables. *Microbes Infect.* **4,** 413–423.

Beuchat, L. R., Adler, B. B., and Lang, M. M. (2004). Efficacy of chlorine and a peroxyacetic acid sanitizer in killing *Listeria monocytogenes* on iceberg and Romaine lettuce using simulated commercial processing conditions. *J. Food Prot.* **67,** 1238–1242.

Beuchat, L. R., and Brackett, R. E. (1990). Survival and growth of *Listeria monocytogenes* on lettuce as influenced by shredding, chlorine treatment, modified atmosphere packaging and temperature. *J. Food Sci.* **55,** 755–758.

Beuchat, L. R., and Ryu, J. H. (1997). Produce handling and processing practices. *Emerg. Infect. Dis.* **3,** 459–465.

Bialka, K. L., and Demirci, A. (2007). Utilization of gaseous ozone for the decontamination of *Escherichia coli* O157:H7 and *Salmonella* on raspberries and strawberries. *J. Food Prot.* **70,** 1093–1098.

Bidawid, S., Farber, J. M., and Sattar, S. A. (2000). Contamination of foods by food handlers: Experiments on hepatitis A virus transfer to food and its interruption. *Appl. Environ. Microbiol.* **66,** 2759–2763.

Black, R. E., Levine, M. M., Clements, M. L., Hughes, T. P., and Blaser, M. J. (1988). Experimental *Campylobacter jejuni* infection in humans. *J. Infect. Dis.* **157,** 472–479.

Boehme, S., Werner, G., Klare, I., Reissbrodt, R., and Witte, W. (2004). Occurrence of antibiotic-resistant enterobacteria in agricultural foodstuffs. *Mol. Nutr. Food Res.* **48,** 522–531.

Bolin, H. R., and Huxsoll, C. C. (1991). Effect of preparation procedures and storage parameters on quality retention of salad-cut lettuce. *J. Food Sci.* **56,** 60.

Bolton, D. J., Byrne, C. M., Sheridan, J. J., McDowell, D. A., and Blair, I. S. (1999). The survival characteristics of a non-toxigenic strain of *Escherichia coli* O157:H7. *J. Appl. Microbiol.* **86,** 407–411.

Brandl, M. T. (2006a). Fitness of human enteric pathogens on plants and implications for food safety. *Annu. Rev. Phytopathol.* **44,** 367–392.

Brandl, M. T. (2006b). Human pathogens and the health threat of the phyllosphere. *In* "Microbial Ecology of Aerial Plant Surfaces" (M. J. Bailey, A. K. Lilley, T. M. Timms-Wilson, and P. T. N. Spencer-Phillips, Eds.), pp. 269–285. CABI International, Wallingford.

Brandl, M. T., and Mandrell, R. E. (2002). Fitness of *Salmonella enterica* serovar Thompson in the cilantro phyllosphere. *Appl. Environ. Microbiol.* **68,** 3614–3621.

BSI (2005). "PAS100 specification for composted materials." British Standards Institute, London.

Burnett, S. L., and Beuchat, L. R. (2000). Human pathogens associated with raw produce and unpasteurized juices, and difficulties in decontamination. *J. Ind. Microbiol. Biotechnol.* **25,** 281–287.

Campbell, J. V., Mohle-Boetani, J., Reporter, R., Abbott, S., Farrar, J., Brandl, M., Mandrell, R., and Werner, S. B. (2001). An outbreak of *Salmonella* serotype Thompson associated with fresh cilantro. *J. Infect. Dis.* **183,** 984–987.

Campo, N. C. D., Pepper, I. L., and Gerba, C. P. (2007). Assessment of *Salmonella typhimurium* growth in class A biosolids and soil/biosolid mixtures. *J. Resid. Sci. Technol.* **4,** 83–88.

Cao, H., Baldini, R. L., and Rahme, L. G. (2001). Common mechanisms for pathogens of plants and animals. *Annu. Rev. Phytopathol.* **39,** 259–284.

Carey-Smith, G. V., Billington, C., Cornelius, A. J., Hudson, J. A., and Heinemann, J. A. (2006). Isolation and characterization of bacteriophages infecting *Salmonella* spp. *FEMS Microbiol. Lett.* **258,** 182–186.

Carmichael, I., Harper, I. S., Coventry, M. J., Taylor, P. W. J., Wan, J., and Hickey, M. W. (1999). Bacterial colonization and biofilm development on minimally processed vegetables. *J. Appl. Microbiol.* **85,** 45S–51S.

CDC (US center for disease control and prevention) (1994). Foodborne outbreaks of enteroxigenic *Escherichia coli* – Rhode Island and New Hampshire, 1993. *Morbidity and Mortality Weekly Report* **43,** 87–89.

CDC (US center for disease control and prevention) (2004). Foodborne outbreaks due to bacterial etiologies, 2003. http://www.cdc.gov/foodborneoutbreaks/us_outb/fbo2003./summary03.htm.

CDC (US center for disease control and prevention) (2006). Update on multi-state outbreak of *E. coli* O157:H7 infections from fresh spinach, October 2006. http://www.cdc.gov/ecoli/2006/September/updates/100606.htm.

CFA (2002). Microbiological guidance for produce suppliers to chilled food manufacturers. Chilled Food Association, Kettering, UK.

Cooley, M. B., Chao, D., and Mandrell, R. E. (2006). *Escherichia coli* O157:H7 survival and growth on lettuce is altered by the presence of epiphytic bacteria. *J. Food Prot.* **69,** 2329–2335.

Cooley, M. B., Miller, W. G., and Mandrell, R. E. (2003). Colonization of *Arabidopsis thaliana* with *Salmonella enterica* and enterohemorrhagic *Escherichia coli* O157:H7 and competition by *Enterobacter asburiae*. *Appl. Environ. Microbiol.* **69,** 4915–4926.

Croci, L., De Medici, D., Scalfaro, C., Fiore, A., and Toti, L. (2002). The survival of hepatitis A virus in fresh produce. *Int. J. Food Microbiol.* **73,** 29–34.

Cummings, K., Barrett, E., Mohle-Boetani, J. C., Brooks, J. T., Farrar, J., Hunt, T., Fiore, A., Komatsu, K., Werner, S. B., and Slutsker, L. (2001). A multistate outbreak of *Salmonella enterica* serotype baildon associated with domestic raw tomatoes. *Emerg. Infect. Dis.* **7,** 1046–1048.

de Freitas, J. R., Schoenau, J. J., Boyetchko, S. M., and Cyrenne, S. A. (2003). Soil microbial populations, community composition, and activity as affected by repeated applications of hog and cattle manure in eastern Saskatchewan. *Can. J. Microbiol.* **49,** 538–548.

de Roever, C. (1998). Microbiological safety evaluations and recommendations on fresh produce. *Food Control* **9,** 321–347.

Delaquis, P. J., Fukumoto, L. R., Toivonen, P. M. A., and Cliff, M. A. (2004). Implications of wash water chlorination and temperature for the microbiological and sensory properties of fresh-cut iceberg lettuce. *Postharvest Biol. Technol.* **31**, 81–91.

Delaquis, P. J., Stewart, S., Toivonen, P. M. A., and Moyls, A. L. (1999). Effect of warm, chlorinated water on the microbial flora of shredded iceberg lettuce. *Food Res. Int.* **32**, 7–14.

Dong, Y. M., Iniguez, A. L., Ahmer, B. M. M., and Triplett, E. W. (2003). Kinetics and strain specificity of rhizosphere and endophytic colonization by enteric bacteria on seedlings of *Medicago sativa* and *Medicago truncatula*. *Appl. Environ. Microbiol.* **69**, 1783–1790.

Duffy, B., Sarreal, C., Ravva, S., and Stanker, L. (2004). Effect of molasses on regrowth of *E. coli* O157:H7 and *Salmonella* in compost teas. *Compost Sci. Util.* **12**, 93–96.

Ercolani, G. L. (1976). Bacteriological quality assessment of fresh marketed lettuce and fennel. *Appl. Environ. Microbiol.* **31**, 847–852.

Farber, J. M., and Peterkin, P. I. (1991). *Listeria monocytogenes*, a food-borne pathogen. *Microbiol. Rev.* **55**, 476–511.

Fenlon, D. R. (1985). Wild birds and silage as reservoirs of *Listeria* in the agricultural environment. *J. Appl. Bact.* **59**, 537–543.

Findlay, C. R. (1972). Persistence of *Salmonella* Dublin in slurry in tanks and on pasture. *Vet. Rec.* **91**, 233–235.

Fonseca, J. M. (2006). Postharvest quality and microbial population of head lettuce as affected by moisture at harvest. *J. Food Sci.* **71**, M45–M49.

Forshell, L. P., and Ekesbo, I. (1993). Survival of Salmonellas in composted and not composted solid animal manures. *J. Vet. Med. B* **40**, 654–658.

Franz, E., van Diepeningen, A. D., de Vos, O. J., and van Bruggen, A. H. C. (2005). Effects of cattle feeding regimen and soil management type on the fate of *Escherichia coli* O157:H7 and *Salmonella enterica* serovar Typhimurium in manure, manure-amended soil, and lettuce. *Appl. Environ. Microbiol.* **71**, 6165–6174.

Franz, E., Visser, A. A., van Diepeningen, A. D., Klerks, M. M., Termorshuizen, A. J., and van Bruggen, A. H. C. (2007). Quantification of contamination of lettuce by GFP-expressing *Escherichia coli* O157:H7 and *Salmonella* enterica serovar Typhimurium. *Food Microbiol.* **24**, 106–112.

Froder, H., Martins, C. C., de Souza, K. L. O., Landgraf, M., Franco, B. D. G. M., and Destro, M. T. (2007). Minimally processed vegetable salads: Microbial quality evaluation. *J. Food. Prot.* **70**, 1277–1280.

Fukushima, H., Hoshina, K., and Gomyoda, M. (1999). Long-term survival of Shiga toxin-producing *Escherichia coli* O26, O113, and O157 in bovine feces. *Appl. Environ. Microbiol.* **65**, 5177–5181.

Gagliardi, J. V., and Karns, J. S. (2002). Persistence of *Escherichia coli* O157:H7 in soil and on plant roots. *Environ. Microbiol.* **4**, 89–96.

Garg, N., Churey, J. J., and Splittstoesser, D. F. (1990). Effect of processing conditions on the microflora of fresh-cut vegetables. *J. Food Prot.* **53**, 701–703.

Garrood, M. J., Wilson, P. D. G., and Brocklehurst, T. F. (2004). Modeling the rate of attachment of *Listeria monocytogenes*, *Pantoea agglomerans*, and *Pseudomonas fluorescens* to, and the probability of their detachment from, potato tissue at 10 °C. *Appl. Environ. Microbiol.* **70**, 3558–3565.

Geldrich, E. E., Fox, K. R., Goodrich, J. A., Rice, E. W., Clark, R. M., and Swerdlow, D. L. (1992). Searching for a water supply connection in the Cabool, Missouri disease outbreak of *Escherichia coli* O157-H7. *Water Res.* **26**, 1127–1137.

Greer, G. G. (2005). Bacteriophage control of foodborne bacteria. *J. Food Prot.* **68**, 1102–1111.

Guan, T. T. Y., Blank, G., and Holley, R. A. (2005). Survival of pathogenic bacteria in pesticide solutions and on treated tomato plants. *J. Food Prot.* **68**, 296–304.

Guan, T. Y., Blank, G., Ismond, A., and van Acker, R. (2001). Fate of foodborne bacterial pathogens in pesticide products. *J. Sci. Food Agric.* **81**, 503–512.

Guo, X., Chen, J. R., Brackett, R. E., and Beuchat, L. R. (2001). Survival of Salmonellae on and in tomato plants from the time of inoculation at flowering and early stages of fruit development through fruit ripening. *Appl. Environ. Microbiol.* **67**, 4760–4764.

Guo, X., Chen, J. R., Brackett, R. E., and Beuchat, L. R. (2002). Survival of *Salmonella* on tomatoes stored at high relative humidity, in soil, and on tomatoes in contact with soil. *J. Food Prot.* **65**, 274–279.

Hamilton, A. J., Stagnitti, F., Premier, R., Boland, A. M., and Hale, G. (2006). Quantitative microbial risk assessment models for consumption of raw vegetables irrigated with reclaimed water. *Appl. Environ. Microbiol.* **72**, 3284–3290.

Handschur, M., Pinar, G., Gallist, B., Lubitz, W., and Haslberger, A. G. (2005). Culture free DGGE and cloning based monitoring of changes in bacterial communities of salad due to processing. *Food Chem. Toxic.* **43**, 1595–1605.

Harp, E., and Gilliland, S. E. (2003). Evaluation of a select strain of *Lactobacillus delbrueckii* subsp *lactis* as a biological control agent for pathogens on fresh-cut vegetables stored at 7 °C. *J. Food Prot.* **66**, 1013–1018.

Hassan, A. N., and Frank, J. F. (2003). Influence of surfactant hydrophobicity on the detachment of *Escherichia coli* O157:H7 from lettuce. *Int. J. Food Microbiol.* **87**, 145–152.

Hassan, A. N., and Frank, J. F. (2004). Attachment of *Escherichia coli* O157:H7 grown in tryptic soy broth and nutrient broth to apple and lettuce surfaces as related to cell hydrophobicity, surface charge, and capsule production. *Int. J. Food Microbiol.* **96**, 103–109.

Hassenberg, K., Idler, C., Molloy, E., Geyer, M., Plochl, M., and Barnes, J. (2007). Use of ozone in a lettuce–washing process: An industrial trial. *J. Food Sci. Agric.* **87**, 914–919.

Heimdal, H., Kuhn, B. F., Poll, L., and Larsen, L. M. (1995). Biochemical changes and sensory quality of shredded and MA-packaged iceberg lettuce. *J. Food Sci.* **60**, 1265–1268.

Hilborn, E. D., Mermin, J. H., Mshar, P. A., Hadler, J. L., Voetsch, A., Wojtkunski, C., Swartz, M., Mshar, R., Lambert-Fair, M. A., Farrar, J. A., Glynn, M. K., and Slutsker, L. (1999). A multistate outbreak of *Escherichia coli* O157:H7 infections associated with consumption of mesclun lettuce. *Arch. Intern. Med.* **159**, 1758–1764.

Himathongkham, S., Bahari, S., Riemann, H., and Cliver, D. (1999). Survival of *Escherichia coli* O157:H7 and *Salmonella typhimurium* in cow manure and cow manure slurry. *FEMS Microbiol. Lett.* **178**, 251–257.

Holah, J. T., Bird, J., and Hall, K. E. (2004). The microbial ecology of high-risk, chilled food factories; evidence for persistent *Listeria* spp. and *Escherichia coli* strains. *J. Appl. Microbiol.* **97**, 68–77.

Hora, R., Warriner, K., Shelp, B. J., and Griffiths, M. W. (2005). Internalization of *Escherichia coli* O157:H7 following biological and mechanical disruption of growing spinach plants. *J. Food Prot.* **68**, 2506–2509.

Horby, P. W., O'Brien, S. J., Adak, G. K., Graham, C., Hawker, J. I., Hunter, P., Lane, C., Lawson, A. J., Mitchell, R. T., Reacher, M. H., Threlfall, E. J., and Ward, L. R. (2003). A national outbreak of multi-resistant *Salmonella enterica* serovar Typhimurium definitive phage type (DT) 104 associated with consumption of lettuce. *Epidemiol. Infect.* **130**, 169–178.

Huang, T. S., Xu, C. L., Walker, K., West, P., Zhang, S. Q., and Weese, J. (2006). Decontamination efficacy of combined chlorine dioxide with ultrasonication on apples and lettuce. *J. Food Sci.* **71**, M134–M139.

Hudson, J. A., Billington, C., Carey-Smith, G., and Greening, G. (2005). Bacteriophages as biocontrol agents in food. *J. Food Prot.* **68**, 426–437.

Ibekwe, A. M., and Grieve, C. M. (2004). Changes in developing plant microbial community structure as affected by contaminated water. *FEMS Microbiol. Ecol.* **48**, 239–248.

Ibekwe, A. M., Watt, P. M., Shouse, P. J., and Grieve, C. M. (2004). Fate of *Escherichia coli* O157:H7 in irrigation water on soils and plants as validated by culture method and real-time PCR. *Can. J. Microbiol.* **50**, 1007–1014.

Ibenyassine, K., AitMhand, R., Karamoko, Y., Cohen, N., and Ennaji, M. M. (2006). Use of repetitive DNA sequences to determine the persistence of enteropathogenic *Escherichia coli* in vegetables and in soil grown in fields treated with contaminated irrigation water. *Lett. Appl. Microbiol.* **43**, 528–533.

Ibenyassine, K., Mhand, R. A., Karamoko, Y., Anajjar, B., Chouibani, M., and Ennaji, M. M. (2007). Bacterial pathogens recovered from vegetables irrigated by wastewater in Morocco. *J. Environ. Health* **69**, 47–51.

Ingham, S. C., Fanslau, M. A., Engel, R. A., Breuer, J. R., Wright, T. H., Reith-Rozelle, J. K., and Zhu, J. (2005). Evaluation of fertilization-to-planting and fertilization-to-harvest intervals for safe use of noncomposted bovine manure in Wisconsin vegetable production. *J. Food Prot.* **68**, 1134–1142.

Ingham, S. C., Losinski, J. A., Andrews, M. P., Breuer, J. E., Breuer, J. R., Wood, T. M., and Wright, T. H. (2004). *Escherichia coli* contamination of vegetables grown in soils fertilized with noncomposted bovine manure: Garden-scale studies. *Appl. Environ. Microbiol.* **70**, 6420–6427.

Ingram, D. T., and Millner, P. D. (2007). Factors affecting compost tea as a potential source of *Escherichia coli* and *Salmonella* on fresh produce. *J. Food Prot.* **70**, 828–834.

Islam, M., Doyle, M. P., Phatak, S. C., Millner, P., and Jiang, X. P. (2004a). Persistence of enterohemorrhagic *Escherichia coli* O157:H7 in soil and on leaf lettuce and parsley grown in fields treated with contaminated manure composts or irrigation water. *J. Food Prot.* **67**, 1365–1370.

Islam, M., Doyle, M. P., Phatak, S. C., Millner, P., and Jiang, X. P. (2005). Survival of *Escherichia coli* O157:H7 in soil and on carrots and onions grown in fields treated with contaminated manure composts or irrigation water. *Food Microbiol.* **22**, 63–70.

Islam, M., Morgan, J., Doyle, M. P., Phatak, S. C., Millner, P., and Jiang, X. P. (2004b). Persistence of *Salmonella enterica* serovar Typhimurium on lettuce and parsley and in soils on which they were grown in fields treated with contaminated manure composts or irrigation water. *Foodborne Path. Dis.* **1**, 27–35.

Itoh, Y., Sugita-Konishi, Y., Kasuga, F., Iwaki, M., Hara-Kudo, Y., Saito, N., Noguchi, Y., Konuma, H., and Kumagai, S. (1998). Enterohemorrhagic *Escherichia coli* O157:H7 present in radish sprouts. *Appl. Environ. Microbiol.* **64**, 1532–1535.

Iwasa, M., Makino, S., Asakura, H., Kobori, H., and Morimoto, Y. (1999). Detection of *Escherichia coli* O157:H7 from *Musca domestica* (Diptera: Muscidae) at a cattle farm in Japan. *J. Med. Entomol.* **36**, 108–112.

Jablasone, J., Warriner, K., and Griffiths, M. (2005). Interactions of *Escherichia coli* 0157:147, *Salmonella typhimurium* and *Listeria monocytogenes* with plants cultivated in a gnotobiotic system. *Int. J. Food Microbiol.* **99**, 7–18.

Janisiewicz, W. J., Conway, W. S., Brown, M. W., Sapers, G. M., Fratamico, P., and Buchanan, R. L. (1999). Fate of *Escherichia coli* O157:H7 on fresh-cut apple tissue and its potential for transmission by fruit flies. *Appl. Environ. Microbiol.* **65**, 1–5.

Jaquette, C. B., Beuchat, L. R., and Mahon, B. E. (1996). Efficacy of chlorine and heat treatment in killing *Salmonella* Stanley inoculated onto alfalfa seeds and growth and survival of the pathogen during sprouting and storage. *Appl. Environ. Microbiol.* **62**, 2212–2215.

Jay, J. M., Loessner, M. J., and Golden, D. A. (2005). "Modern Food Microbiology,". 7th edn. Springer Science, New York.

Jiang, X. P., Morgan, J., and Doyle, M. P. (2002). Fate of *Escherichia coli* O157:H7 in manure-amended soil. *Appl. Environ. Microbiol.* **68**, 2605–2609.

Johannessen, G. S., Bengtsson, G. B., Heier, B. T., Bredholt, S., Wasteson, Y., and Rorvik, L. M. (2005). Potential uptake of *Escherichia coli* O157:H7 from organic manure into crisphead lettuce. *Appl. Environ. Microbiol.* **71**, 2221–2225.

Johannessen, G. S., Froseth, R. B., Solemdal, L., Jarp, J., Wasteson, Y., and Rorvik, L. M. (2004). Influence of bovine manure as fertilizer on the bacteriological quality of organic Iceberg lettuce. *J. Appl. Microbiol.* **96**, 787–794.

Johannessen, G. S., Loncarevic, S., and Kruse, H. (2002). Bacteriological analysis of fresh produce in Norway. *Int. J. Food Microbiol.* **77,** 199–204.

Johnston, L. M., Jaykus, L. A., Moll, D., Martinez, M. C., Anciso, J., Mora, B., and Moe, C. L. (2005). A field study of the microbiological quality of fresh produce. *J. Food Prot.* **68,** 1840–1847.

Jones, P., and Martin, M. (2003). "The occurrence and survival of pathogens of animals and humans in green compost." The Waste and Resources Action Programme, Banbury, UK.

Jones, P. W. (1980). Health hazards associated with the handling of animal wastes. *Vet. Rec.* **106,** 4–7.

Kapperud, G., Rorvik, L. M., Hasseltvedt, V., Hoiby, E. A., Iversen, B. G., Staveland, K., Johnsen, G., Leitao, J., Herikstad, H., Andersson, Y., Langeland, G., Gondrosen, B., *et al.* (1995). Outbreak of *Shigella sonnei* infection traced to imported iceberg lettuce. *J. Clin. Microbiol.* **33,** 609–614.

Kearney, T. E., Larkin, M. J., and Levett, P. N. (1993). The effect of slurry storage and anaerobic digestion on survival of pathogenic bacteria. *J. Appl. Bacteriol.* **74,** 86–93.

Keeling, S., Moutafis, G., Hayman, B., and Coloe, P. (2007). Application of a pathogenicity marker found in *Escherichia coli* for the assessment of irrigation water quality. *Water Environ. Res.* **79,** 561–566.

Kim, H., Ryu, J. H., and Beuchat, L. R. (2006). Survival of *Enterobacter sakazakii* on fresh produce as affected by temperature, and effectiveness of sanitizers for its elimination. *Int. J. Food Microbiol.* **111,** 134–143.

King, A. D., Magnuson, J. A., Torok, T., and Goodman, N. (1991). Microbial flora and storage quality of partially processed lettuce. *J. Food Sci.* **56,** 459–461.

Kondo, N., Murata, M., and Isshiki, K. (2006). Efficiency of sodium hypochlorite, fumaric acid, and mild heat in killing native microflora and *Escherichia coli* O157:H7, *Salmonella typhimurium* DT104, and *Staphylococcus aureus* attached to fresh-cut lettuce. *J. Food Prot.* **69,** 323–329.

Koseki, S., and Isobe, S. (2005a). Growth of *Listeria monocytogenes* on iceberg lettuce and solid media. *Int. J. Food Microbiol.* **101,** 217–225.

Koseki, S., and Isobe, S. (2005b). Prediction of pathogen growth on iceberg lettuce under real temperature history during distribution from farm to table. *Int. J. Food Microbiol.* **104,** 239–248.

Koseki, S., Isobe, S., and Itoh, K. (2004). Efficacy of acidic electrolyzed water ice for pathogen control on lettuce. *J. Food Prot.* **67,** 2544–2549.

Kostrzynska, M., and Bachand, A. (2006). Use of microbial antagonism to reduce pathogen levels on produce and meat products: A review. *Can. J. Microbiol.* **52,** 1017–1026.

Kudva, I. T., Blanch, K., and Hovde, C. J. (1998). Analysis of *Escherichia coli* O157:H7 survival in ovine or bovine manure and manure slurry. *Appl. Environ. Microbiol.* **64,** 3166–3174.

Kudva, I. T., Jelacic, S., Tarr, P. I., Youderian, P., and Hovde, C. J. (1999). Biocontrol of *Escherichia coli* O157 with O157-specific bacteriophages. *Appl. Environ. Microbiol.* **65,** 3767–3773.

Kutter, S., Hartmann, A., and Schmid, M. (2006). Colonization of barley (*Hordeum vulgare*) with *Salmonella enterica* and *Listeria* spp. *FEMS Microbiol. Ecol.* **56,** 262–271.

Lang, M. M., Harris, L. J., and Beuchat, L. R. (2004). Survival and recovery of *Escherichia coli* O157:H7, *Salmonella*, and *Listeria monocytogenes* on lettuce and parsley as affected by method of inoculation, time between inoculation and analysis, and treatment with chlorinated water. *J. Food Prot.* **67,** 1092–1103.

Leverentz, B., Conway, W. S., Alavidze, Z., Janisiewicz, W. J., Fuchs, Y., Camp, M. J., Chighladze, E., and Sulakvelidze, A. (2001). Examination of bacteriophage as a biocontrol method for *Salmonella* on fresh-cut fruit: A model study. *J. Food Prot.* **64,** 1116–1121.

Leverentz, B., Conway, W. S., Camp, M. J., Janisiewicz, W. J., Abuladze, T., Yang, M., Saftner, R., and Sulakvelidze, A. (2003). Biocontrol of *Listeria monocytogenes* on fresh-cut produce by treatment with lytic bacteriophages and a bacteriocin. *Appl. Environ. Microbiol.* **69**, 4519–4526.

Li, Y., Brackett, R. E., Chen, J., and Beuchat, L. R. (2002). Mild heat treatment of lettuce enhances growth of *Listeria monocytogenes* during subsequent storage at 5 °C or 15 °C. *J. Appl. Microbiol.* **92**, 269–275.

Li, Y., Brackett, R. E., Chen, J. R., and Beuchat, L. R. (2001a). Survival and growth of *Escherichia coli* O157:H7 inoculated onto cut lettuce before or after heating in chlorinated water, followed by storage at 5 °C or 15 °C. *J. Food Prot.* **64**, 305–309.

Li, Y., Brackett, R. E., Shewfelt, R. L., and Beuchat, L. R. (2001b). Changes in appearance and natural microflora on iceberg lettuce treated in warm, chlorinated water and then stored at refrigeration temperature. *Food Microbiol.* **18**, 299–308.

Liao, C. H. (2007). Inhibition of foodborne pathogens by native microflora recovered from fresh peeled baby carrot and propagated in cultures. *J. Food Sci.* **72**, M134–M139.

Liao, C. S., and Fett, W. F. (2001). Analysis of native microflora and selection of strains antagonistic to human pathogens on fresh produce. *J. Food Prot.* **64**, 1110–1115.

Lin, C. M., Fernando, S. Y., and Wei, C. I. (1996). Occurrence of *Listeria monocytogenes*, *Salmonella* spp, *Escherichia coli* and *E. coli* O157:H7 in vegetable salads. *Food Control* **7**, 135–140.

Lindow, S. E., and Brandl, M. T. (2003). Microbiology of the phyllosphere. *Appl. Environ. Microbiol.* **69**, 1875–1883.

Little, C., Roberts, D., Youngs, E., and De Louvois, J. (1999). Microbiological quality of retail imported unprepared whole lettuces: A PHLS food working group study. *J. Food Prot.* **62**, 325–328.

Loaiza-Velarde, J. G., Tomas-Barbera, F. A., and Saltveit, M. E. (1997). Effect of intensity and duration of heat-shock treatments on wound-induced phenolic metabolism in iceberg lettuce. *J. Amer. Soc. Hort. Sci.* **122**, 873–877.

Loncarevic, S., Johannessen, G. S., and Rorvik, L. M. (2005). Bacteriological quality of organically grown leaf lettuce in Norway. *Lett. Appl. Microbiol.* **41**, 186–189.

Long, S. M., Adak, G. K., O'Brien, S. J., and Gillespie, I. A. (2002). General outbreaks of infectious intestinal disease linked with salad vegetables and fruit, England and Wales, 1992–2002. *Commun. Dis. Public Health* **5**, 101–105.

López-Gálvez, G., Saltveit, M., and Cantwell, M. (1996). Wound-induced phenylalanine ammonia lyase activity: Factors affecting its induction and correlation with the quality of minimally processed lettuces. *Postharvest Biol. Tech.* **9**, 223–233.

Losinger, W. C., Wells, S. J., Garber, L. P., Hurd, H. S., and Thomas, L. A. (1995). Management factors related to *Salmonella* shedding by dairy heifers. *J. Dairy Sci.* **78**, 2464–2472.

Lung, A. J., Lin, C. M., Kim, J. M., Marshall, M. R., Nordstedt, R., Thompson, N. P., and Wei, C. I. (2001). Destruction of *Escherichia coli* O157:H7 and *Salmonella enteritidis* in cow manure composting. *J. Food Prot.* **64**, 1309–1314.

Machado, D. C., Maia, C. M., Carvalho, I. D., da Silva, N. F., Andre, M. C. D. B. P., and Serafini, A. B. (2007). Microbiological quality of organic vegetables produced in soil treated with different types of manure and mineral fertilizer. *Brazilian J. Microbiol.* **37**, 538–544.

Magnuson, J. A., King, A. D., and Torok, T. (1990). Microflora of partially processed lettuce. *Appl. Environ. Microbiol.* **56**, 3851–3854.

Matos, A., and Garland, J. L. (2005). Effects of community versus single strain inoculants on the biocontrol of *Salmonella* and microbial community dynamics in alfalfa sprouts. *J. Food Prot.* **68**, 40–48.

Maule, A. (1997). Survival of verotoxigenic strains of *E. coli* O157:H7 in laboratory-scale microcosms. *In* "Coliforms and *E. coli*: Problem or solution?" (D. Kay and C. Fricker, Eds.), pp. 61–65. Royal Society of Chemistry, London.

Maule, A. (1999). Environmental aspects of *E. coli* O157. *Int. Food Hygiene* **9**, 21–23.

Maule, A. (2000). Survival of verocytotoxigenic *Escherichia coli* O157 in soil, water and on surfaces. *J. Appl. Microbiol.* **88,** 71S–78S.

Maxcy, R. B. (1978). Lettuce salad as a carrier of microorganisms of public health significance. *J. Food Prot.* **41,** 435–438.

McMahon, M. A. S., and Wilson, I. G. (2001). The occurrence of enteric pathogens and *Aeromonas* species in organic vegetables. *Int. J. Food Microbiol.* **70,** 155–162.

Melloul, A. A., Hassani, L., and Rafouk, L. (2001). *Salmonella* contamination of vegetables irrigated with untreated wastewater. *World J. Microb. Biot.* **17,** 207–209.

Miyamoto, T., Kawahara, M., and Minamisawa, K. (2004). Novel endophytic nitrogen-fixing clostridia from the grass *Miscanthus sinensis* as revealed by terminal restriction fragment length polymorphism analysis. *Appl. Environ. Microbiol.* **70,** 6580–6586.

Moganedi, K. L. M., Goyvaerts, E. M. A., Venter, S. N., and Sibara, M. M. (2007). Optimisation of the PCR-invA primers for the detection of *Salmonella* in drinking and surface waters following a pre-cultivation step. *Water S A.* **33,** 195–202.

Monaghan, J. M. (2006). United Kingdom and European approach to fresh produce food safety and security. *HortTechnol.* **16,** 559–562.

Mukherjee, A., Cho, S., Scheftel, J., Jawahir, S., Smith, K., and Diez-Gonzalez, F. (2006). Soil survival of *Escherichia coli* O157:H7 acquired by a child from garden soil recently fertilized with cattle manure. *J. Appl. Microbiol.* **101,** 429–436.

Mukherjee, A., Speh, D., Dyck, E., and Diez-Gonzalez, F. (2004). Preharvest evaluation of coliforms, *Escherichia coli*, *Salmonella*, and *Escherichia coli* O157:H7 in organic and conventional produce grown by Minnesota farmers. *J. Food Prot.* **67,** 894–900.

Naimi, T. S., Wicklund, J. H., Olsen, S. J., Krause, G., Wells, J. G., Bartkus, J. M., Boxrud, D. J., Sullivan, M., Kassenborg, H., Besser, J. M., Mintz, E. D., Osterholm, M. T., et al. (2003). Concurrent outbreaks of *Shigella sonnei* and enterotoxigenic *Escherichia coli* infections associated with parsley: Implications for surveillance and control of foodborne illness. *J. Food Prot.* **66,** 535–541.

Nascimento, M. S., Silva, N., Catanozi, M., and Silva, K. C. (2003). Effects of different disinfection treatments on the natural microbiota of lettuce. *J. Food Prot.* **66,** 1697–1700.

Natvig, E. E., Ingham, S. C., Ingham, B. H., Cooperband, L. R., and Roper, T. R. (2002). *Salmonella enterica* serovar Typhimurium and *Escherichia coli* contamination of root and leaf vegetables grown in soils with incorporated bovine manure. *Appl. Environ. Microbiol.* **68,** 2737–2744.

Nguyen-the, C., and Carlin, F. (1994). The microbiology of minimally processed fresh fruits and vegetables. *Crit. Rev. Food Sci.* **34,** 371–401.

Nicholl, P., McInerney, S., and Prendergast, M. (2004). Growth dynamics of indigenous microbial populations on vegetables after decontamination and during refrigerated storage. *J. Food Process. Pres.* **28,** 442–459.

Nicholson, F. A., Groves, S. J., and Chambers, B. J. (2005). Pathogen survival during livestock manure storage and following land application. *Bioresource Technol.* **96,** 135–143.

Nuorti, J. P., Niskanen, T., Hallanvuo, S., Mikkola, J., Kela, E., Hatakka, M., Fredriksson-Ahomaa, M., Lyytikainen, O., Siitonen, A., Korkeala, H., and Ruutu, P. (2004). A widespread outbreak of *Yersinia pseudotuberculosis* O:3 infection from iceberg lettuce. *J. Infect. Dis.* **189,** 766–774.

O'Brien, R. D., and Lindow, S. E. (1989). Effect of plant species and environmental conditions on epiphytic population sizes of *Pseudomonas syringae* and other bacteria. *Phytopathology* **79,** 619–627.

Odumeru, J. A., Mitchell, S. J., Alves, D. M., Lynch, J. A., Yee, A. J., Wang, S. L., Styliadis, S., and Farber, J. M. (1997). Assessment of the microbiological quality of ready-to-use vegetables for health-care food services. *J. Food Prot.* **60,** 954–960.

Oh, S. W., Dancer, G. I., and Kang, D. H. (2005). Efficacy of aerosolized peroxyacetic acid as a sanitizer of lettuce leaves. *J. Food Prot.* **68,** 1743–1747.

Ohara, T., and Itoh, K. (2003). Significance of *Pseudomonas aeruginosa* colonization of the gastrointestinal tract. *Internal Med.* **42,** 1072–1976.

O'Mahony, M., Cowden, J., Smyth, B., Lynch, D., Hall, M., Rowe, B., Teare, E. L., Tettmar, R. E., Rampling, A. M., Coles, M., Gilbert, R. J., Kingcott, E., *et al.* (1990). An outbreak of *Salmonella* Saint-Paul infection associated with bean sprouts. *Epidemiol. Infect.* **104,** 229–235.

Parke, J. L., and Gurian-Sherman, D. (2001). Diversity of the *Burkholderia cepacia* complex and implications for risk assessment of biological control strains. *Annu. Rev. Phytopathol.* **39,** 225–258.

Pell, A. N. (1997). Manure and microbes: Public and animal health problem? *J. Dairy Sci.* **80,** 2673–2681.

Pieczarka, D. J., and Lorbeer, J. W. (1975). Bacterial populations on basal lettuce leaves and in soil from under lettuce plants. *Phytopathology* **65,** 509–513.

Perry, L., Heard, P., Kane, M., Kim, H., Savikhin, S., Dominguez, W., and Applegate, B. (2007). Application of multiplex polymerase chain reaction to the detection of pathogens in food. *J. Rapid Meth. Aut. Mic.* **15,** 176–198.

Porter, J., Mobbs, K., Hart, C. A., Saunders, J. R., Pickup, R. W., and Edwards, C. (1997). Detection, distribution and probable fate of *Escherichia coli* O157 from asymptomatic cattle on a dairy farm. *J. Appl. Microbiol.* **83,** 297–306.

Rangel, J. M., Sparling, P. H., Crowe, C., Griffin, P. M., and Swerdlow, D. L. (2005). Epidemiology of *Escherichia coli* O157:H7 outbreaks, United States, 1982–2002. *Emerg. Infect. Dis.* **11,** 603–609.

Rankin, J. D., and Taylor, R. J. (1969). A study of some disease hazards which could be associated with the system of applying cattle slurry on pasture. *Vet. Rec.* **85,** 587–591.

Rasmussen, T. B., and Givskov, M. (2006). Quorum sensing inhibitors: A bargain of effects. *Microbiology* **152,** 895–904.

Rice, D. H., Hancock, D. D., and Besser, T. E. (1995). Verotoxigenic *Escherichia coli* O157 colonization of wild deer and range cattle. *Vet. Rec.* **137,** 524.

Rodgers, S. L., Cash, J. N., Siddiq, M., and Ryser, E. T. (2004). A comparison of different chemical sanitizers for inactivating *Escherichia coli* O157:H7 and *Listeria monocytogenes* in solution and on apples, lettuce, strawberries, and cantaloupe. *J. Food Prot.* **67,** 721–731.

Rodriguez, C., Lang, L., Wang, A., Altendorf, K., Garcia, F., and Lipski, A. (2006). Lettuce for human consumption collected in Costa Rica contains complex communities of culturable oxytetracycline- and gentamicin-resistant bacteria. *Appl. Environ. Microbiol.* **72,** 5870–5876.

Rudi, K., Flateland, S. L., Hanssen, J. F., Bengtsson, G., and Nissen, H. (2002). Development and evaluation of a 16S ribosomal DNA array-based approach for describing complex microbial communities in ready-to-eat vegetable salads packed in a modified atmosphere. *Appl. Environ. Microbiol.* **68,** 1146–1156.

Sagoo, S. K., Little, C. L., and Mitchell, R. T. (2001). The microbiological examination of ready-to-eat organic vegetables from retail establishments in the United Kingdom. *Lett. Appl. Microbiol.* **33,** 434–439.

Sagoo, S. K., Little, C. L., Ward, L., Gillespie, I. A., and Mitchell, R. T. (2003). Microbiological study of ready-to-eat salad vegetables from retail establishments uncovers a national outbreak of salmonellosis. *J. Food Prot.* **66,** 403–409.

Santamaria, J., and Toranzos, G. A. (2003). Enteric pathogens and soil: A short review. *Int. Microbiol.* **6,** 5–9.

Schlech, W. F., Lavigne, P. M., Bortolussi, R. A., Allen, A. C., Haldane, E. V., Wort, A. J., Hightower, A. W., Johnson, S. E., King, S. H., Nicholls, E. S., and Broome, C. V. (1983). Epidemic Listeriosis – Evidence for transmission by food. *New Engl. J. Med.* **308,** 203–206.

Schuenzel, K. M., and Harrison, M. A. (2002). Microbial antagonists of foodborne pathogens on fresh, minimally processed vegetables. *J. Food Prot.* **65,** 1909–1915.

Scolari, G., and Vescovo, M. (2004). Microbial antagonism of *Lactobacillus casei* added to fresh vegetables. Ital. *J. Food Sci.* **16**, 465–475.

Sela, S., Nestel, D., Pinto, R., Nemny-Lavy, E., and Bar-Joseph, M. (2005). Mediterranean fruit fly as a potential vector of bacterial pathogens. *Appl. Environ. Microbiol.* **71**, 4052–4056.

Seo, K. H., and Frank, J. F. (1999). Attachment of *Escherichia coli* O157:H7 to lettuce leaf surface and bacterial viability in response to chlorine treatment as demonstrated by using confocal scanning laser microscopy. *J. Food Prot.* **62**, 3–9.

Singh, B. R., Chandra, M., Agarwal, R., and baju, N. (2007). Interactions between *Salmonella enterica* subspecies *enterica* serovar Typhimurium and cowpea (*Vigna unguiculata* variety sinsensis) seeds, plants and persistence in hay. *J. Food Safety* **27**, 169–187.

Singh, N., Singh, R. K., Bhunia, A. K., and Stroshine, R. L. (2002). Efficacy of chlorine dioxide, ozone, and thyme essential oil or a sequential washing in killing *Escherichia coli* O157:H7 on lettuce and baby carrots. *Food Sci. Technol.* **35**, 720–729.

Sivapalasingam, S., Friedman, C. R., Cohen, L., and Tauxe, R. V. (2004). Fresh produce: A growing cause of outbreaks of foodborne illness in the United States, 1973 through 1997. *J. Food Prot.* **67**, 2342–2353.

Sjogren, R. E. (1995). 13-year survival study of an environmental *Escherichia coli* in field mini-plots. *Water Air Soil Poll.* **81**, 315–335.

Solomon, E. B., and Matthews, K. R. (2006). Interaction of live and dead *Escherichia coli* O157:H7 and fluorescent microspheres with lettuce tissue suggests bacterial processes do not mediate adherence. *Lett. Appl. Microbiol.* **42**, 88–93.

Solomon, E. B., Pang, H. J., and Matthews, K. R. (2003). Persistence of *Escherichia coli* O157:H7 on lettuce plants following spray irrigation with contaminated water. *J. Food Prot.* **66**, 2198–2202.

Solomon, E. B., Potenski, C. J., and Matthews, K. R. (2002a). Effect of irrigation method on transmission to and persistence of *Escherichia coli* O157:H7 on lettuce. *J. Food Prot.* **65**, 673–676.

Solomon, E. B., Yaron, S., and Matthews, K. R. (2002b). Transmission of *Escherichia coli* O157:H7 from contaminated manure and irrigation water to lettuce plant tissue and its subsequent internalization. *Appl. Environ. Microbiol.* **68**, 397–400.

Starr, M. P., and Chatterjee, A. K. (1972). Genus *Erwinia* - Enterobacteria pathogenic to plants and animals. *Annu. Rev. Microbiol.* **26**, 389–426.

Steele, M., Mahdi, A., and Odumeru, J. (2005). Microbial assessment of irrigation water used for production of fruit and vegetables in Ontario, Canada. *J. Food Prot.* **68**, 1388–1392.

Steele, M., and Odumeru, J. (2004). Irrigation water as source of foodborne pathogens on fruit and vegetables. *J. Food Prot.* **67**, 2839–2849.

Stine, S. W., Song, I., Choi, C. Y., and Gerba, C. P. (2005). Effect of relative humidity on preharvest survival of bacterial and viral pathogens on the surface of cantaloupe, lettuce, and bell peppers. *J. Food Prot.* **68**, 1352–1358.

Strauch, D., and Ballarini, G. (1994). Hygienic aspects of the production and agricultural use of animal wastes. *J. Vet. Med. B* **41**, 176–228.

Suslow, T. V. (2002). Production practices affecting the potential for persistent contamination of plants by microbial foodborne pathogens. *In* "Phyllosphere Microbiology" (S. E. Lindow, E. I. Hecht-Poinar, and V. J. Elliott, Eds.) pp. 241–256. APS Press, St. Paul, MN.

Takeuchi, K., and Frank, J. F. (2000). Penetration of *Escherichia coli* O157:H7 into lettuce tissues as affected by inoculum size and temperature and the effect of chlorine treatment on cell viability. *J. Food Prot.* **63**, 434–440.

Takeuchi, K., and Frank, J. F. (2001). Quantitative determination of the role of lettuce leaf structures in protecting *Escherichia coli* O157:H7 from chlorine disinfection. *J. Food Prot.* **64**, 147–151.

Takeuchi, K., Matute, C. M., Hassan, A. N., and Frank, J. F. (2000). Comparison of the attachment of *Escherichia coli* O157:H7, *Listeria monocytogenes, Salmonella typhimurium,* and *Pseudomonas fluorescens* to lettuce leaves. *J. Food Prot.* **63,** 1433–1437.

Taormina, P. J., Beuchat, L. R., and Slutsker, L. (1999). Infections associated with eating seed sprouts: An international concern. *Emerg. Infect. Dis.* **5,** 626–634.

Tauxe, R., Kruse, H., Hedberg, C., Potter, M., Madden, J., and Wachsmuth, K. (1997). Microbial hazards and emerging issues associated with produce – A preliminary report to the National Advisory Committee on Microbiologic Criteria for Foods. *J. Food Prot.* **60,** 1400–1408.

Tauxe, R. V. (1997). Emerging foodborne diseases: An evolving public health challenge. *Emerg. Infect. Dis.* **3,** 425–434.

Taylor, R. J., and Burrows, M. R. (1971). Survival of *Escherichia coli* and *Salmonella* Dublin in slurry on pasture and infectivity of S-Dublin for grazing calves. *Brit. Vet. J.* **127,** 536–538.

Toth, I. K., Pritchard, L., and Birch, P. R. J. (2006). Comparative genomics reveals what makes an enterobacterial plant pathogen. *Annu. Rev. Phytopathol.* **44,** 305–336.

Turner, C. (2002). The thermal inactivation of *E. coli* in straw and pig manure. *Bioresource Tech.* **84,** 57–61.

Tyrrel, S. F., Knox, J. W., and Weatherhead, E. K. (2006). Microbiological water quality requirements for salad irrigation in the United Kingdom. *J. Food Prot.* **69,** 2029–2035.

Unc, A., and Goss, M. J. (2004). Transport of bacteria from manure and protection of water resources. *Appl. Soil Ecol.* **25,** 1–18.

Vaz da Costa-Vargas, S. M., Mara, D. D., and Vargas-Lopez, C. E. (1991). Residual fecal contamination on effluent-irrigated lettuces. *Water Sci. Tech.* **24,** 89–94.

Wachsmuth, K., and Morris, G. K. (1989). Shigella. In "Foodborne Bacterial Pathogens" (M. P. Doyle, Ed.), pp. 447–462. Marcel Dekker, New York.

Wachtel, M. R., Whitehand, L. C., and Mandrell, R. E. (2002a). Association of *Escherichia coli* O157:H7 with preharvest leaf lettuce upon exposure to contaminated irrigation water. *J. Food Prot.* **65,** 18–25.

Wachtel, M. R., Whitehand, L. C., and Mandrell, R. E. (2002b). Prevalence of *Escherichia coli* associated with a cabbage crop inadvertently irrigated with partially treated sewage wastewater. *J. Food Prot.* **65,** 471–475.

Wang, G. D., and Doyle, M. P. (1998). Survival of enterohemorrhagic *Escherichia coli* O157:H7 in water. *J. Food Prot.* **61,** 662–667.

Wang, G. D., Zhao, T., and Doyle, M. P. (1996). Fate of enterohemorrhagic *Escherichia coli* O157:H7 in bovine feces. *Appl. Environ. Microbiol.* **62,** 2567–2570.

Warriner, K., Ibrahim, F., Dickinson, M., Wright, C., and Waites, W. M. (2003a). Interaction of *Escherichia coli* with growing salad spinach plants. *J. Food Prot.* **66,** 1790–1797.

Warriner, K., Ibrahim, F., Dickinson, M., Wright, C., and Waites, W. M. (2003b). Internalization of human pathogens within growing salad vegetables. *Biotech. Gen. Engin. Rev.* **20,** 117–134.

Warriner, K., Ibrahim, F., Dickinson, M., Wright, C., and Waites, W. M. (2005). Seed decontamination as an intervention step for eliminating *Escherichia coli* on salad vegetables and herbs. *J. Sci. Food Agric.* **85,** 2307–2313.

Warriner, K., Spaniolas, S., Dickinson, M., Wright, C., and Waites, W. M. (2003c). Internalization of bioluminescent *Escherichia coli* and *Salmonella* Montevideo in growing bean sprouts. *J. Appl. Microbiol.* **95,** 719–727.

Weissinger, W. R., Chantarapanont, W., and Beuchat, L. R. (2000). Survival and growth of *Salmonella* baildon in shredded lettuce and diced tomatoes, and effectiveness of chlorinated water as a sanitizer. *Int. J. Food Microbiol.* **62,** 123–131.

Wells, J. M., and Butterfield, J. E. (1997). *Salmonella* contamination associated with bacterial soft rot of fresh fruits and vegetables in the marketplace. *Plant Dis.* **81,** 867–872.

Wells, S. J., Fedorka-Cray, P. J., Dargatz, D. A., Ferris, K., and Green, A. (2001). Fecal shedding of *Salmonella* spp. by dairy cows on farm and at cull cow markets. *J. Food Prot.* **64,** 3–11.

Whipps, J. M. (2001). Microbial interactions and biocontrol in the rhizosphere. *J. Exp. Bot.* **52,** 487–511.

Yuk, H. G., Yoo, M. Y., Yoon, J. W., Moon, K. D., Marshall, D. L., and Oh, D. H. (2006). Effect of combined ozone and organic acid treatment for control of *Escherichia coli* O157:H7 and *Listeria monocytogenes* on lettuce. *J. Food Sci.* **71,** M83–M87.

Zschock, M., Hamann, H. P., Kloppert, B., and Wolter, W. (2000). Shiga-toxin-producing *Escherichia coli* in faeces of healthy dairy cows, sheep and goats: prevalence and virulence properties. *Lett. Appl. Microbiol.* **31,** 203–208.

CHAPTER 8

Microbial Retention on Open Food Contact Surfaces and Implications for Food Contamination

Joanna Verran,[1] Paul Airey, Adele Packer, and Kathryn A. Whitehead

Contents			
	I.	Introduction	224
	II.	Microbial Attachment, Biofilm Formation, and Cell Retention	225
	III.	Surfaces (substrata) Encountered	226
	IV.	Factors Affecting Retention	227
		A. Surface topography	227
		B. Surface chemistry	230
		C. Presence of organic material	231
	V.	Characterization of Surfaces	232
		A. Topography	232
		B. Chemistry	234
	VI.	Measuring Retention and Assessing Cleaning and Disinfection	234
		A. Substratum preparation	234
		B. Amount of retention	235
		C. Strength of retention	238
		D. Quantifying organic material	240
	VII.	Conclusions	241
		References	241

School of Biology, Chemistry and Health Science, Manchester Metropolitan University, Manchester M1 5GD, United Kingdom
[1] Corresponding author: School of Biology, Chemistry and Health Science, Manchester Metropolitan University, Chester St., Manchester M1 5GD, United Kingdom

I. INTRODUCTION

In the food industry, food contact surfaces are generally described as being "open" or "closed". Closed surfaces are primarily pipework, where wet product, or ingredients are contained within a flowing, liquid system. Open surfaces are exposed, with moist or dry food passing along conveyors; thus liquid does not necessarily enclose the food, or cover the surface, and consequently flow is absent.

Closed systems present any contaminating microorganisms with a solid–liquid interface for attachment and colonization. Access to these surfaces for cleaning is difficult, thus the opportunity exists for the development of biofilm. Open surfaces present a solid–air–, or a solid–liquid–air interface, where microorganisms attached on the surface may encounter an environment less conducive to growth, encompassing opportunity for dehydration during drying, lack of moisture, and exposure to regular cleaning and disinfection. Thus for hygienic, open, food contact surfaces, the retention and survival of viable microorganisms pre- and postcleaning and disinfection is of key concern. True biofilm is unlikely to be present, thus recognition of the particular properties of microbial cells in biofilm, and the use of biofilm in the assessment of cleaning and disinfection protocols, may not be relevant.

The presence of viable, if not necessarily multiplying, cells on open surfaces poses a biotransfer potential. Transfer of microorganisms from an inert, nonnutritive surface to a different environment (e.g., food) may result in multiplication and colonization and accompanying potential for infection, spoilage, and/or disruption to quality assurance/quality control procedures. Focus on retention rather than on initial attachment or subsequent colonization is therefore relevant when considering cleaning and disinfection processes, where the aim is removal, and/or inactivation, of viable microorganisms. Thus any viable cells that remain, that is, are retained, are of concern. Several factors affect this retention: properties associated with the microorganism (identity, physiology, and viability), the surface (topography, chemistry, orientation, and design), the presence of inorganic and organic soil on the surface (e.g., from food and cleaning formulations), and interactions between these parameters. The methods used to assess the effectiveness of cleaning and disinfection should take account of these parameters. For example, considering the microorganisms, it is not necessarily the *amount* of microbial attachment in terms of cell numbers per unit area that is important in aspects of food hygiene, rather the *strength* of attachment, and hence removal. For surfaces, those that are more worn are less easy to clean. Similarly, the presence of organic material can affect cell viability, and surface cleanability, thus the differential response of these two components (microbial cell and organic soil) to cleaning and disinfection should be considered.

This paper reviews some of the scenarios encountered, addresses factors affecting retention, and suggests some methods for characterization of the interactions between cells, soil and inert substrata encountered in the food processing industry, enabling evaluation and maintenance of effective cleaning and disinfection procedures in a given environment.

II. MICROBIAL ATTACHMENT, BIOFILM FORMATION, AND CELL RETENTION

Food and food contact surfaces can become contaminated with pathogenic and nonpathogenic microorganisms through contact with soil, water, fertilizers, equipment, humans, aerosols, and animals. The identity of the microorganism will vary depending on the product of concern: thus in the closed systems of the dairy industry, thermophilic streptococci are considered a problem in terms of product spoilage and cleanability of surfaces fouled with heat-treated, milk-based soil (Flint et al., 2000). *Listeria monocytogenes* is a chill-tolerant pathogen of concern in the cheese industry, but also in the meat and fish industries (Gibbons et al., 2006; Jeong and Frank, 1994; Miettinen et al., 2003; Navratilova et al., 2004; Tresse et al., 2007). In these three different industrial scenarios, the food substrates differ considerably, thus the interactions occurring between cells and soil will also differ, and the cleaning agents of choice will vary. Indeed, with the exception perhaps of aerial contamination, it is unlikely that cells or spores will be transferred to a surface in the absence of organic material, be it sweat/sebum from human fingers, blood and other body fluids, or other food components such as fats, oils, and proteins. Organic material and microbial cells may be deposited together, sequentially, and/or unevenly across the surface (Verran and Whitehead, 2006), providing a scenario difficult to model *in vitro*, and one which is very different from the more familiar initial stages of biofilm formation at the solid–liquid interface (Kumar and Anand, 1998; Verran, 2002). Additionally, cells may be transferred to a surface via a large contamination event, or a few residual cells may have survived a cleaning procedure. Cells may also adhere more strongly to surface layers when the suspension is allowed to dry than when moist (Moore and Griffith, 2002), as would be likely on an open surface

The ability of bacteria to attach to, and be retained on, surfaces will influence their persistence during manufacturing and retail, as well as their ability to cause infection (Kumar and Anand, 1998). The physiological state of the microorganisms [dormant, multiplying, presence of exopolymeric substances (EPS), stressed] is also pertinent to the consideration of any given scenario. Microorganisms contaminating food processing surfaces may be actively multiplying and colonizing the surface or may be merely surviving, retaining viability but being unable to multiply because of adverse environmental conditions (Verran, 2002).

At a solid–liquid interface, before a cell can bind to a surface, both the surface and the cell are conditioned by adsorbing molecules from the surrounding fluid. At a solid–air interface, the surface conditioning may be less "specific", resulting from the passive transfer of food material from substrate to substratum, as the food passes through the processing plant, or is handled in the domestic environment. Similarly, microbial cells may be deposited onto the surface via food-substratum contact, again without any specificity other than that which caused the cell to interact with the food material in the first place. The strength of attachment of this cell-soil layer to the surface will vary depending on time elapsed prior to cleaning (i.e., drying), temperature (of food, surface, process), and the interactions occurring between the soil, cell, and surface. All these properties impact on cleaning.

In the food industry, biofilms are rarely quantified. For closed systems, access prevents appropriate sampling. For environmental sampling, strategically placed coupons are assessed for contamination postexposure. Coverage of surfaces by "biofilm" is not uniform, with cell density less than 10^4 cells cm^{-2}. Apparently, the term biofilm is usually used when the density of 10^6 cells cm^{-1} is exceeded (Flint *et al.*, 1997; Mettler and Carpentier, 1998; Verran and Boyd, 2001). When assessing open food contact surfaces for contamination, the lack of uniform coverage poses considerable problems in terms of monitoring cleaning, but is useful for identifying critical control points.

III. SURFACES (SUBSTRATA) ENCOUNTERED

Hygienic surfaces are hard, inert, and easy to clean. Stainless steel remains the surface of choice in most instances in the food industry, since it is stable at a variety of temperatures, inert, relatively resistant to corrosion and may be treated electrolytically or mechanically to achieve functionally and aesthetically improved surfaces (Verran *et al.*, 2000). Wear of steel tends to result in linear features (scratches), with occasional pits. Glass/ceramics are similarly "hygienic" and ceramic tiles may be used on walls or floors. "Wear" of such surfaces tends to present as fractures. Plastics, epoxy resins, and rubbers are softer, more flexible, and much more problematic in terms of hygienic status and cleanability, although they are essential for some appliances such as gaskets and conveyors in some parts of the food processing plant. Loss of flexibility through excess wear, and resultant cracking increases the potential for penetration of contaminating microorganisms into the material.

Most "open" surfaces tend to be exposed to liquid only intermittently, for example during cleaning, thus the attached cells do not form a true biofilm as would be seen at a solid–liquid interface, although microcolonies might be observed if conditions are suitable for any multiplication, and areas of poor accessibility and surfaces of increased porosity/flexibility would facilitate the accumulation of microorganisms. Surface hydrophobicity will affect drying kinetics (Whitehead and Verran, 2007), with moisture droplets being more prominent on hydrophobic surfaces, taking longer time to dry, thus promoting cell deposition at the solid–liquid–air interface. Vertically orientated hydrophobic surfaces will facilitate liquid run-off, leaving a relatively clean surface—a phenomenon employed in part by the "Lotus effect" (Furstner et al., 2005).

IV. FACTORS AFFECTING RETENTION

A. Surface topography

One factor that significantly affects microbial retention is substratum topography. Stainless steels often present a "finish", for example a brush finish, where parallel linear features of defined dimensions give a pleasing aesthetic to the naked eye, and reduce the visibility of fingerprints (Fig. 8.1). It has been shown that the finish of stainless steels used in food processing does not affect their cleanability or hygienic status (Airey and Verran, 2007; Hilbert et al., 2003), although bacterial attachment (not retention) on electropolished surfaces has been shown to be less than that on rougher stainless steel finishes (Arnold et al., 2004). However, wear of such surfaces inevitably results in the production of linear features, of differing dimensions and length, randomized across a surface (i.e., scratches), and "pits" (Fig. 8.2). Simulation of worn surfaces to assess cleanability *in vitro* has demonstrated that hygienic status of stainless steel and ceramic was not affected in terms of microbial retention, but cleanability, in terms of removal of organic (food) soil was reduced (Boyd et al., 2001a,b, 2002; Frank and Chmieliewski, 1997; Korber et al., 1997; Verran et al., 2001a; Verran and Whitehead, 2006).

It is generally acknowledged that an increased substratum surface roughness affects the retention of microorganisms on that surface (Verran et al., 2004). The R_a measurement provides an indication of surface roughness, usually given in micrometers, describing the average departure of the surface profile from a constructed "center line"—in effect a two-dimensional measure of a three-dimensional parameter. R_a is usually chosen to describe surface topography, since it is the most universally used roughness parameter for general quality control (Anon, 1988). An R_a value of less than 0.8 µm is generally accepted as indicative of

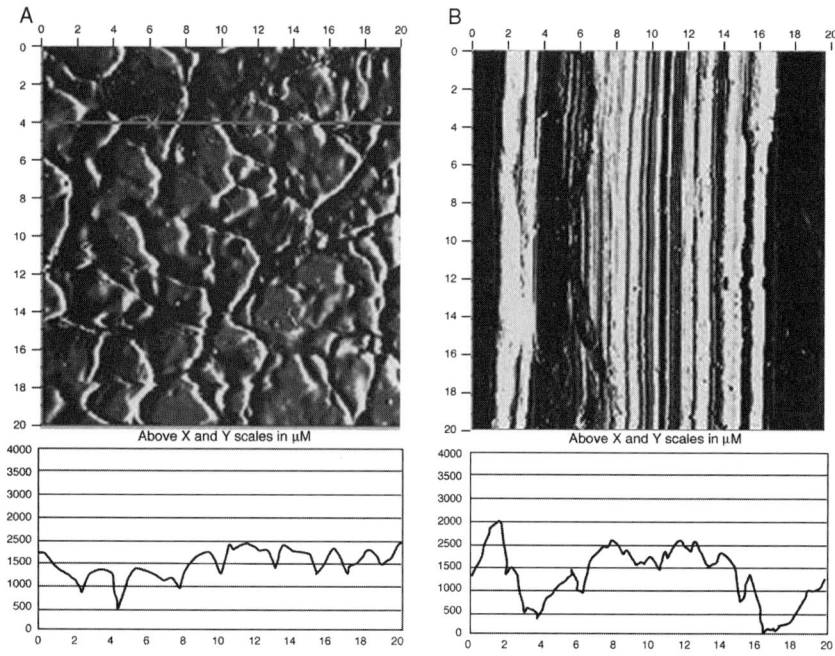

FIGURE 8.1 AFM images of 304 stainless steel with (A) 2B and (B) brushed finish.

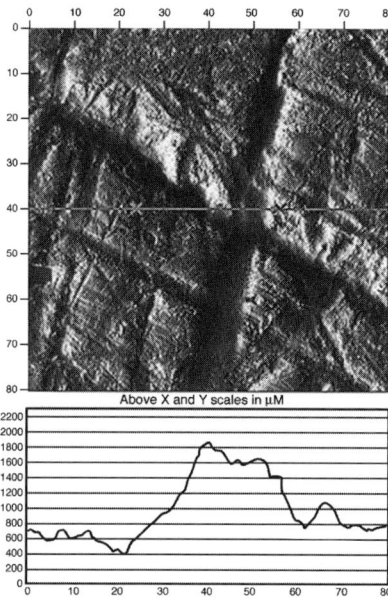

FIGURE 8.2 AFM images of replicas of a worn surface demonstrating patterns of differing dimensions and length, randomized across a surface (i.e., scratches), and "pits" produced by wear on a stainless steel surface.

a hygienic surface. However, R_a values can only be really representative if there is a regular surface topography, as for polished and brushed stainless steel, but not for worn surfaces whose aberrations tend to be of a random nature (Verran et al., 2000). Thus, in the food processing industry, wear of food contact surfaces through abrasion and impact damage will affect topography, but may not necessarily alter the key parameters used to measure surface roughness (Boulange-Petermann, 1996; Boyd et al., 2000, 2001b; Holah and Thorpe, 1990; Packer et al., 2007; Verran et al., 2000, 2001a; Verran and Boyd, 2001). It is therefore important to visualize the surface as well as deriving an R_a value, or any other statistically derived parameter.

Attempts to reexamine the effect of surface topography on microbial retention have revealed apparently conflicting data. Some have observed no relationship between surface roughness (in terms of R_a) and the ability of bacteria to attach to host (Boulange-Petermann et al., 1997; Flint, 1996, 2000; Langeveld et al., 1972; Tide et al., 1999; Vanhaecke et al., 1990; Verran and Boyd, 2001). Others have suggested that the greater the degree of surface roughness, the greater the retention of microorganisms (Arnold and Bailey, 2000; Bollen et al., 1997; Characklis and Wilderer, 1990; Holah et al., 1990; Medilanski et al., 2002; Verran and Maryan, 1997). However, these apparent contradictions arise primarily from the use of different perspectives of scale (Verran and Boyd, 2001). If the surface irregularities are much larger than the microorganisms, passive retention will be minimal (Verran et al., 1991) unless within these macroflaws, micro or even nanosize features existed (Verran and Boyd, 2001). If the features are of microbial dimension or slightly smaller, then retention will be enhanced. However, surfaces with a regular nanotopography have been shown to reduce microbial attachment, due to the lack of sufficient area for contact between the cell and the substratum (Cousins et al., 2007; Li et al., 2004; Whitehead and Verran, 2006). On worn, hygienic food contact surfaces, all of these topographical features are likely to be present. Their separate impact on surface cleanability and cell retention has yet to be investigated, requiring the fabrication of surfaces with defined topography. Thus, microbial retention assays have been carried out in our laboratories on a range of engineered surfaces with controlled topographical features of dimensions comparable to those of microbial cells [e.g., pits (Whitehead et al., 2005) and grooves (Packer et al., 2007; Scheuerman et al., 1998)]. Whitehead et al. (2005) demonstrated that with a range of differently sized unrelated microorganisms, the size of circular surface defects was important with respect to the size of the cell, and its subsequent retention. Using defined linear features, the same phenomenon was observed (Fig. 8.3), and an additional factor of directionality of rinsing comes into play, where one might anticipate improved cleaning parallel to the features, rather than perpendicular.

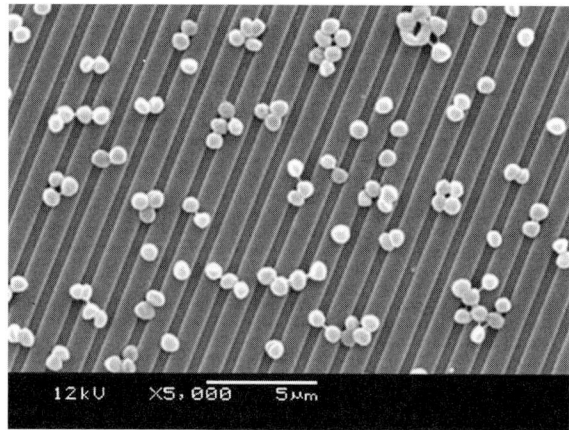

FIGURE 8.3 SEM of *Staphylococcus aureus* retained on titanium coated surface presenting linear features of 1 μm width. The vast majority of individual cells or colony forming units (96.9%) is associated within surface features.

B. Surface chemistry

In food hygiene, the contact surface is required to be inert, so that transfer of any potential chemical contaminant from substratum to food does not occur. The chemistry of stainless steel, the material of choice, is complex, being modified in order to produce materials with properties relating to conditions of use: chromium for example enhances corrosion resistance. Other elements can be added, for example, nickel, manganese, molybdenum (Maller, 1998). Chromium reacts with the atmosphere to form a protective oxide layer (passivation) and it is this oxide layer that gives stainless steel its enhanced corrosion resistance. Because of the speed of re-passivation (sec or min) in normal ambient conditions it is difficult to determine the exact chemical makeup of the surface of stainless steel and a review of the experiments and *in situ* techniques concludes that the passive film is never static and changes across its environment (Olsson and Landolt, 2003). The effect of variations in the surface chemistry of stainless steel on microbial retention has not been explored to date: for example, differences between grain boundary and the bulk surface properties may affect cell–substratum interactions over and above those of simple topography. However, in food processing, the substratum employed is rarely that of mill finish 2B (Fig. 8.2a). Brushed or polished surfaces are in common use, and these should present a more homogeneous surface chemistry.

There have been studies assessing the effect of surface chemistry on microbial attachment, using surfaces of comparable topography. For example, the adhesion of *L. monocytogenes*, *Salmonella typhimurium*, *Staphylococcus aureus*, and *Escherichia coli* have been assessed using surfaces with varying chemical groups grafted onto the surfaces (hydrophilicity, hydrophobicity, chain length, and chemical functionality) on glass substrata where it was shown that the bacterial attachment of L. *monocytogenes* and E. *coli* was affected by the chemistry of the underlying substratum (Cunliffe et al., 1999). The underlying surface chemistry has also been shown to affect biofilm formation (Teughels et al., 2006).

Wilks et al. (2005) found that the persistence and survival of E. *coli* O157 was greatly reduced on copper alloys than on stainless steel. However, the toxicity of copper would preclude its use on food contact surfaces (Airey and Verran, 2007). When a 2B finished stainless steel was coated with a titanium grid, *Pseudomonas aeruginosa* (Verran et al., 2003) and S. *aureus* (Verran and Whitehead, 2005) were found to preferentially bind to the raised titanium features. In contrast, E. *coli* would not attach to titanium coated surfaces (unpublished results). This different behavior of cells on surfaces might provide an opportunity for coating surfaces at critical control points on equipment with a material less conducive to retention, enabling targeted hygiene control strategies.

C. Presence of organic material

At the solid–liquid interface, the biological surface presented to microorganisms by the conditioning layer provides an element of specificity via receptors for the adhesion of pioneer organisms (Verran and Boyd, 2001). On an open surface (solid–air interface), this might instead occur more nonspecifically via direct contact with food (Verran, 2002). Attached microorganisms may be retained in surface features mixed with organic material such as fats, carbohydrates or proteins, or detergent residue (Fig. 8.4) (Verran, 2002). Thus, the term "conditioning film" may not be appropriate especially where a more significant transfer of organic matter occurs (Verran, 2002),

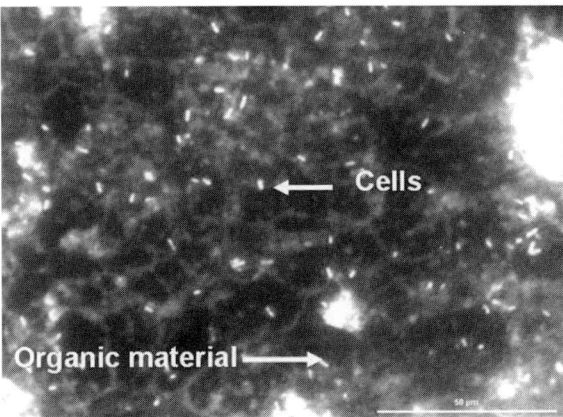

FIGURE 8.4 Microorganisms (L. *monocytogenes*) with organic material (fish extract) on stainless steel stained with acridine orange.

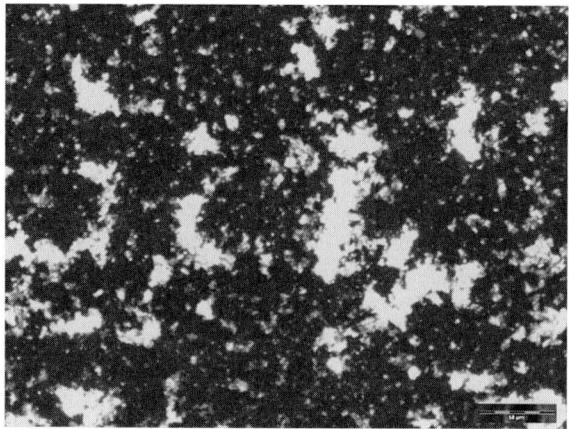

FIGURE 8.5 Uneven coverage of organic material (cottage cheese soil) on substrata, stained with acridine orange and image using epifluorescence microscopy.

and surface coverage is uneven (Fig. 8.5): "soiling" may be a more appropriate term. Relatively large deposits of organic soil may fill larger defects prior to microbial surface contamination (Frank and Chmielewski, 1997; Kumar and Anand, 1998; Milledge and Jowitt, 1980; Verran and Jones, 2000), and mask the underlying topography, while the formation of a thinner conditioning film (Carpentier and Cerf, 1993) on any surface in an aqueous environment may mask small topographical features (Boyd et al., 2000). There is also evidence that nanoscale roughness enhances the adhesion of the conditioning layer to the substratum (Hanarp et al., 1999).

Thus the presence of organic (and inorganic) material on a surface affects its cleanability, and also, potentially, its hygienic status since the "soil" can interfere with the activity of cleaning and disinfecting agents, by physically and "chemically" protecting microorganisms. This organic material may also potentially provide nutrients for the residual microorganisms, enabling multiplication and an increase in contamination of the surface. It has been shown that following continued cleaning and fouling cycles, stainless steel grain boundaries become progressively more contaminated with organic soil (Fig. 8.6) (Verran et al., 2001a; Verran and Whitehead, 2006). This cumulative soiling will inherently affect surface conditions and thus microbial attachment and retention.

V. CHARACTERIZATION OF SURFACES

A. Topography

There are several methods by which surface topography can be characterized, but the choice of method is dependent on the size of the samples and surface features. Most currently available instrumentation claims to

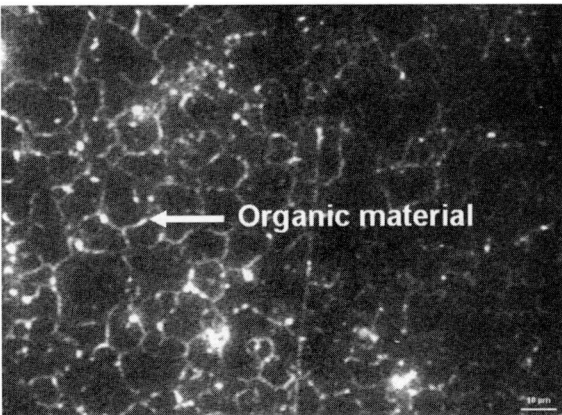

FIGURE 8.6 Following 15 cycles increased cleaning/soiling cycles organic material becomes trapped in grain boundaries on stainless steel.

resolve surface features at the lower nanoscale, depending on the equipment and the sample type. At the nanoscale and for surface with a z height < 8 μm surface topography is usually characterized using atomic force microscopy (AFM). The AFM has a sharp tip that can vary in shape but is around 10 nm in size, and can be used in a contact or noncontact mode when scanning a surface. The force applied to the tip and the tip shape can also be varied depending on whether hard or biological samples need to be imaged. Although the surface can be visualized in three-dimensions with excellent resolution, the area analyzed is small, thus irregularities in the surface can greatly skew the results unless a large number of measurements are made (Verran and Boyd, 2001). Another disadvantage of this instrument is that many samples may be too rough to image.

White light interferometry may also be used at the nanoscale level, but can scan larger surface areas. It utilizes the deflection of light from surface irregularities to produce a three-dimensional noncontact image of the surface. The advantage of using a white light interferometer over AFM is that it is easy to operate and is relatively quick; however it is not always easy to image some polymers and translucent materials. In laser profilometers, again noncontact instruments use highly precise stages to create profiles and three-dimensional topographies.

The solid stylus (probe radius 2–10 μm) profilometer traces across the surface producing a two-dimensional trace from which roughness values are calculated. However if an entire surface is to be mapped, many lines must be scanned. There are also limits in lateral resolution, set by the size of the probe tip, which can range from 10–25 μm diameter.

For both the solid stylus profilometer and the AFM, probe dimensions affect observed results. If the probe is physically incapable of reaching the bottom of narrow troughs, features are recorded as being more shallow than they are. It is also difficult to accurately chart highly curved, undercut, or convoluted surfaces with steep slopes.

B. Chemistry

Sophisticated analytical techniques have been employed to characterize surface properties before and after contamination. These can provide information on the nature of the elements and their associated chemical bonds on a surface [X-ray photoelectron spectroscopy (XPS)], their distribution [imaging secondary ion mass spectrometry (iSIMS)], and the elemental and molecular identity [time of flight secondary ion mass spectrometry (ToFSIMS)]. XPS spectra measure components in the top 1 to 10 nm of the material being analyzed under ultra high vacuum. XPS has been used to assess the cleanability of stainless steel fouled with starch and bacteria (Boyd *et al.*, 2001a) where it was shown that starch and bacteria were removed from surfaces at different rates following cleaning. In our laboratories, other work using XPS has further shown that stainless steel surfaces were rapidly conditioned with organic material following one cycle of soiling and cleaning (Verran *et al.*, 2001a). iSIMS is a sensitive surface analysis technique, which has been used to show the heterogeneity of conditioning film coverage on surfaces (Verran *et al.*, 2001a). Unlike XPS, ToFSIMS is not quantitative but has been used to analyze stainless steel surfaces fouled with starch and milk powder, pre and postcleaning, showing that surface cleanability was affected by the cleaning regime and the shape of the surface defects (Boyd *et al.*, 2001b). However, because of their sensitivity, signals produced by both XPS and ToFSIMS are not easily interpreted signals for complex soil/microorganism mixes (Verran *et al.*, 2001a). Similar unpublished observations have been made regarding the use of Fourier Transform Infrared Spectroscopy (FTIR).

VI. MEASURING RETENTION AND ASSESSING CLEANING AND DISINFECTION

A. Substratum preparation

Substrata used in *in vitro* cleaning and disinfection assays should provide a realistic and reproducible challenge. The use of recycled substrata is acceptable, since the surfaces would be repeatedly cleaned and soiled *in situ*. Reused substrata were stained with acridine orange and visualized using epifluorescence microscopy. Stained (organic) material was evident within grain boundaries despite rigorous cleaning between assays (Fig. 8.6). Thus it is unrealistic to use pristine, unused surfaces, since the more natural situation is for the substrata to have been conditioned.

Additionally, the use of simulated worn surfaces is preferable. Characterization of in-use worn surfaces has been undertaken using imprint technology such as dental impression materials (Verran et al., 2003) and cellulose acetate strips, softened by the application of acetone and pushed onto surfaces enabling reproduction of surface topography on the removable strip (results not presented). Transfer of impression to the laboratory enables examination by AFM, measurement of feature dimensions, and reproduction of these features on test surfaces *in vitro* (Fig. 8.2).

B. Amount of retention

1. Total viable count

Measurement of the amount of attachment (rather than strength) has traditionally relied on cell counting methods. Methods include removal of cells from the surface by swabbing or sonication with subsequent plate counts (Boyd et al., 2002; Wirtanen et al., 2000). Problems include clumping of cells, and low recovery rates, resulting in underestimation of the level of contamination. It has been suggested that to use these techniques accurately the numbers of cells on surfaces should be $>10^5$ cells cm^{-2} (Holah et al., 1988). Furthermore, swabbing has been shown to be only approximately 10% efficient at removing a known population of cells when unabsorbed (wet) and 4% when adsorbed (dry) onto surfaces (Obee et al., 2007). When stainless steel surfaces were inoculated with a persistent *L. monocytogenes* strain and surfaces were re-inoculated with cells, dried and washed with water for five cycles, cells were evident on the surface (Fig. 8.7A). In the absence of organic soil, or a cleaning agent, coupled with the effect of drying between soiling episodes, a marked increase in the amount of microbial contamination on the surface was evident, as measured by substratum surface coverage (Fig. 8.7B). However the number of cells recoverable in the absence of soil using the swab technique was less than 4%, perhaps also due to the drying process making cell removal difficult. Contact plates/dipslides are more efficient (50–70%) in terms of cell recovery; but the sample area is small. The effect of the presence of organic material on the removal of attached cells has not been explored although work is currently being carried out in our laboratories on the effect of organic food soil on microbial attachment and retention.

Direct enumeration of viable bacteria may be carried out on cells that have been dried onto surfaces. The surfaces are over layered with agar, and once incubated colonies are stained with nitroblue tetrazolium solution (Airey and Verran, 2006; Barnes et al., 1996). This method was shown

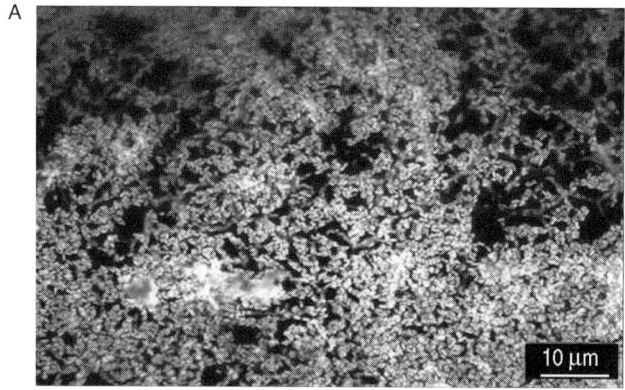
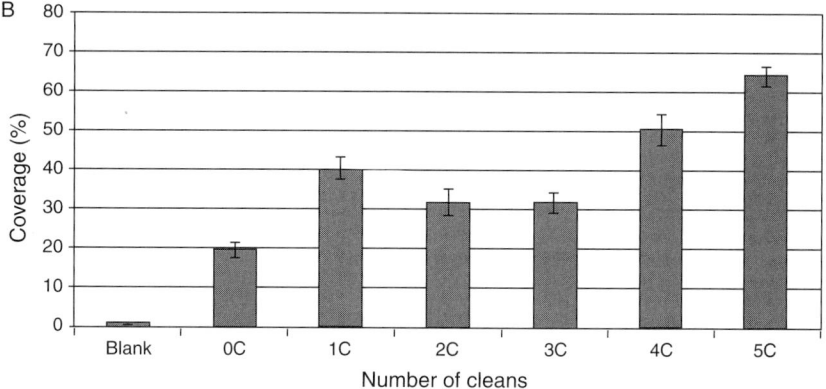

FIGURE 8.7 (A) *L. monocytogenes* retained on stainless steel surface following five soiling and cleaning cycles (cleaning with water), stained with acridine orange (B) Percentage coverage of cells left on surfaces with re-soiling and drying of the surfaces as measured using acridine orange staining and epifluorescence microscopy ($n = 40$).

to recover five times more bacteria than a conventional swabbing procedure, but cannot be used if an antimicrobial element is diffusing from the surface.

2. Microscopic counts

Imaging cells *in situ* has a significant advantage over culture, since results are immediate, and higher numbers of microorganisms are obtained, because there is no interim step in which cells can be lost. Cells are viewed directly on the surface, and may be differentially stained to determine cell viability. Microscopy methods are more reliable at detecting low levels of surface contamination (Hood and Zottola, 1995; Verran and Whitehead, 2006), but in all cases the area examined is relatively small, necessitating the use of a number of replicates. Cell counts can also be replaced by surface coverage data, via image analysis. Visualization of cells on the surfaces also gives some indication of their distribution, and interaction with organic material and any flaws in the substratum surface.

The LIVE/DEAD™® stain (Molecular probes) enables determination of cell viability on surfaces. The kits provide two-color fluorescence staining on both live (green) and dead (red) cells. The green-fluorescent nucleic acid dye stains both live and dead bacteria with intact and damaged cell membranes while the red-fluorescent nucleic acid dye stains only dead bacteria with damaged cell membranes (Fig. 8.8). The method enables comparison between surfaces, but should not be taken as infallible, since some bacteria with damaged membranes may be able to recover and reproduce while others with intact membranes may be unable to reproduce in nutrient medium (Anonymous, 2004).

It has been suggested that acridine orange can be used as an indicator of cell viability. When the concentration of acridine orange is kept relatively low, bacteria growing at high growth rates will fluoresce red-orange because of the predominance of RNA while inactive bacteria fluoresce green because of DNA. However the color can be affected by growth medium, contact time, and species (Back and Kroll, 1991; Yu *et al.*, 1995), thus the stain is more typically used as a total stain. The DNA specific dye DAPI has been considered to be superior to acridine orange for the enumeration of bacteria (Hobbie *et al.*, 1977) since acridine orange may also detect other organic components and in some cases also the surfaces of polymers thus making enumeration of cells using acridine orange difficult.

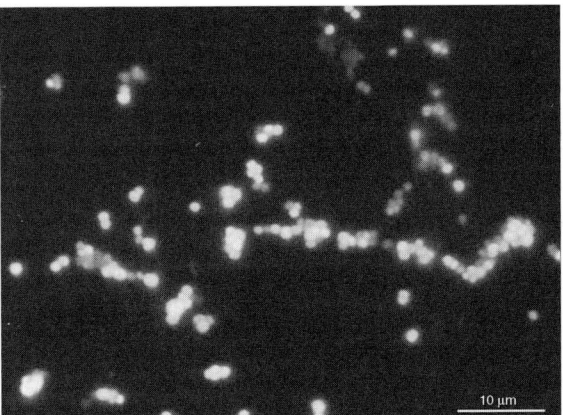

FIGURE 8.8 LIVE/DEAD™® staining of cells. The green-fluorescent nucleic acid dye stains both live and dead bacteria with intact and damaged cell membranes while the red-fluorescent nucleic acid dye stains only dead bacteria with damaged cell membranes.

However, this nonselective staining by acridine orange may be used to advantage, by providing a relatively simple means for visualizing both cells and soil on surfaces (Fig. 8.4).

Fluorescent *in situ* hybridization (FISH) uses fluorescent probes to detect specific DNA sequences on surfaces, thus enabling targetting of pathogens within microbial consortia (Sunde *et al.*, 2003) and cell/soil mixtures.

C. Strength of retention

The few studies that have been concerned with quantifying the strength of microbial attachment to a surface have tended to focus on flow cells where a known shear force is applied across the test surface and cell removal is monitored (Callow and Fletcher, 1994; Sjollema *et al.*, 1989), or via the passing of an air–liquid interface, which has the ability to displace attached bacterial cells by applying large shear forces (Boyd *et al.*, 2002; Gomez-Suarez *et al.*, 2001). At solid–air interfaces, the blot succession technique has been used. Bacteria that have been dried onto surfaces are transferred onto a sequence of agar plates, under the premise that less strongly attached cells are removed at earlier "blots" (Allison *et al.*, 2000; Eginton *et al.*, 1995; Kim and Silva, 2005). For example, *E. coli* O157:H7 and *S. typhimurium* had higher removal exponents than *Listeria* spp.

AFM is now being used to quantify, in nanoNewtons, the force of retention of bacteria on surfaces (Boonaert and Rouxhet, 2000; Boyd *et al.*, 2002; Méndez-Vilas *et al.*, 2007; Whitehead *et al.*, 2006). Methods include coating the AFM tip with cells (Razatos *et al.*, 2000), immobilizing a single cell onto the tip of the AFM cantilever (Bowen *et al.*, 2001) applying the AFM tip perpendicularly to the attached cell, or using the tip to push cells from the surface (Boyd *et al.*, 2002; Whitehead *et al.*, 2006).

It is at first difficult to reconcile measurements taken at the cellular level in terms of strength of attachment with measurements on the microscopic scale. Using microscopy and counts, it appears that small changes in surface topography due to surface wear do not affect the *amount* of attachment, but when AFM is used to measure the lateral force required to move retained cells, the location of cells with respect to surface features has a significant effect. However, cell counts are assessing cells which are retained; AFM is measuring the force required to remove these cells. The *strength* of attachment seems to be particularly related to the area of cell-substratum surface in contact: rod shaped cells straddling a linear feature with a relatively small percentage of their area in contact with a surface are more easily removed than cocci which nestle within the feature itself (Whitehead *et al.*, 2006). The application of detergent to this micro (nano) scale interaction demonstrates an improved ease or removal (Fig. 8.9); differences between strains and environmental features can also be investigated.

FIGURE 8.9 AFM force studies demonstrating the removal of *S. aureus* from brushed stainless steel surfaces under water and detergent. When detergent is applied cells are more easily removed from the surfaces (Images courtesy of R. Boyd).

D. Quantifying organic material

Quantification of organic material on a surface is often difficult because of the small amounts present and the complexity of the soil. Some chemical analysis has been attempted (e.g., XPS (Boyd et al., 2001a), TOFSIMS (Boyd et al., 2001b) or FTIR; however, because of the complexity of the organic material the results are not the easiest to elucidate. However it is of importance to be able to determine the amount of residual material retained on a surface following cleaning or removal of organic material. One of the easiest methods is staining the surface and visualization using epifluorescence microscopy. Using differential staining techniques it is possible to visualize the distribution of cells and organic material on a surface before and following cleaning assays (Fig. 8.10), and to measure coverage via image analysis. In addition to the more commonly used "non-specific" stains such as acridine orange, stains are also available that are specific for particular organic material (Verran et al., 2006), and/or for microorganisms (Declerck et al., 2003).The strength of attachment of the soil film to the surface has been measured via micromanipulation techniques, where the force required to remove the film is measured (Hooper et al., 2006; Saikhwan et al., 2007).

Methods for monitoring the presence of organic fouling material rather than microorganisms are available, but in general find limited use in the microbiology laboratory (Holah, 2000; Kulkarni et al., 1975; Verran et al., 2002). For the assessment of cleaning in place (CIP), detection of adenosine triphosphate (ATP) by bioluminescence provides an indication of surface

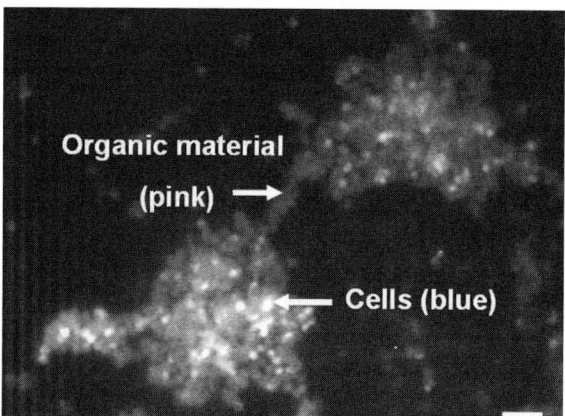

FIGURE 8.10 Differential staining techniques of cells and organic material allow visualization of cell and organic material distribution on a surface before and following cleaning assays (scale bar 10 μm). Cells are stained with DAPI, while organic material is stained with a molecular probes dye L-1908.

contamination (Redsvena *et al.*, 2007). The method measures ATP not only from microorganisms but also from product residues. It has been argued that ATP bioluminescence can only detect cells when nonbacterial ATP-soil is removed, following which the limit of detection is only <100 CFU cm^{-2} (Moore and Griffith, 2002). This method is not intended to indicate the presence of microorganisms, but only the cleanliness of a surface. Likewise, surfaces irradiated with ultraviolet light fluoresce in the presence of residual soil (http://www.bactoforce.co.uk/).

VII. CONCLUSIONS

By defining precisely the interactions occurring between microorganisms and organic soil on well characterized surfaces, modifications may be made to ensure minimal adhesion and/or maximum cleanability (minimal retention). For the food industry, major changes to the chemistry of food contact surfaces are unlikely, for reasons of consumer safety, effective evaluation, and installation cost. Targeted hygiene, via the careful use of selected surfaces at critical control points, might be more feasible, on a site-specific basis. Plant design and personnel training likewise should address issues of optimum surface flow in closed systems, easy access for cleaning wherever possible, and effective cleaning protocols and monitoring procedures. The use of inert, hygienic surfaces, with regular topography whose features are optimized to minimize retention, coupled with effective cleaning and disinfection procedures, remains the least invasive approach.

Microbial and organic fouling of surfaces in the food processing industry is of key importance in terms of hygiene and cleanability. A multidisciplinary and multifactorial approach to the topic, addressing interactions occurring between the substratum, the microbial cell and the food material, should enable an improved understanding of factors contributing to surface cleanability, and facilitate the development of strategies to minimize the retention of microorganisms and organic material. The consequences of poor hygienic practice are significant. Unacceptable outcomes include impact on food quality and food safety, and the reputation of the industry in general, and the food producer in particular, thus safe, cost effective and efficient functioning of equipment (and personnel) is essential (Verran, 2002).

REFERENCES

Airey, P., and Verran, J. (2006). A method for monitoring substratum hygiene using a complex soil: The human fingerprint. *In* "Fouling, Cleaning and Disinfection" (D. I. Wilson, Y. M. Chew, P. J. Fryer, and A. P. M. Hasting, Eds.) pp. 92–96. University of Cambridge, Cambridge, UK.

Airey, P., and Verran, J. (2007). Potential use of copper as a hygienic surface; problems associated with cumulative soiling and cleaning. *J. Hosp. Infect.* **67**, 271–277.

Allison, D. G., Cronin, M. A., Hawker, J., and Freeman, S. (2000). Influence of cranberry juice on attachment of *Escherichia coli* to glass. *J. Basic Microbiol.* **40**, 3–6.

Anonymous (1988). "BS1134-1: Assessment of Surface Texture—Part 1: Methods and Instrumentation." British Standards Institute, Milton Keynes, UK.

Anonymous (2004). Live/dead BacLight viability kits. Invitrogen, Carlsbad, CA http://probes.invitrogen.com/media/publications/147.pdf.

Arnold, J. W., and Bailey, G. W. (2000). Surface finishes on stainless steel reduce bacterial attachment and early biofilm formation: Scanning electron and atomic force microscopy study. *Poult. Sci.* **79**, 1839–1845.

Arnold, J. W., Boothe, D. H., Suzuki, O., and Bailey, G. W. (2004). Multiple imaging techniques demonstrate the manipulation of surfaces to reduce bacterial contamination and corrosion. *J. Microsc.-Oxford* **216**, 215–221.

Back, J. P., and Kroll, R. G. (1991). The differential fluorescence of bacteria stained with acridine-orange and the effects of heat. *J. Appl. Bacteriol.* **71**, 51–58.

Barnes, B. I., Cassar, C. A., Halablab, M. A., Parkinson, M. H., and Miles, R. J. (1996). An *in situ* method for determining bacterial survival on food preparation surfaces using a redox dye. *Lett. Appl. Microbiol.* **23**, 325–328.

Bollen, C. M. L., Lambrechts, P., and Quirynen, M. (1997). Comparison of surface roughness of oral hard materials to the threshold surface roughness for bacterial plaque retention: A review of the literature. *Dent. Mater.* **13**, 258–269.

Boonaert, C. J. P., and Rouxhet, P. G. (2000). Surface of lactic acid bacteria: Relationships between chemical composition and physicochemical properties. *Appl. Environ. Microbiol.* **66**, 2548–2554.

Boulange-Petermann, L. (1996). Processes of bioadhesion on stainless steel surfaces and cleanability: A review with special reference to the food industry. *Biofouling* **10**, 275–300.

Boulange-Petermann, L., Rault, J., and Bellon-Fontaine, M.-N. (1997). Adhesion of *Streptococcus thermophilus* to stainless steel with different surface topography and roughness. *Biofouling* **11**, 201–216.

Bowen, W. R., Lovitt, R. W., and Wright, C. J. (2001). Atomic force microscopy study of the adhesion of *Saccharomyces cerevisiae*. *J. Coll. Inter. Sci.* **237**, 54–61.

Boyd, R. D., Cole, D., Rowe, D., Verran, J., Coultas, S. J., Paul, A. J., West, R. H., and Goddard, D. T. (2000). Surface characterization of glass and poly(methyl methacrylate) soiled with a mixture of fat, oil, and starch. *J. Adh. Sci. Technol.* **14**, 1195–1207.

Boyd, R. D., Verran, J., Hall, K. E., Underhill, C., Hibbert, S., and West, R. (2001a). The cleanability of stainless steel as determined by X-ray photoelectron spectroscopy. *Appl. Surf. Sci.* **172**, 135–143.

Boyd, R. D., Cole, D., Rowe, D., Verran, J., Paul, A. J., and West, R. H. (2001b). Cleanability of soiled stainless steel as studied by atomic force microscopy and time of flight secondary ion mass spectrometry. *J. Food Prot.* **64**, 87–93.

Boyd, R. D., Verran, J., Jones, M. V., and Bhakoo, M. (2002). Use of the atomic force microscope to determine the effect of substratum surface topography on bacterial adhesion. *Langmuir* **18**, 2343–2346.

Callow, M. E., and Fletcher, R. L. (1994). The influence of low surface-energy materials on bioadhesion—A review. *Inter. Biodeter. Biodeg.* **34**, 333–348.

Carpentier, B., and Cerf, O. (1993). Biofilms and their consequences, with particular reference to hygiene in the food-industry. *J. Appl. Bacteriol.* **75**, 499–511.

Characklis, W. G., and Wilderer, P. A. (1990). "Structure and Function of Biofilms." John Wiley and Sons, New York.

Cousins, B. G., Allison, H. E., Doherty, P. J., Edwards, C., Garvey, M. J., Martin, D. S., and Williams, R. L. (2007). Effects of a nanoparticulate silica substrate on cell attachment of *Candida albicans*. *J. Appl. Microbiol.* **102**, 757–765.

Cunliffe, D., Smart, C. A., Alexander, C., and Vulfson, E. N. (1999). Bacterial adhesion at synthetic surfaces. *Appl. Environ. Microbiol.* **65,** 4995–5002.

Declerck, P., Verelst, L., Duvivier, L., Van Damme, A., and Ollevier, F. (2003). A detection method for *Legionella spp.* in (cooling) water: Fluorescent *in situ* hybridisation (FISH) on whole bacteria. *Water Sci. Technol.* **47,** 143–146.

Eginton, P. J., Gibson, H., Holah, J., Handley, P. S., and Gilbert, P. (1995). Quantification of the ease of removal of bacteria from surfaces. *J. Indus. Microbiol.* **15,** 305–310.

Flint, S. (1996). Report on the evaluation of seven plasma weld samples. *In* "Stainless Steel Weld Surface Finish and Biofilm Development. A Round Robin Test" (W. Scholz, Ed.), (Report R8–14), pp. 1–7. New Zealand Welding Centre, Manukau City, New Zealand.

Flint, S. H., Bremer, P. J., and Brooks, J. D. (1997). Biofilms in dairy manufacturing plant—Description, current concerns and methods of control. *Biofouling* **11,** 81–97.

Flint, S. H., Brooks, J. D., and Bremer, P. J. (2000). Properties of the stainless steel substrate, influencing the adhesion of thermo-resistant streptococci. *J. Food Eng.* **43,** 235–242.

Frank, J. F., and Chmielewski, R. A. N. (1997). Effectiveness of sanitation with quaternary ammonium compound or chlorine on stainless steel and other domestic food-preparation surfaces. *J. Food Prot.* **60,** 43–47.

Furstner, R., Barthlott, W., Neinhuis, C., and Walzel, P. (2005). Wetting and self-cleaning properties of artificial superhydrophobic surfaces. *Langmuir* **21,** 956–961.

Gibbons, I., Adesiyun, A., Seepersadsingh, N., and Rahaman, S. (2006). Investigation for possible source(s) of contamination of ready-to-eat meat products with *Listeria* spp. and other pathogens in a meat processing plant in Trinidad. *Food Microbiol.* **23,** 359–366.

Gomez-Suarez, C., Busscher, H. J., and van der Mei, H. C. (2001). Analysis of bacterial detachment from substratum surfaces by the passage of air–liquid interfaces. *Appl. Environ. Microbiol.* **67,** 2531–2537.

Hanarp, P., Sutherland, D., Gold, J., and Kasemo, B. (1999). Nanostructured model biomaterial surfaces prepared by colloidal lithography. *Nanostructured Mat.* **12,** 429–432.

Hilbert, L. R., Bagge-Ravn, D., Kold, J., and Gram, L. (2003). Influence of surface roughness of stainless steel on microbial adhesion and corrosion resistance. *Inter. Biodet. Biodeg.* **52,** 175–185.

Hobbie, J. E., Daley, R. J., and Jasper, S. (1977). Use of nuclepore filters for counting bacteria by fluorescence microscopy. *Appl. Environ. Microbiol.* **33,** 1225–1228.

Holah, J. T. (2000). Food processing equipment design and cleanability. *Flair-Flow Eur. Tech. Man.* F-FE 377 A/00.

Holah, J. T., Betts, R. P., and Thorpe, R. H. (1988). The use of direct epifluorescent microscopy (DEM) and the direct epifluorescent filter technique (DEFT) to assess microbial-populations on food contact surfaces. *J. Appl. Bacteriol.* **65,** 215–221.

Holah, J. T., Higgs, C., Robinson, S., Worthington, D., and Spenceley, H. (1990). A conductance-based surface disinfection test for food hygiene. *Lett. Appl. Microbiol.* **11,** 255–259.

Holah, J. T., and Thorpe, R. H. (1990). Cleanability in relation to bacterial retention on unused and abraded domestic sink materials. *J. Appl. Bacteriol.* **69,** 599–608.

Hood, S. K., and Zottola, E. A. (1995). Biofilms in food processing. *Food Control* **6,** 9–18.

Hooper, R. J., Liu, W., Fryer, P. J., Paterson, W. R., Wilson, D. I., and Zhang, Z. (2006). Comparative studies of fluid dynamic gauging and a micromanipulation probe for strength measurements. *Food Bioprod. Proc.* **84,** 353–358.

Jeong, D. K., and Frank, J. F. (1994). Growth of *Listeria-monocytogenes* at 10-degrees-C in biofilms with microorganisms isolated from meat and dairy processing environments. *J. Food Prot.* **57,** 576–586.

Kim, T., and Silva, J. L. (2005). Quantification of attachment strength of selected foodborne pathogens by the blot succession method. *J. Rapid Meth. Auto. Microbiol.* **13,** 127–133.

Korber, D. R., Choi, A., Wolfaardt, G. M., Ingham, S. C., and Caldwell, D. E. (1997). Substratum topography influences susceptibility of *Salmonella enteritidis* biofilms to trisodium phosphate. *Appl. Environ. Microbiol.* **63,** 3352–3358.

Kulkarni, S. M., Arnold, R. G., and Maxcy, R. B. (1975). Reuse limits and regeneration of solutions for cleaning dairy equipment. *J. Dairy Sci.* **58**, 1095–1100.

Kumar, C. G., and Anand, S. K. (1998). Significance of microbial biofilms in food industry: A review. *Inter. J. Food Microbiol.* **42**, 9–27.

Langeveld, L. P. M., Bolle, A. C., and Vegter, J. E. (1972). The cleanability of stainless steel with different degrees of surface roughness. *Neth. Milk Dairy J.* **42**, 149–154.

Li, X., Liu, T., and Chen, Y. (2004). The effects of the nanotopography of biomaterial surfaces on *Pseudomonas fluorescens* cell adhesion. *Biochem. Eng. J.* **22**, 11–17.

Maller, R. R. (1998). Passivation of stainless steel. *Trends Food Sci. Tech.* **9**, 28–32.

Medilanski, E., Kaufmann, K., Wick, L. Y., Wanner, O., and Harms, H. (2002). Influence of the surface topography of stainless steel on bacterial adhesion. *Biofouling* **18**, 193–203.

Méndez-Vilas, A., Bruque, J. M., and González-Martín, M. L. (2007). Sensitivity of surface roughness parameters to changes in the density of scanning points in multi-scale AFM studies. Application to a biomaterial surface. *Ultramicroscopy* **107**, 617–625.

Mettler, E., and Carpentier, B. (1998). Variations over time of microbial load and physico-chemical properties of floor materials after cleaning in food industry premises. *J. Food Prot.* **61**, 57–65.

Miettinen, H., Arvola, A., Luoma, T., and Wirtanen, G. (2003). Prevalence of *Listeria monocytogenes* in, and microbiological and sensory quality of, rainbow trout, whitefish, and vendance roes from Finnish retail markets. *J. Food Prot.* **66**, 1832–1839.

Milledge, J. J., and Jowitt, R. (1980). The cleanability of stainless steel used as a food contact surface. *Inst. Food Sci. Technol. Proc.* **13**, 57–62.

Moore, G., and Griffith, C. (2002). A comparison of surface sampling methods for detecting coliforms on food contact surfaces. *Food Microbiol.* **19**, 65–73.

Navratilova, P., Schlegelova, J., Sustackova, A., Napravnikova, E., Lukasova, J., and Klimova, E. (2004). Prevalence of *Listeria monocytogenes* in milk, meat and foodstuff of animal origin and the phenotype of antibiotic resistance of isolated strains. *Vet. Med.* **49**, 243–252.

Obee, P., Griffith, C. J., Cooper, R. A., and Bennion, N. E. (2007). An evaluation of different methods for the recovery of methicillin-resistant *Staphylococcus aureus* from environmental surfaces. *J. Hosp. Infect.* **65**, 35–41.

Olsson, C. O. A., and Landolt, D. (2003). Passive films on stainless steels—Chemistry, structure and growth. *Electrochim. Acta* **48**, 1093–1104.

Packer, A., Kelly, P., Whitehead, K., and Verran, J. (2007). Effects of defined linear features on surface hygiene and cleanability *In* "Society of Vacuum Coaters 50th Annual Technical Conference Proceedings." (See pp. 90–93). Kentucky.

Razatos, A., Ong, Y. L., Boulay, F., Elbert, D. L., Hubbell, J. A., Sharma, M. M., and Georgiou, G. (2000). Force measurements between bacteria and poly(ethylene glycol)-coated surfaces. *Langmuir* **16**, 9155–9158.

Redsvena, I., Kymäläinena, H.-R., Pesonen-Leinonena, E., Kuismaa, R., Ojala-Paloposkia, T., Hautalaa, M., and Sjöberg, A.-M. (2007). Evaluation of a bioluminescence method, contact angle measurements and topography for testing the cleanability of plastic surfaces under laboratory conditions. *Appl. Surf. Sci.* **253**, 5536–5543.

Saikhwan, P., Chew, Y., Paterson, W., and Wilson, D. (2007). Fluid dynamic gauging: A technique for studying the cleaning of food processing surfaces. *Food Manuf. Efficiency* **1**, 35–41.

Scheuerman, T. R., Camper, A. K., and Hamilton, M. A. (1998). Effects of substratum topography on bacterial adhesion. *J. Coll. Inter. Sci.* **208**, 23–33.

Sjollema, J., Busscher, H. J., and Weerkamp, A. H. (1989). Experimental approaches for studying adhesion of microorganisms to solid substrata: Applications and mass transport. *J. Microbiol. Meth.* **9**, 73–78.

Sunde, P. T., Olsen, I., Gobel, U. B., Theegarten, D., Winter, S., Debelian, G. J., Tronstad, L., and Moter, A. (2003). Fluorescence *in situ* hybridization (FISH) for direct visualization of bacteria in periapical lesions of asymptomatic root-filled teeth. *Microbiol. SGM* **149**, 1095–1102.

Teughels, W., Van Assche, N., Sliepen, I., and Quirynen, M. (2006). Effect of material characteristics and/or surface topography on biofilm development. *Clin. Oral Impl. Res.* **17**, 68–81.

Tide, C., Harkin, S., Geesey, G. G., Bremer, J., and Scholz, W. (1999). The influence of welding procedures on bacterial colonization of stainless steel weldments. *J. Food Eng.* **42**, 85–96.

Tresse, O., Shannon, K., Pinon, A., Malle, P., Vialette, M., and Midelet-Bourdin, G. (2007). Variable adhesion of *Listeria monocytogenes* isolates from food-processing facilities and clinical cases to inert surfaces. *J. Food Prot.* **70**, 1569–1578.

Vanhaecke, E., Remon, J.-P., Moors, M., Raes, F., de Rudder, D., and van Peteghem, A. (1990). Kinetics of *Pseudomonas aeruginosa* adhesion to 304 and 316-L stainless steel: Role of cell surface hydrophobicity. *Appl. Environ. Microbiol.* **56**, 788–795.

Verran, J. (2002). Biofouling in food processing—Biofilm or biotransfer potential? *Food Bioprod. Proc.* **80**, 292–298.

Verran, J., and Maryan, C. (1997). Retention of Candida albicans on acrylic resin and silicone of different surface topography. *J. Prosthet. Dent.* **77**, 535–539.

Verran, J., and Jones, M. (2000). Problems of biofilms in the food and beveridge industry. *In* "Industrial Biofouling Detection, Prevention and Control" (J. Walker, S. Surmann, and J. Jass, Eds.) pp. 145–173. Wiley, Chichester, UK.

Verran, J., and Boyd, R. D. (2001). The relationship between substratum surface roughness and microbiological and organic soiling: A review. *Biofouling* **17**, 59.

Verran, J., and Whitehead, K. (2005). Factors affecting microbial adhesion to stainless steel and other materials used in medical devices. *Inter. J. Artif. Organs* **28**, 1138–1145.

Verran, J., and Whitehead, K. A. (2006). Assessment of organic materials and microbial components on hygienic surfaces. *Food Bioprod. Proc.* **84**, 260–264.

Verran, J., Lees, G., and Shakespeare, A. P. (1991). The effect of surface roughness on the adhesion of *Candida albicans* to acrylic. *Biofouling* **3**, 183–192.

Verran, J., Rowe, D. L., Cole, D., and Boyd, R. D. (2000). The use of the atomic force microscope to visualise and measure wear of food contact surfaces. *Inter. Biodet. Biodeg.* **46**, 99–105.

Verran, J., Boyd, R. D., Hall, K., and West, R. H. (2001a). Microbiological and chemical analyses of stainless steel and ceramics subjected to repeated soiling and cleaning treatments. *J. Food Prot.* **64**, 1377–1387.

Verran, J., Boyd, R. D., Hall, K. E., and West, R. (2002). The detection of microorganisms and organic material on stainless steel food contact surfaces. *Biofouling* **18**, 167–176.

Verran, J., Rowe, D. L., and Boyd, R. D. (2003). Visualization and measurement of nanometer dimension surface features using dental impression materials and atomic force microscopy. *Inter. Biodet. Biodeg.* **51**, 221–228.

Verran, J., Boyd, R., Whitehead, K. A., and Hall, K. (2004). Surface topography and the retention of organic soil and microorganisms on hygienic surfaces. *In* "Hygienic Coatings and Surfaces Conference Papers." Paint Research Association, Orlando, Florida.

Verran, J., Airey, P., and Whitehead, K. A. (2006). Assessment of organic material and microbial componenets on hygienic surfaces. *In* "Fouling, Cleaning and Disinfection" (D. I. Wilson, Y. M. Chew, P. J. Fryer, and A. P. M. Hasting, Eds.) pp. 1–7. University of Cambridge, Cambridge, UK.

Whitehead, K. A., and Verran, J. (2006). The effect of surface topography on the retention of microorganisms. A review. *Food Bioprod. Proc.* **84**, 253–259.

Whitehead, K. A., and Verran, J. (2007). The effect of application method on the retention of *Pseudomonas aeruginosa* on stainless steel, and titanium coated stainless steel of differing topographies. *Inter. Biodet. Biodeg.* **60,** 74–80.

Whitehead, K. A., Colligon, J., and Verran, J. (2005). Retention of microbial cells in substratum surface features of micrometer and sub-micrometer dimensions. *Colloids Surf. B-Biointerfaces* **41,** 129–138.

Whitehead, K. A., Rogers, D., Colligon, J., Wright, C., and Verran, J. (2006). Use of the atomic force microscope to determine the effect of substratum surface topography on the ease of bacterial removal. *Colloids Surf. B-Biointerfaces* **51,** 44–53.

Wilks, S. A., Michels, H., and Keevil, C. W. (2005). The survival of *Escherichia coli* O157 on a range of metal surfaces. *Int. J. Food Microbiol.* **105,** 445–454.

Wirtanen, G., Saarela, M., and Mattila-Sandholm, T. (2000). Biofilms—impact of hygiene in food industries. *In* "Biofilms II: Process Analysis and Applications" (J. Bryers, ed.). Wiley-Liss, New York.

Yu, F. P., Dodds, W. K., Banks, M. K., Skalsky, J., and Strauss, E. A. (1995). Optimal staining and sample storage time for direct microscopic enumeration of total and active bacteria in soil with two fluorescent dyes. *Appl. Environ. Microbiol.* **61,** 3367–3372.

INDEX

A

Aequorea victoria, 138, 146
Aerobic bacteria, toluene degradation gene clusters, 9
AFM. *See* Atomic force microscopy
AFP. *See* Autofluorescent proteins
Anaerobic toluene pathway
　advantages of, 47
　Azoarcus sp, 51
　Thauera aromatica k172
　　enzymes role, 47–51
　　genetic studies, 52
　Thauera aromatica T1, 51
Arabidopsis thaliana, 195
Atomic force microscopy
　retention strength measurement, 238
　for surface topography, 233
Autofluorescent proteins, 138, 146–148
Azoarcus sp EbN1and T, 51

B

Bacillus megaterium, 148
Bacterial artificial chromosomes (BAC), 130
　for mapping and analysis, 130
　mini- F plasmid pMB0131, 130
Bacterial community assembly
　niche- assembly perspective, 177
　probability- based similarity index (S_{RC}), 179
　random assembly, 178–179
　stochastic assembly, 178
Bacterial ecology, 168
Beta diversity
　distance–decay relationships in, 177–178
　smaller islands *vs.* larger islands, 175
　species- time relationship in, 175–176
BFP. *See* Blue fluorescent protein
Biofilm formation, 83, 100, 225–226
　catecholamine- induced, 101
　surface chemistry affecting, 231
Bioluminescence, 144–145
　detection of adenosine triphosphate (ATP), 240

Biopreservatives, 207
Biosensors
　definition of, 138
　host cell specifications and working principles, 139
　molecular biosensors
　　electrical, 159–160
　　optical, 150–158
Blue fluorescent protein, 154
Burkholderia sp. strain JS150
　biological nature of, 25
　multiple toluene degradation pathways
　　genes and coding regions, 26–27
　　toluene conversion, 25
Burkholderia vietnamiensis G4
　genome of, 15
　toluene 2- monooxygenase pathway in
　　enzymes involved in, 12–14
　　genes involved in, 14–15
Burkholderia xenovorans, 23

C

Campylobacter sp., 83–84, 185
Catecholamine
　enteric bacteria, specificity
　　in animals, 88
　　epinephrine *vs.* norepinephrine, 85–87
　　in mammalian systems, 87–88
　immune system and infectious agent, 77–79
　molecular analyses, 88–89
　spectrum of, 84–85
　stress- related neuroendocrine hormones, 79–84
Cell retention, 225–226
Chloramphenicol acetyltransferase (Cat), 140
Cladophialophora, 44
Cladosporium sphaerospermum
　isolation of, 44
　in toluene degradation pathway
　　enzyme assay study, 44
　　oxidation, 46
Cleaning in place (CIP), 240

247

Closed surface systems
  biofilm formation in, 224, 226
  microbial attachment, 225
  optimum surface flow in, 241
Conditioning film, 231–232
Cosmid vector, 110–111, 114

D

Dimethylsulfoxide (DMSO), 109, 116–117
Dioxygenase mediated toluene
    degradation pathway
  enzymes involved in, 4–8
  toluene degradation (tod) genes in, 8–11
Discosoma striata, 147
Distance–decay relationships, 177–178
DsRed, fluorescent protein, 147

E

Electron transport chain, 53, 109, 117–119
Enhanced cyan fluorescent protein (ECFP), 155, 157–158
Enhanced yellow fluorescent protein (EYFP), 155, 157–158
Enteric nervous system (ENS), catecholamine
  in animals, 88
  epinephrine *vs.* norepinephrine, 85–87
  in mammalian systems, 87–88
  spectrum of, 84–85
Enterobacter cloacae SLD1a-1, in selenate reduction
  biochemical characteristics, 109
  direct cloning, 110–111
  genetic analysis, 115–117
  growth, 110
  molecular model, 118–119
  physiology of, 108–109
  targeted gene knock-out procedure, 112–113
  transposon mutagenesis in, 113–115
Erwinia carotovora, 203, 207
Escherichia coli
  cysteine synthesis, 148–149
  in food poisoning, 185–186
  as human pathogens, 187–188
  K-12, 117
  maltose biosensor, 155
  O157, 231
  O157:H7, 84–85, 87–89, 95, 188, 195, 202, 207–208, 238

  in phyllosphere contamination, 205–208
  plasmid replication, 139
  S17-1, 110–111, 114–115
  S17-λpir (mini-Tn5-*lac*Z1), 114
  surface chemistry effects, 230–231
Exophiala jeanselmei, 44

F

Fluorescent indicator proteins (FLIPs), molecular biosensors
  FlAsH and ReAsH, 151
  fluorescent dyes, 152
  FRET, 153–159
  site directed mutagenesis (SDM), 150–151
Fluorescent *in situ* hybridization (FISH), 238
Fluorescent resonance energy transfer (FRET) biosensors
  Ca^{2+}-dependent change, 154
  fluorescent proteins, use of, 154
  glucose nanosensor, 156, 159
  maltose biosensor, 155–156
  microscopic ruler, use as, 153
  SBP-based, 157–158
  signal to noise ratio, 159
fnr (Fumurate Nitrate Reduction regulator) gene, 111, 113, 115–116
Food poisoning outbreaks, 185–186
Fosmids, 126, 130
Fourier transform infrared (FTIR) spectroscopy, 234
Fungal toluene degradation pathway
  vs. bacterial pathways, 46
  Cladosporium sphaerospermum
    isolation and growth, 44
    oxidation, 46
  Phanerochaete chrysosporium, 43

G

Geobacter metallireducens, 51
Green fluorescent protein (GFP), 138, 154

H

Haemolytic uremic syndrome, 185
Hazard analysis critical control point (HACCP), 189
Human pathogens, ecology and survival in association with plants
  E. coli and Salmonella, 195
  horizontal gene transfer, 203

Index 249

population dynamics, 195–201
transmission process, 195, 202
during processing
 biocontrol agents in, 206–207
 decontamination treatments in, 205
 E. coli O157:H7 in, 207–208
 in lettuce, 204–205
 lower ability plants, 208
 mesophilic bacteria, 204
 psychrophiles, 206
pathogens sources
 fecal contamination, 186–187
 irrigation water, 187–188
 preharvest period, 188–189
 raw animal manures, 187
research prospects, 208–209
in sewage and manure, 189–192
in soil, 193–194
in water, 190
Hygienic surfaces, 226

I

Ice nucleation proteins, 149–150
Imaging secondary ion mass spectrometry (iSIMS), 234
InaZ proteins, 149

L

Lactobacillus casei, 207
Ligand
 detection for (*see* Biosensors)
 dinuclear iron binding sites, TouA, 30
 p- toluic acid, for affinity chromatography, 6
Listeria monocytogenes
 chill- tolerant pathogen, 225
 in food poisoning, 185–186
 in phyllosphere contamination, 207–208
LIVE/DEAD$^{TM®}$, 237
Luria- Broth (LB) agar, 110–114

M

Maltose binding protein (MBP), 151, 155, 159–160
Menaquinone
 cellular production of, 119
 menD gene, 115–117
 in selenate reduction process, 118
Metagenomic DNA
 cloning preparation
  mechanical shearing of, 129
  partial restriction digestion, 129

current techniques used, 127–128
extraction of, 128
human DNA removal
 differential centrifugation, 130
 lysis method, 129
isolation of, 126
library construction
 large insert libraries, 130–131
 novel characterizations in, 131
 phage display technologies, 132–133
 small insert libraries, 131–132
reduced transport fluid (RTF) addition, 126
sample collection and processing, 126, 128
Metagenomics
 future perspectives, 134
 gene expression modes, 134
 limitations and challenges, 133–134
Methanobacterium ivanovii, 148
Microbial ecology
 beta diversity in
  distance–decay relationships of, 177
  smaller islands *vs.* larger islands, 175
  species–time relationships of, 175–176
 community assemblies
  niche- assembly perspective, 177
  probability- based similarity index (SRC) in, 179
  randomly assembled, 178–179
  stochastic or niche considerations, 178
 island systems
  biogeographic islands, 169–170
  immigration/extinction balance, 170
 species–area relationship (SAR)
  bacterial endemicity, 172
  colonization and extinction of species, 171
  metal- cutting fluids, 172
  wastewater treatment plant (WWTP), 173–174
  waterfilled tree holes, 172–173
Microbial endocrinology
 bacterial catecholamine
  enteric nervous system (ENS), 85–88
  molecular analyses of, 88–89
  spectrum of, 84–85
  stress- related neuroendocrine hormones, 79–84
 description of, 76–77
 immune system and infectious agent, 77–79

Microbial endocrinology (cont.)
  medium comparability
    neuroendocrine–bacterial interactions, 90
    serum- SAPI medium, 91
Microbial induction biosensors, 139
Microbial retention
  factors affecting
    organic material, 231–232
    surface chemistry, 230–231
    surface topography, 227–230
  measurement, on surfaces
    cell counting methods, 235–236
    cell visualization, 237
    fluorescent *in situ* hybridization (FISH), 238
    organic material, 240–241
    strength quantification, 238
    substratum preparation, 235–236
  on open food contact surfaces, 226
mini- Tn5- *lacZ*1, 113–114
Molecular biosensors
  electrical biosensors, 159–160
  optical biosensors
    multiple fluorophores, 153–159
    single fluorophores, 150–153
Mutant strain(s)
  Burkholderia strain JS150, 25
  derivatives DsRedT.3 and DsRedT.4, 147
  E. cloacae SLD1a- 1, knock- out *fnr* gene, 113
  E. coli IHF, 43
  E. coli O157:H7 luxS, 82
  P. mendocina KR1, T4MO cloned, 22
  P. putida DOT–T1E todST, 24
  P. putida F1, cis- dihydrodiol dehydrogenase mutant, 4
  P. putida F39/D, for toluene *cis*- dihydrodiol, 7
  P. stutzeri OX1, 28
  S. enterica, with reduced adhesion to alfalfa, 205

N

Nanosensors, 138, 154–156, 159
Neuroendocrine stress neurohormone assay methodologies
  bacterial transferrin iron uptake, 98–99
  catecholamine- mediated biofilm production, 99–100
  serum- SAPI medium preparation, 97–98
  transferrin- iron removal, 98
bacterial catecholamine
  and autoinducer activity, 82–83
  enhance bacterial attachment, 83–84
  induced host sequestered iron, 79, 82
bacterial inoculum size, 92–94
bacterial population density, 93
catecholamine specificity, 93–94
dose–response effect of, 95–96
Non specific toulene monooxygenase pathway. *See* Toulene/o- xylene monooxygenase pathway
Norepinephrine- induced autoinducer (NE- AI), 82

O

Open food contact surfaces
  biofilm coverage of, 226
  cell retention on, 225–226
  CIP assessment on, 240–241
  food- substratum contact, 226
  microbial attachment on, 225
  organic fouling of, 240
  preparation of, 234–235
  substrata used in, 226–227
Optical biosensors
  FRET in, 154–159
  multiple fluorophores
    improvement modifications, 156, 159
    SBP- based FRET biosensor, 154–158
  single fluorophores
    biosensor development, 151
    fluorescent dyes, 152
    modified SBPs, 151–153
    site directed mutagenesis (SDM) in, 150–151
    specificity modification, 151, 153

P

Phage display technologies
  DNA fragment insertion
    hyperphage, use of, 133
    phagemid vector, use of, 132
    process stages, 133
  DNA fragment purification, 132
Phanerochaete chrysosporium, 43
Phenyl ammonia lyase, 204
Photinus pyralis, 145

Phyllosphere, in human pathogens
  in association with plants
    E.coli and Salmonella, 195
    horizontal gene transfer, 203
    population dynamics, 195–201
    transmission process, 195, 202
  during processing
    biocontrol agents, 206–207
    decontamination treatment, 205
    E. coli O157:H7, 207–208
    in lettuce, 204–205
    lower ability plants, 208
    mesophilic bacteria, 204
    psychrophiles, 206
    research prospects, 208–209
  in sewage and manure, 189–192
  in soil, 193–194
  in water, 190
Plasmid vector, 112
Propionibacterium freudenreichii, 148
Pseudomonas aeruginosa, 186, 231
Pseudomonas denitrificans, 148
Pseudomonas mendocina KR1
  isolation of, 20
  PcuR protein, 15
  toluene degradation pathway, 21
    genes involved, 24–25
    toluene 4- monooxygenase, 22–23
Pseudomonas putida F1
  cis- dihydroxylation in, 6
  isolation of, 4
  toluene degradation (tod) genes, 8, 11
Pseudomonas putida mt- 2
  isolation of, 33
  TOL pathway
    genes involved, 40–41
    plasmid role in lower pathway, 37–39
    plasmid role in upper pathway, 36–37
    regulatory proteins, 41–43
Pseudomonos stutzeri OX1
  toluene/o- xylene monooxygenase
      pathway
    catechol production, 32
    enzyme activity, gene cloning for, 29–31
    gene expression, 32–33
    isolation of, 28
    oxidation of, 29

Q

Quantification, organic material on surface.
  See also Microbial retention

CIP assessment and bioluminescence,
  240–241
differential staining techniques, 240

R

Ralstonia pickettii PKO1
  isolation of, 15
  toluene- 3- monooxygenase pathway
    enzymes involved, 16, 18
    genes involved, 18–20
Reporter genes. See also Biosensors
  advantages and disadvantages, 141–143
  autofluorescent proteins (AFP), 146–148
  β- gal, 140, 144
  bioluminescence, 144–145
  Cat, 140
  GUS, 144
  ice nucleation proteins, 149–150
  lux genes, 145
  uroporphyrinogen III methyltransferase
    (UMT), 148–149

S

sacB gene, 112
Salmonella
  in food poisoning, 185–186
  as human pathogens, 187–188
  in phyllosphere contamination, 202–203
Salmonella Newport PT33, 185
Salmonella Saint- Paul, 185
Salmonella typhimurium, 230, 238
SBP. See Solute binding proteins
Selenomonas ruminantium, 148
S. enterica, 83, 85, 88, 92, 185, 205, 208
Shewanella oneidensis MR- 1, 117
Shigella, 185–186
Sinorhizobium meliloti, 147–148
Solid–air interface, biofilm formation, 226
Solid–liquid interface
  in biofilm formation, 225–226
  hygienic surfaces, 227
Solute binding proteins, 154–156, 159
Species–area relationship (SAR)
  for bacterial colonization
    in metal- cutting fluids, 172
    in wastewater treatment plant (WWTP)
      bioreactors, 173–174
    in waterfilled tree- holes, 172–173
  bacterial endemicity, 172
  colonization and extinction of species, 171
Species–time relationship (STR), 175–176

Staphylococcus aureus, 230
Substrata, 235–236
Surface
 characterisation
  surface chemistry, 234
  surface topography, 233
 chemistry
  in microbial retention, 230–231
  in surface characterization, 234
 hydrophobicity, 227
 topography
  in microbial retention, 227–229
  in surface characterization, 233–234

T

tatC gene, 115–117
TCE. *See* Trichloroethylene
TDO. *See* Toluene dioxygenase
Thauera aromatica k172
 anaerobic toluene pathway
  enzymatic functions, 47–51
  genetic studies, 52
 isolation of, 47
Thauera aromatica T1
 genetic studies, 52
 isolation of, 51
Thermus thermophilus, 23
TMAO. *See* Trimethylamine N- oxide
T2MO. *See* Toluene 2- monooxygenase
T3MO. *See* Toluene 3- monooxygenase
T4MO. *See* Toluene 4- monooxygenase
tod genes
 cis- dishydroxylation in *putida*, 6
 cloning and sequencing, 8, 10
 vs. dioxygenase genes in *putida*, 54
 promoters, 6
 proteins and operons, 11
TOL pathway
 pseudomonos putida mt- 2
  benzoate formation, 34
  genes in, 40–41
  isolation of, 33
  lower pathway, plasmid role in, 37–39
  regulatory proteins, 41–43
  upper pathway, plasmid role in, 36–37
Toluene- degrading fungal isolates, 44
Toluene dioxygenase
 cis- hydroxylation of toluene, 6
 oxidation role of, 7

Toluene 2- monooxygenase, 13
Toluene 3- monooxygenase
 genes encoding, 19
 oxidation by, 16
 subunits of, 18
Toluene 4- monooxygenase
 genes in, 22
 structural analysis of, 23
Tom genes, 15
ToMO pathway. *See* Toulene/*o*- xylene monooxygenase pathway
Toulene
 degradation pathways, 2–4
 production of, 2
Toluene 2- monooxygenase pathway
 Burkholderia vietnamiensis G4
  enzymes involved in, 12–14
  genes involved in, 14–15
Toulene- 3- monooxygenase pathway
 Ralstonia pickettii PKO1
  enzymes involved, 16, 18
  genes involved, 18–20
  isolation of, 15
  pathway overview, 17
  transport of toluene, 20
Toluene 4- monooxygenase pathway
 Pseudomonos mendocina KR1
  genes involved, 24–25
  isolation of, 20
  overview of, 21
  toluene 4- monooxygenase, role of, 22–23
Toluene/*o*- xylene monooxygenase pathway
 Pseudomonos stutzeri OX1
  catechol production, 32
  gene cloning for enzyme activity, 29–31
  gene expression, 32–33
  isolation of, 28
  oxidation of, 29
Transposon mutagenesis
 E. cloacae mutants, 114–115
 mini- Tn5 transposon system, 113–114
Trichloroethylene, 7
Trimethylamine N- oxide, 109, 117

U

ubiE gene, 119
Uroporphyrinogen III methyltransferase (UMT), 148–149

X

Xanthomonas, 149
X- ray photoelectron spectroscopy (XPS), 234
o- Xylene
  conversion of, 16, 28
  genes for degradation, 30
  oxidation of, 29
  P. stutzeri OX1, inducible growth, 32
  as sole carbon and energy sources, 30

Y

Yersinia enterocolitica, 92

CONTENTS OF PREVIOUS VOLUMES

Volume 40

Microbial Cellulases: Protein Architecture, Molecular Properties, and Biosynthesis
Ajay Singh and Kiyoshi Hayashi

Factors Inhibiting and Stimulating Bacterial Growth in Milk: An Historical Perspective
D. K. O'Toole

Challenges in Commercial Biotechnology. Part I. Product, Process, and Market Discovery
Aleš Prokop

Challenges in Commercial Biotechnology. Part II. Product, Process, and Market Development
Aleš Prokop

Effects of Genetically Engineered Microorganisms on Microbial Populations and Processes in Natural Habitats
Jack D. Doyle, Guenther Stotzky, Gwendolyn McClung, and Charles W. Hendricks

Detection, Isolation, and Stability of Megaplasmid-Encoded Chloroaromatic Herbicide-Degrading Genes within *Pseudomonas* Species
Douglas J. Cork and Amjad Khalil

Index

Volume 41

Microbial Oxidation of Unsaturated Fatty Acids
Ching T. Hou

Improving Productivity of Heterologous Proteins in Recombinant *Saccharomyces cerevisiae* Fermentations
Amit Vasavada

Manipulations of Catabolic Genes for the Degradation and Detoxification of Xenobiotics
Rup Lal, Sukanya Lal, P. S. Dhanaraj, and D. M. Saxena

Aqueous Two-Phase Extraction for Downstream Processing of Enzymes/Proteins
K. S. M. S. Raghava Rao, N. K. Rastogi, M. K. Gowthaman, and N. G. Karanth

Biotechnological Potentials of Anoxygenic Phototrophic Bacteria. Part I. Production of Single Cell Protein, Vitamins, Ubiquinones, Hormones, and Enzymes and Use in Waste Treatment
Ch. Sasikala and Ch. V. Ramana

Biotechnological Potentials of Anoxygenic Phototrophic Bacteria. Part II. Biopolyesters, Biopesticide, Biofuel, and Biofertilizer
Ch. Sasikala and Ch. V. Ramana

Index

Volume 42

The Insecticidal Proteins of *Bacillus thuringiensis*
P. Ananda Kumar, R. P. Sharma, and V. S. Malik

Microbiological Production of Lactic Acid
John H. Litchfield

Biodegradable Polyesters
Ch. Sasikala

The Utility of Strains of Morphological Group II *Bacillus*
Samuel Singer

Phytase
Rudy J. Wodzinski and A. H. J. Ullah

Index

Volume 43

Production of Acetic Acid by *Clostridium thermoaceticum*
Munir Cheryan, Sarad Parekh, Minish Shah, and Kusuma Witjitra

Contact Lenses, Disinfectants, and *Acanthamoeba Keratitis*
Donald G. Ahearn and Manal M. Gabriel

Marine Microorganisms as a Source of New Natural Products
V. S. Bernan, M. Greenstein, and W. M. Maiese

Stereoselective Biotransformations in Synthesis of Some Pharmaceutical Intermediates
Ramesh N. Patel

Microbial Xylanolytic Enzyme System: Properties and Applications
Pratima Bajpai

Oleaginous Microorganisms: An Assessment of the Potential
Jacek Leman

Index

Volume 44

Biologically Active Fungal Metabolites
Cedric Pearce

Old and New Synthetic Capacities of Baker's Yeast
P. D'Arrigo, G. Pedrocchi-Fantoni, and S. Servi

Investigation of the Carbon- and Sulfur-Oxidizing Capabilities of Microorganisms by Active-Site Modeling
Herbert L. Holland

Microbial Synthesis of D-Ribose: Metabolic Deregulation and Fermentation Process
P. de Wulf and E. J. Vandamme

Production and Application of Tannin Acyl Hydrolase: State of the Art
P. K. Lekha and B. K. Lonsane

Ethanol Production from Agricultural Biomass Substrates
Rodney J. Bothast and Badal C. Saha

Thermal Processing of Foods, A Retrospective, Part I: Uncertainties in Thermal Processing and Statistical Analysis
M. N. Ramesh, S. G. Prapulla, M. A. Kumar, and M. Mahadevaiah

Thermal Processing of Foods, A Retrospective, Part II: On-Line Methods for Ensuring Commercial Sterility
M. N. Ramesh, M. A. Kumar, S. G. Prapulla, and M. Mahadevaiah

Index

Volume 45

One Gene to Whole Pathway: The Role of Norsolorinic Acid in Aflatoxin Research
J. W. Bennett, P.-K. Chang, and D. Bhatnagar

Formation of Flavor Compounds in Cheese
P. F. Fox and J. M. Wallace

The Role of Microorganisms in Soy Sauce Production
Desmond K. O'Toole

Gene Transfer Among Bacteria in Natural Environments
Xiaoming Yin and G. Stotzky

Breathing Manganese and Iron:
   Solid-State Respiration
*Kenneth H. Nealson and
   Brenda Little*

Enzymatic Deinking
Pratima Bajpai

Microbial Production of Docosahexaenoic
   Acid (DHA, C22:6)
Ajay Singh and Owen P. Word

Index

Volume 46

Cumulative Subject Index

Volume 47

Seeing Red: The Story of Prodigiosin
J. W. Bennett and Ronald Bentley

Microbial/Enzymatic Synthesis of Chiral
   Drug Intermediates
Ramesh N. Patel

Recent Developments in the
   Molecular Genetics of the
   Erythromycin-Producing Organism
   Saccharopolyspora erythraea
Thomas J. Vanden Boom

Bioactive Products from Streptomyces
Vladisalv Behal

Advances in Phytase Research
*Edward J. Mullaney, Catherine B. Daly,
   and Abdul H. J. Ullah*

Biotransformation of Unsaturated
   Fatty Acids of industrial Products
Ching T. Hou

Ethanol and Thermotolerance in
   the Bioconversion of Xylose
   by Yeasts
Thomas W. Jeffries and Yong-Su Jin

Microbial Degradation of the
   Pesticide Lindane
   (γ-Hexachlorocyclohexane)
*Brajesh Kumar Singh, Ramesh Chander
   Kuhad, Ajay Singh, K. K. Tripathi, and
   P. K. Ghosh*

Microbial Production of
   Oligosaccharides: A Review
*S. G. Prapulla, V. Subhaprada, and
   N. G. Karanth*

Index

Volume 48

Biodegredation of Nitro-Substituted
   Explosives by White-Rot Fungi:
   A Mechanistic Approach
*Benoit Van Aken and
   Spiros N. Agathos*

Microbial Degradation of Pollutants in
   Pulp Mill Effluents
Pratima Bajpai

Bioremediation Technologies
   for Metal-Containing
   Wastewaters Using Metabolically
   Active Microorganisms
*Thomas Pumpel and
   Kishorel M. Paknikar*

The Role of Microorganisms in
   Ecological Risk Assessment
   of Hydrophobic Organic
   Contaminants in Soils
*C. J. A. MacLeod, A. W. J. Morriss,
   and K. T. Semple*

The Development of Fungi: A New
   Concept Introduced By Anton de Bary
Gerhart Drews

Bartolomeo Gosio, 1863–1944:
   An Appreciation
Ronald Bentley

Index

Volume 49

Biodegredation of Explosives
*Susan J. Rosser, Amrik Basran,
   Emmal R. Travis, Christopher E. French,
   and Neil C. Bruce*

Biodiversity of Acidophilic Prokaryotes
*Kevin B. Hallberg and
   D. Barrie Johnson*

Laboratory Birproduction of Paralytic Shellfish Toxins in Dinoflagellates
Dennis P. H. Hsieh, Dazhi Wang, and Garry H. Chang

Metal Toxicity in Yeasts and the Role of Oxidative Stress
S. V. Avery

Foodbourne Microbial Pathogens and the Food Research Institute
M. Ellin Doyle and Michael W. Pariza

Alexander Flemin and the Discovery of Penicillin
J. W. Bennett and King-Thom Chung

Index

Volume 50

Paleobiology of the Archean
Sherry L. Cady

A Comparative Genomics Approach for Studying Ancestral Proteins and Evolution
Ping Liang and Monica Riley

Chromosome Packaging by Archaeal Histones
Kathleen Sandman and John N. Reeve

DNA Recombination and Repair in the Archaea
Erica M. Seitz, Cynthia A. Haseltine, and Stephen C. Kowalczykowski

Basal and Regulated Transcription in Archaea
Jörg Soppa

Protein Folding and Molecular Chaperones in Archaea
Michel R. Leroux

Archaeal Proteasomes: Proteolytic Nanocompartments of the Cell
Julie A. Maupin-Furlow, Steven J. Kaczowka, Mark S. Ou, and Heather L. Wilson

Archaeal Catabolite Repression: A Gene Regulatory Paradigm
Elisabetta Bini and Paul Blum

Index

Volume 51

The Biochemistry and Molecular Biology of Lipid Accumulation in Oleaginous Microorganisms
Colin Ratledge and James P. Wynn

Bioethanol Technology: Developments and Perspectives
Owen P. Ward and Ajay Singh

Progress of *Aspergillus oryzae* Genomics
Masayuki Machida

Transmission Genetics of *Microbotryum violaceum (Ustilago violacea)*: A Case History
E. D. Garber and M. Ruddat

Molecular Biology of the *Koji* Molds
Katsuhiko Kitamoto

Noninvasive Methods for the Investigation of Organisms at Low Oxygen Levels
David Lloyd

The Development of the Penicillin Production Process in Delft, The Netherlands, During World War II Under Nazi Occupation
Marlene Burns and Piet W. M. van Dijck

Genomics for Applied Microbiology
William C. Nierman and Karen E. Nelson

Index

Volume 52

Soil-Based Gene Discovery: A New Technology to Accelerate and Broaden Biocatalytic Applications
Kevin A. Gray, Toby H. Richardson, Dan E. Robertson, Paul E. Swanson, and Mani V. Subramanian

The Potential of Site-Specific Recombinases as Novel Reporters in Whole-Cell Biosensors of Pollution
Paul Hinde, Jane Meadows, Jon Saunders, and Clive Edwards

Microbial Phosphate Removal and
  Polyphosphate Production from
  Wastewaters
John W. McGrath and John P. Quinn

Biosurfactants: Evolution and Diversity
  in Bacteria
Raina M. Maier

Comparative Biology of Mesophilic and
  Thermophilic Nitrile Hydratases
*Don A. Cowan, Rory A. Cameron, and
  Tsepo L. Tsekoa*

From Enzyme Adaptation to Gene
  Regulation
William C. Summers

Acid Resistance in *Escherichia coli*
Hope T. Richard and John W. Foster

Iron Chelation in Chemotherapy
Eugene D. Weinberg

Angular Leaf Spot: A Disease Caused
  by the Fungus *Phaeoisariopsis griseola*
  (Sacc.) Ferraris on *Phaseolus vulgaris* L.
*Sebastian Stenglein, L. Daniel Ploper,
  Oscar Vizgarra, and
  Pedro Balatti*

The Fungal Genetics Stock Center: From
  Molds to Molecules
Kevin McCluskey

Adaptation by Phase Variation in
  Pathogenic Bacteria
*Laurence Salaün, Lori A. S. Snyder, and
  Nigel J. Saunders*

What Is an Antibiotic? Revisited
Ronald Bentley and J. W. Bennett

An Alternative View of the Early History
  of Microbiology
Milton Wainwright

The Delft School of Microbiology, from
  the Nineteenth to
  the Twenty-first Century
Lesley A. Robertson

Index

Volume 53

Biodegradation of Organic Pollutants in
  the Rhizosphere
Liz J. Shaw and Richard G. Burns

Anaerobic Dehalogenation of
  Organohalide Contaminants in the
  Marine Environment
*Max M. Häggblom, Young-Boem Ahn,
  Donna E. Fennell, Lee J. Kerkhof, and
  Sung-Keun Rhee*

Biotechnological Application of
  Metal-Reducing Microorganisms
*Jonathan R. Lloyd, Derek R. Lovley, and
  Lynne E. Macaskie*

Determinants of Freeze Tolerance in
  Microorganisms, Physiological
  Importance, and Biotechnological
  Applications
*An Tanghe, Patrick Van Dijck, and Johan
  M. Thevelein*

Fungal Osmotolerance
*P. Hooley, D. A. Fincham,
  M. P. Whitehead, and N. J. W. Clipson*

Mycotoxin Research in South Africa
M. F. Dutton

Electrophoretic Karyotype Analysis in
  Fungi
*J. Beadle, M. Wright, L. McNeely, and
  J. W. Bennett*

Tissue Infection and Site-Specific
  Gene Expression in *Candida albicans*
Chantal Fradin and Bernard Hube

LuxS and Autoinducer-2: Their
  Contribution to Quorum Sensing and
  Metabolism in Bacteria
*Klaus Winzer, Kim R. Hardie, and
  Paul Williams*

Microbiological Contributions to the
  Search of Extraterrestrial Life
Brendlyn D. Faison

Index

Volume 54

Metarhizium spp.: Cosmopolitan Insect-
  Pathogenic Fungi – Mycological
  Aspects
*Donald W. Roberts and
  Raymond J. St. Leger*

Molecular Biology of the
  Burkholderia cepacia Complex
Jimmy S. H. Tsang

Non-Culturable Bacteria in Complex
    Commensal Populations
William G. Wade

λ Red-Mediated Genetic
    Manipulation of
    Antibiotic-Producing
    Streptomyces
*Bertolt Gust, Govind Chandra,
    Dagmara Jakimowicz, Tian Yuqing,
    Celia J. Bruton, and
    Keith F. Chater*

Colicins and Microcins: The Next
    Generation Antimicrobials
*Osnat Gillor, Benjamin C. Kirkup, and
    Margaret A. Riley*

Mannose-Binding Quinone Glycoside,
    MBQ: Potential Utility and Action
    Mechanism
*Yasuhiro Igarashi and
    Toshikazu Oki*

Protozoan Grazing of
    Freshwater Biofilms
Jacqueline Dawn Parry

Metals in Yeast Fermentation Processes
Graeme M. Walker

Interactions between Lactobacilli
    and Antibiotic-Associated
    Diarrhea
Paul Naaber and Marika Mikelsaar

Bacterial Diversity in the Human Gut
*Sandra MacFarlane and
    George T. MacFarlane*

Interpreting the Host-Pathogen Dialogue
    Through Microarrays
*Brian K. Coombes, Philip R. Hardwidge,
    and B. Brett Finlay*

The Inactivation of Microbes
    by Sunlight: Solar Disinfection
    as a Water Treatment Process
Robert H. Reed

Index

Volume 55

Fungi and the Indoor Environment:
    Their Impact on Human Health
*J. D. Cooley, W. C. Wong, C. A. Jumper,
    and D. C. Straus*

Fungal Contamination as a Major
    Contributor to Sick Building
    Syndrome
De-Wei LI and Chin S. Yang

Indoor Moulds and Their Associations
    with Air Distribution Systems
*Donald G. Ahearn, Daniel L. Price,
    Robert Simmons,
    Judith Noble-Wang, and
    Sidney A. Crow, Jr.*

Microbial Cell Wall Agents and Sick
    Building Syndrome
Ragnar Rylander

The Role of *Stachybotrys* in the
    Phenomenon Known as Sick
    Building Syndrome
Eeva-Liisa Hintikka

Moisture-Problem Buildings with Molds
    Causing Work-Related Diseases
Kari Reijula

Possible Role of Fungal Hemolysins in
    Sick Building Syndrome
*Stephen J. Vesper and
    Mary Jo Vesper*

The Roles of *Penicillium* and *Aspergillus* in
    Sick Building Syndrome (SBS)
*Christopher J. Schwab and
    David C. Straus*

Pulmonary Effects of *Stachybotrys
    chartarum* in Animal Studies
Iwona Yike and Dorr G. Dearborn

Toxic Mold Syndrome
Michael B. Levy and Jordan N. Fink

Fungal Hypersensitivity:
    Pathophysiology, Diagnosis, Therapy
Vincent A. Marinkovich

Indoor Molds and Asthma in Adults
*Maritta S. Jaakkola and
    Jouni J. K. Jaakkola*

Role of Molds and Mycotoxins in
    Being Sick in Buildings:
    Neurobehavioral and Pulmonary
    Impairment
Kaye H. Kilburn

The Diagnosis of Cognitive Impairment Associated with Exposure to Mold
Wayne A. Gordon and Joshua B. Cantor

Mold and Mycotoxins: Effects on the Neurological and Immune Systems in Humans
Andrew W. Campbell, Jack D. Thrasher, Michael R. Gray, and Aristo Vojdani

Identification, Remediation, and Monitoring Processes Used in a Mold-Contaminated High School
S. C. Wilson, W. H. Holder, K. V. Easterwood, G. D. Hubbard, R. F. Johnson, J. D. Cooley, and D. C. Straus

The Microbial Status and Remediation of Contents in Mold-Contaminated Structures
Stephen C. Wilson and Robert C. Layton

Specific Detection of Fungi Associated With SBS When Using Quantitative Polymerase Chain Reaction
Patricia Cruz and Linda D. Stetzenbach

Index

Volume 56

Potential and Opportunities for Use of Recombinant Lactic Acid Bacteria in Human Health
Sean Hanniffy, Ursula Wiedermann, Andreas Repa, Annick Mercenier, Catherine Daniel, Jean Fioramonti, Helena Tlaskolova, Hana Kozakova, Hans Israelsen, Søren Madsen, Astrid Vrang, Pascal Hols, Jean Delcour, Peter Bron, Michiel Kleerebezem, and Jerry Wells

Novel Aspects of Signaling in *Streptomyces* Development
Gilles P. van Wezel and Erik Vijgenboom

Polysaccharide Breakdown by Anaerobic Microorganisms Inhabiting the Mammalian Gut
Harry J. Flint

Lincosamides: Chemical Structure, Biosynthesis, Mechanism of Action, Resistance, and Applications
Jaroslav Spížek, Jitka Novotná, and Tomáš Řezanka

Ribosome Engineering and Secondary Metabolite Production
Kozo Ochi, Susumu Okamoto, Yuzuru Tozawa, Takashi Inaoka, Takeshi Hosaka, Jun Xu, and Kazuhiko Kurosawa

Developments in Microbial Methods for the Treatment of Dye Effluents
R. C. Kuhad, N. Sood, K. K. Tripathi, A. Singh, and O. P. Ward

Extracellular Glycosyl Hydrolases from Clostridia
Wolfgang H. Schwarz, Vladimir V. Zverlov, and Hubert Bahl

Kernel Knowledge: Smut of Corn
María D. García-Pedrajas and Scott E. Gold

Bacterial ACC Deaminase and the Alleviation of Plant Stress
Bernard R. Glick

Uses of *Trichoderma* spp. to Alleviate or Remediate Soil and Water Pollution
G. E. Harman, M. Lorito, and J. M. Lynch

Bacteriophage Defense Systems and Strategies for Lactic Acid Bacteria
Joseph M. Sturino and Todd R. Klaenhammer

Current Issues in Genetic Toxicology Testing for Microbiologists
Kristien Mortelmans and Doppalapudi S. Rupa

Index

Volume 57

Microbial Transformations of Mercury: Potentials, Challenges, and Achievements in Controlling Mercury Toxicity in the Environment
Tamar Barkay and Irene Wagner-Döbler

Interactions Between Nematodes and Microorganisms: Bridging Ecological and Molecular Approaches
Keith G. Davies

Biofilm Development in Bacteria
Katharine Kierek-Pearson and Ece Karatan

Microbial Biogeochemistry of Uranium Mill Tailings
Edward R. Landa

Yeast Modulation of Wine Flavor
Jan H. Swiegers and Isak S. Pretorius

Moving Toward a Systems Biology Approach to the Study of Fungal Pathogenesis in the Rice Blast Fungus *Magnaporthe grisea*
Claire Veneault-Fourrey and Nicholas J. Talbot

The Biotrophic Stages of Oomycete–Plant Interactions
Laura J. Grenville-Briggs and Pieter van West

Contribution of Nanosized Bacteria to the Total Biomass and Activity of a Soil Microbial Community
Nicolai S. Panikov

Index

Volume 58

Physiology and Biotechnology of *Aspergillus*
O. P. Ward, W. M. Qin, J. Dhanjoon, J. Ye, and A. Singh

Conjugative Gene Transfer in the Gastrointestinal Environment
Tine Rask Licht and Andrea Wilcks

Force Measurements Between a Bacterium and Another Surface *In Situ*
Ruchirej Yongsunthon and Steven K. Lower

Actinomycetes and Lignin Degradation
Ralph Kirby

An ABC Guide to the Bacterial Toxin Complexes
Richard ffrench-Constant and Nicholas Waterfield

Engineering Antibodies for Biosensor Technologies
Sarah Goodchild, Tracey Love, Neal Hopkins, and Carl Mayers

Molecular Characterization of Ochratoxin A Biosynthesis and Producing Fungi
J. O'Callaghan and A. D. W. Dobson

Index

Volume 59

Biodegradation by Members of the Genus *Rhodococcus*: Biochemistry, Physiology, and Genetic Adaptation
Michael J. Larkin, Leonid A. Kulakov, and Christopher C. R. Allen

Genomes as Resources for Biocatalysis
Jon D. Stewart

Process and Catalyst Design Objectives for Specific Redox Biocatalysis
Daniel Meyer, Bruno Bühler, and Andreas Schmid

The Biosynthesis of Polyketide Metabolites by Dinoflagellates
Kathleen S. Rein and Richard V. Snyder

Biological Halogenation has Moved far Beyond Haloperoxidases
Karl-Heinz van Pée, Changjiang Dong, Silvana Flecks, Jim Naismith, Eugenio P. Patallo, and Tobias Wage

Phage for Rapid Detection and Control of Bacterial Pathogens in Food
Catherine E. D. Rees and Christine E. R. Dodd

Gastrointestinal Microflora: Probiotics
S. Kolida, D. M. Saulnier, and G. R. Gibson

The Role of Helen Purdy Beale in the Early Development of Plant Serology and Virology
Karen-Beth G. Scholthof and Paul D. Peterson

Index

Volume 60

Microbial Biocatalytic Processes and Their Development
John M. Woodley

Occurrence and Biocatalytic Potential of Carbohydrate Oxidases
Erik W. van Hellemond, Nicole G. H. Leferink, Dominic P. H. M. Heuts, Marco W. Fraaije, and Willem J. H. van Berkel

Microbial Interactions with Humic Substances
J. Ian Van Trump, Yvonne Sun, and John D. Coates

Significance of Microbial Interactions in the Mycorrhizosphere
Gary D. Bending, Thomas J. Aspray, and John M. Whipps

Escherich and *Escherichia*
Herbert C. Friedmann

Index

Volume 61

Unusual Two-Component Signal Transduction Pathways in the Actinobacteria
Matthew I. Hutchings

Acyl-HSL Signal Decay: Intrinsic to Bacterial Cell–Cell Communications
Ya-Juan Wang, Jean Jing Huang, and Jared Renton Leadbetter

Microbial Exoenzyme Production in Food
Peggy G. Braun

Biogenetic Diversity of Cyanobacterial Metabolites
Ryan M. Van Wagoner, Allison K. Drummond, and Jeffrey L. C. Wright

Pathways to Discovering New Microbial Metabolism for Functional Genomics and Biotechnology
Lawrence P. Wackett

Biocatalysis by Dehalogenating Enzymes
Dick B. Janssen

Lipases from Extremophiles and Potential for Industrial Applications
Moh'd Salameh and Juergen Wiegel

In Situ Bioremediation
Kirsten S. Jørgensen

Bacterial Cycling of Methyl Halides
Hendrik Schäfer, Laurence G. Miller, Ronald S. Oremland, and J. Colin Murrell

Index

Volume 62

Anaerobic Biodegradation of Methyl *tert*-Butyl Ether (MTBE) and Related Fuel Oxygenates
Max M. Häggblom, Laura K. G. Youngster, Piyapawn Somsamak, and Hans H. Richnow

Controlled Biomineralization by and Applications of Magnetotactic Bacteria
Dennis A. Bazylinski and Sabrina Schübbe

The Distribution and Diversity of *Euryarchaeota* in Termite Guts
Kevin J. Purdy

Understanding Microbially Active Biogeochemical Environments
Deirdre Gleeson, Frank McDermott, and Nicholas Clipson

The Scale-Up of Microbial Batch and Fed-Batch Fermentation Processes
Christopher J. Hewitt and Alvin W. Neinow

Production of Recombinant Proteins in *Bacillus subtilis*
Wolfgang Schumann

Quorum Sensing: Fact, Fiction, and
 Everything in Between
 *Yevgeniy Turovskiy, Dimitri Kashtanov,
 Boris Paskhover, and
 Michael L. Chikindas*

Rhizobacteria and Plant Sulfur Supply
 *Michael A. Kertesz, Emma Fellows,
 and Achim Schmalenberger*

Antibiotics and Resistance Genes:
 Influencing the Microbial Ecosystem
 in the Gut
 *Katarzyna A. Kazimierczak and
 Karen P. Scott*

Index

Volume 63

A Ferment of Fermentations: Reflections
 on the Production of Commodity
 Chemicals Using Microorganisms
 Ronald Bentley and Joan W. Bennett

Submerged Culture Fermentation of
 "Higher Fungi": The Macrofungi
 *Mariana L. Fazenda, Robert Seviour,
 Brian McNeil, and Linda M. Harvey*

Bioprocessing Using Novel Cell Culture
 Systems
 *Sarad Parekh, Venkatesh Srinivasan, and
 Michael Horn*

Nanotechnology in the Detection and
 Control of Microorganisms
 Pengju G. Luo and Fred J. Stutzenberger

Metabolic Aspects of Aerobic Obligate
 Methanotrophy
 Yuri A. Trotsenko and John Colin Murrell

Bacterial Efflux Transport in
 Biotechnology
 Tina K. Van Dyk

Antibiotic Resistance in the Environment,
 with Particular Reference to MRSA
 *William Gaze, Colette O'Neill, Elizabeth
 Wellington, and Peter Hawkey*

Host Defense Peptides in the Oral Cavity
 Deirdre A. Devine and Celine Cosseau

Index

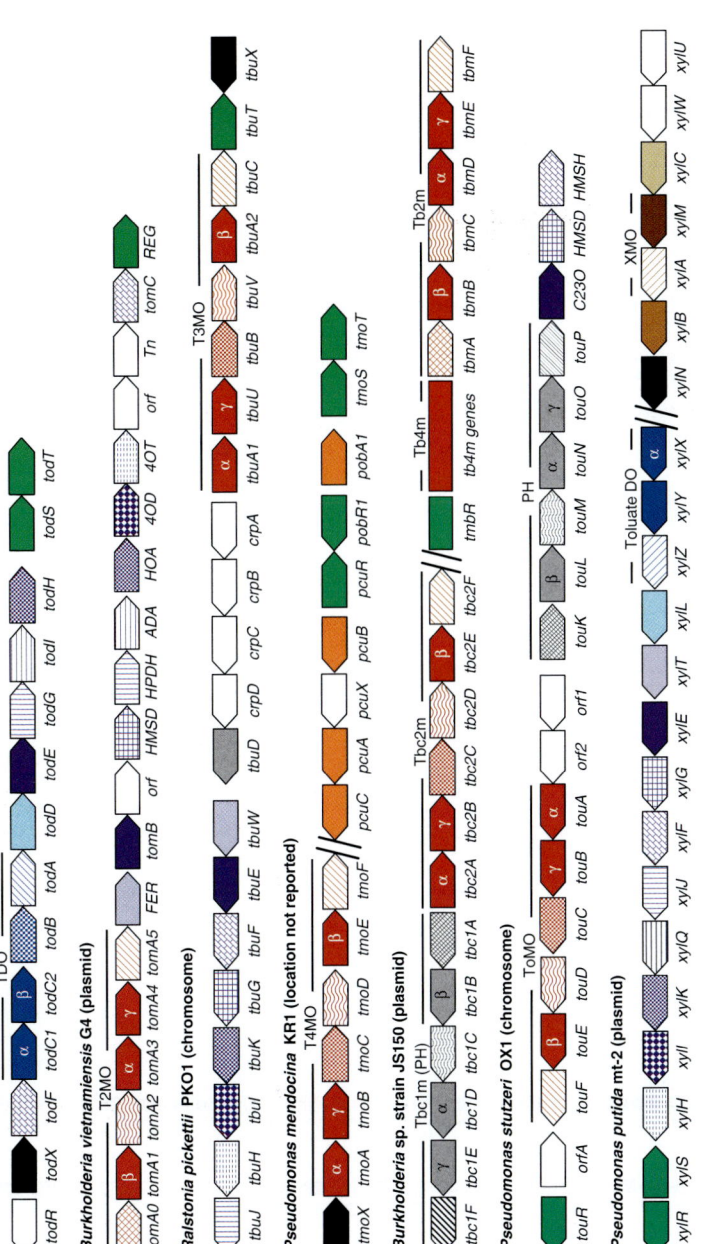

Please refer to Fig. 1.3 in text for figure legend.

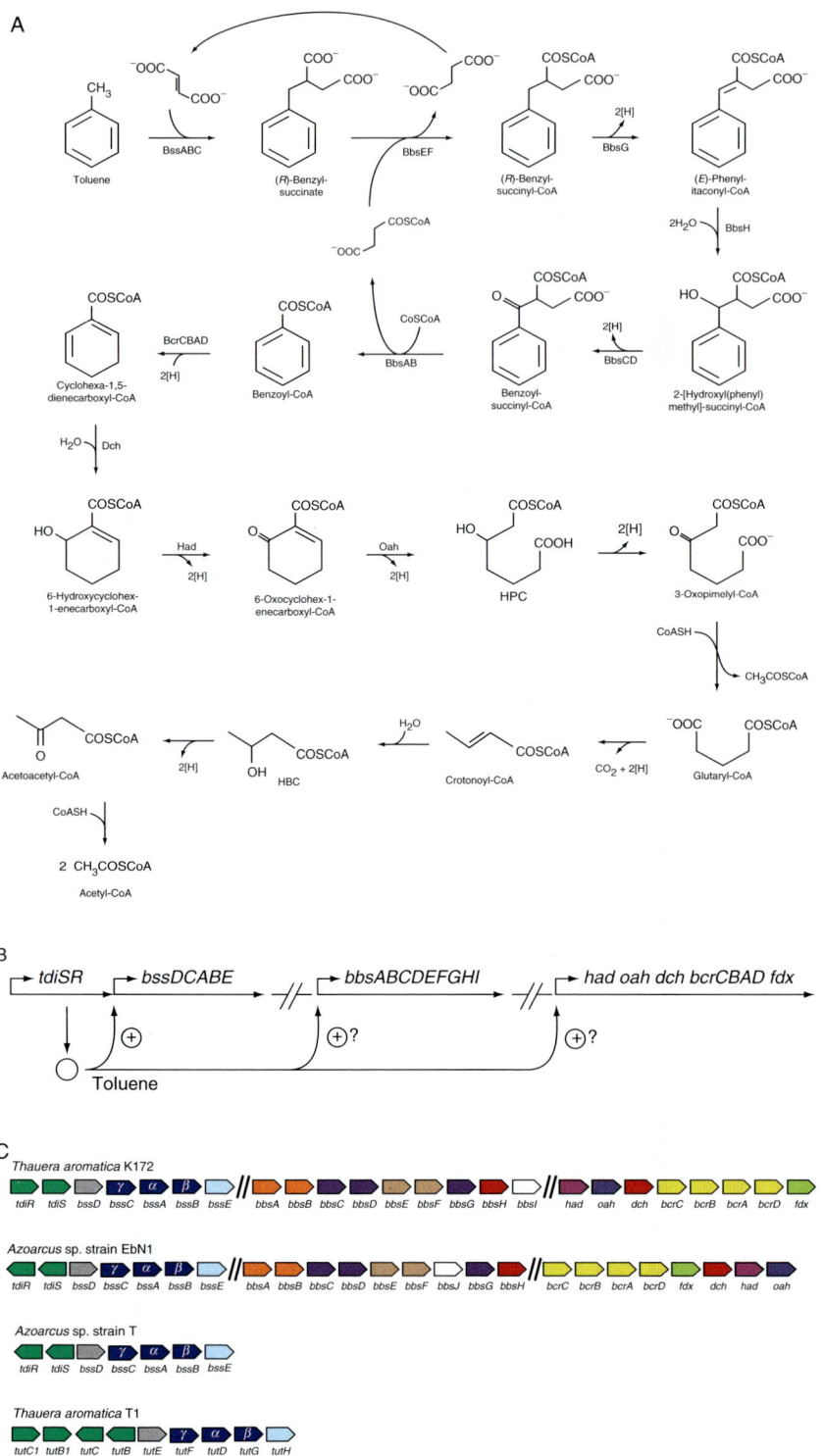

Please refer to Fig. 1.10 in text for figure legend.

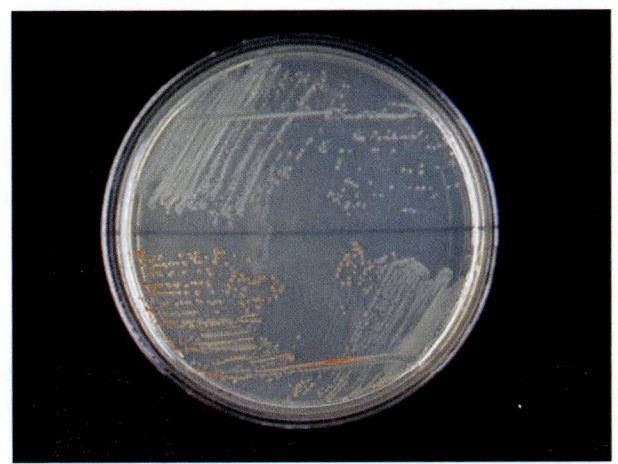

Please refer to Fig. 3.1 in text for figure legend.

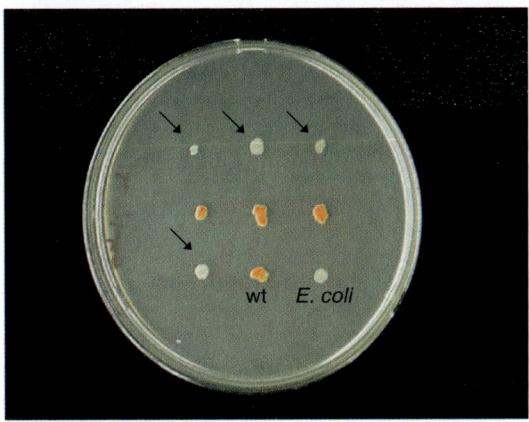

Please refer to Fig. 3.2 in text for figure legend.

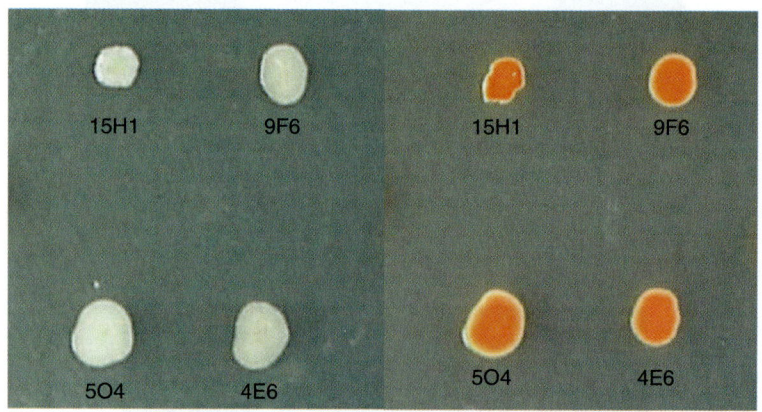

Please refer to Fig. 3.3 in text for figure legend.